LINEAR ALGEBRA |

LINEAR
ALGEBRA

D. C. MURDOCH

Professor of Mathematics

The University of British Columbia

John Wiley & Sons, Inc.

NEW YORK | LONDON | SYDNEY | TORONTO

PREFACE

Linear Algebra for Undergraduates, first published in 1957, was planned and largely written between 1950 and 1952. At that time, undergraduate courses in linear algebra were rare in North American universities and most students were not exposed to an abstract approach to algebra until their first year of graduate studies. The major changes in the undergraduate curriculum in the past fifteen years have necessitated not only substantial rewriting but also a change in title. In the new version I have taken a somewhat more abstract approach to the subject and have treated a number of topics more completely. This will, I hope, better meet the needs of today's student and take advantage of his more adequate preparation for abstract mathematics. Nevertheless, the book is still a quite elementary introduction designed for all students who need some knowledge of linear algebra, matrices, and their applications and not primarily for mathematics specialists.

Several features of the original book have been retained. The field of scalars is still the real or the complex field and real and complex inner product spaces are still treated separately in different chapters. Matrix algebra has been emphasized, and properties of linear transformations and operators are usually derived from the theorems about matrices. I believe this to be the best approach for an elementary course, especially for students who are not planning to specialize in mathematics. On the other hand, linear transformations and the geometric aspects of the subject have by no means been neglected, and the applications to systems of linear differential equations and to the classification of quadric surfaces have been retained.

The first three chapters have been completely rewritten. The principal additions are a discussion of the solution of linear systems by Gaussian elimination, added emphasis on the reduction of a matrix to row-echelon form, and the exploitation of this process both for numerical solution of systems of linear equations and for easy derivation of theorems about matrices and linear systems. In addition, the section on determinants has been expanded to include proofs of their basic properties. Real inner product spaces are treated in Chapter 4 and unitary spaces in Chapter 8. An instructor who so wishes could easily do Chapter 8, except for Section 8.10, along with Chapter 4. Linear transformations are treated in Chapter 5 more completely than in the original, and operators have been written on the left instead of on the right throughout the book. Chapters 6 and 7 have approximately the same content as Chapters 7 and 8 of *Linear Algebra for Undergraduates*.

An appendix on three-dimensional analytic geometry, included in the original edition for the benefit of students who lacked this background, has been partly rewritten and now includes an introduction to vectors in three-dimensional space. Moreover, it has been incorporated into the text as Chapter 0. It is expected that many classes, having covered this material previously, will start with Chapter 1. Others may find it convenient to do Sections 0.1 to 0.9 before starting Chapter 1 but to postpone Sections 0.10 to 0.15 to do in conjunction with Chapter 7.

I would like to thank all those who helped directly or indirectly in the making of this book. Special thanks are due Mr. William Celmaster who read most of the manuscript with the critical eye of a student and provided invaluable assistance in checking the problems and their solutions. I would also like to thank the publishers and their staff and especially their consultants whose helpful criticism of early versions of the manuscript substantially improved the final product.

David C. Murdoch

CONTENTS

Chapter 1

| VECTOR SPACES 44

Chapter 2

| MATRICES AND SYSTEMS OF LINEAR EQUATIONS 82

LINEAR ALGEBRA |

THREE-DIMENSIONAL ANALYTIC GEOMETRY

0.1 Coordinates in Space

Let l be a straight line, O any point on l, and P_1 any point on l different from O. We assume that the reader is familiar with the fact that coordinates can be assigned to the points of l in such a way that the point O, the *origin*, has coordinate 0, the point P_1, the *unit point*, has coordinate 1. If P_x is the point whose coordinate is x (thus $P_0 = O$), there is a one-to-one correspondence $x \leftrightarrow P_x$ between the real numbers x and the points P_x of l. If $x > 0$ then P_x is on the same side of the origin as P_1 and if $x < 0$, P_x is on the opposite side of the origin to P_1. If we think of l as horizontal with O to the left of P_1, then P_x is to the left of P_y if and only if $x < y$. Finally, for arbitrary x and y, the length of the segment $P_x P_y$ in terms of the segment OP_1 as unit is $|y - x|$.

Now let l_1, l_2, and l_3 be any three lines that intersect in a common point O but do not lie in one plane. We assign coordinates to the points of each line choosing O as origin in each case, and unit points P_1 on l_1, Q_1 on l_2, and R_1 on l_3. Now let P be any point in space. A plane through P parallel to the plane of l_2 and l_3 will cut l_1 in a point P_x with coordinate x (Figure 0.1). Similarly the plane through P parallel to the plane of l_1 and l_3 will cut l_2 in Q_y with coordinate y and the plane through P parallel to the plane of l_1 and l_2 will cut l_3 in R_z with coordinate z. The

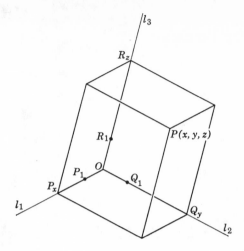

Figure 0.1

three numbers x, y, z are uniquely determined by P and are called Cartesian coordinates of P. We refer to P as the point $(x, y, z,)$ or $P(x, y, z)$. Conversely, any ordered triple (x, y, z) of real numbers determines a unique point whose coordinates they are.

Coordinates on l_1, l_2, and l_3 are designated by the letters x, y, and z, respectively. For this reason these lines will be referred to as the x-, y-, and z-axes, respectively, and collectively as the coordinate axes. The origin O divides each axis into two half lines or *rays*. The *positive rays* of the coordinate axes are those which contain the unit points P_1, Q_1, and R_1 and these will be designated by OX, OY, and OZ. The plane containing the x- and y-axes will be called the xy-plane with similar defini-tions for the yz-plane and the xz-plane. We shall adopt the following convention for the choice of the unit points on the coordinate axes. The unit points P_1, Q_1, and R_1 on the x-, y-, and z-axes will be chosen so that to an observer at R_1 looking at the xy-plane, OX can be rotated into OY counterclockwise through an angle less than $180°$. Such a set of coordinate axes is called a right-hand system.

So far we have placed no restriction on the coordinate axes except that they have the point O in common and do not lie in one plane. Neither have we imposed any restriction on the unit points P_1, Q_1, and R_1 except that they do not coincide with O. For many purposes, no additional restrictions are necessary, but for the derivation of the distance formula in the next section we need some additional assumptions. The coordinate system is said to be *rectangular* if each of the coordi-nate axes is perpendicular to the other two axes, and is *isometric* if the units are the same on each axis, that is, if $OP_1 = OQ_1 = OR_1$. We shall assume throughout this chapter that our coordinate systems are rectangular and isometric.

0.2 The Distance Formula

Assume a rectangular isometric coordinate system and let $A_1(x_1, y_1, z_1)$ and $A_2(x_2, y_2, z_2)$ be any two points in space. Through A_1 and A_2 pass planes parallel to the yz-plane. These planes are perpendicular to the x-axis and will cut it in points P_{x_1} and P_{x_2}, whose linear coordinates are x_1 and x_2. It follows that the perpendicular distance between these parallel planes is $|x_1 - x_2|$. Similarly, if planes parallel to the xz-plane are passed through A_1 and A_2, they will cut the y-axis in Q_{y_1} and Q_{y_2} whose distance apart is $|y_1 - y_2|$, and planes through A_1 and A_2 parallel to the xy-plane will cut the z-axis in R_{z_1} and R_{z_2} whose distance apart is $|z_1 - z_2|$. From Figure 0.2 it is seen that these six planes contain the six faces of a rectangular parallelepiped, or box. The segment A_1A_2 is a diagonal of this box, and the lengths of the three edges A_1B, A_1C, and A_1D are, respectively, $|x_1 - x_2|$, $|y_1 - y_2|$, and $|z_1 - z_2|$. By drawing the diagonal A_1A_2 of the box and the diagonal A_1E of the face A_1BEC, we see from the right-angled triangle A_1A_2E that

$$A_1A_2^2 = A_1E^2 + EA_2^2,$$

and, from the right triangle A_1CE,

$$A_1E^2 = A_1C^2 + CE^2.$$

From these two equations we deduce that

$$A_1A_2^2 = A_1C^2 + CE^2 + EA_2^2,$$

or that the square of the diagonal of a rectangular box is the sum of the squares of its three edges. Since it was shown above that the edges of this box are the absolute

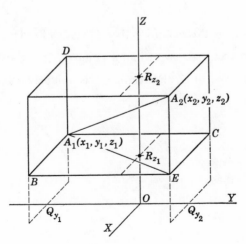

Figure 0.2

values of the differences of the coordinates of A_1 and A_2, we have the distance formula

(1) $$A_1A_2 = \sqrt{(x_1 - x_2)^2 + (y_1 - y_2)^2 + (z_1 - z_2)^2}.$$

0.3 Vectors

In order to specify directions in space, we introduce the concept of a *vector*. Although this term will be given a more general meaning in Chapter 1, for the purposes of this introductory chapter a vector is an ordered pair (A, B) of points in space. The first point A is called the *initial point* of the vector and the second point B is called its *terminal point*. The vector with initial point A and terminal point B will be designated by the notation $\overrightarrow{AB}$ and is represented geometrically (Figure 0.3) by the line segment AB with an arrowhead at B to indicate the terminal point.

Associated with a vector $\overrightarrow{AB}$ is a *length*, namely, the distance between the initial and terminal points, and a *direction*, which is the direction from the initial to the terminal point.

If two vectors lie in the same straight line, it is clear what is meant by saying they have the same or opposite directions. If two vectors $\overrightarrow{AB}$ and $\overrightarrow{CD}$ are parallel, they have the same direction if the segments AC and BD do not intersect and opposite direction if these segments do intersect (Figure 0.4).

Definition.   *Two vectors $\overrightarrow{AB}$ and $\overrightarrow{CD}$ are said to be equal if and only if they have the same length and the same direction.*

It follows from this definition that if O is any fixed point every vector $\overrightarrow{AB}$ is equal to a unique vector $\overrightarrow{OP}$ with initial point O. If a Cartesian coordinate system is chosen with origin at O, the vector $\overrightarrow{OP}$ is uniquely determined by the coordinates

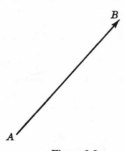

Figure 0.3

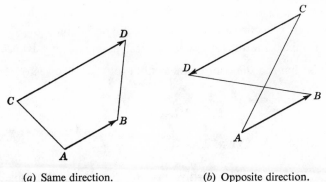

(*a*) Same direction. (*b*) Opposite direction.

Figure 0.4.

(x, y, z) of the point P. These are also called coordinates of the vector $\overrightarrow{AB}$ (or of $\overrightarrow{OP}$) relative to the given coordinate system. The segments AB and OP are opposite sides of a parallelogram (Figure 0.5).

Definition. *The coordinates of a vector $\overrightarrow{AB}$, relative to a Cartesian coordinate system with origin O, are the coordinates of the point P such that $\overrightarrow{OP} = \overrightarrow{AB}$. This definition implies that equal vectors have the same coordinates.*

If a vector $\overrightarrow{AB}$ has coordinates x, y, z relative to a given coordinate system, we write

$$\overrightarrow{AB} = [x, y, z],$$

using square brackets to distinguish the vector from the point $P(x, y, z)$. We also

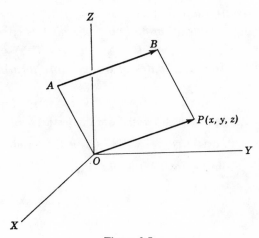

Figure 0.5

frequently use a single capital letter such as V for the vector $[x, y, z]$. It must be emphasized, of course, that when a vector is denoted by a coordinate symbol $[x, y, z]$, a fixed coordinate system has been designated in advance. If the coordinate system is changed the coordinate symbol for the same vector $\overrightarrow{AB}$ will, in general, change.

0.4 Translation of Axes

Let OX, OY, OZ be the positive rays of the x-, y-, and z-axes of a rectangular Cartesian coordinate system. Through an arbitrary point O' with coordinates (a, b, c) draw rays $O'X'$, $O'Y'$, $O'Z'$ having the same direction as OX, OY, and OZ. This means, for example, that any vector with initial point O and terminal point in the ray OX has the same direction as a vector with initial point O' and terminal point in $O'X'$. It is clear that $O'X'$, $O'Y'$, and $O'Z'$ can be used as the positive rays of a new set of rectangular coordinate axes with the same unit of length and with origin at O'. If a point P has coordinates (x, y, z) with respect to the x-, y-, and z-axes and coordinates (x', y', z') with respect to the x'-, y'-, and z'-axes, it is easy to verify by reference to Figure 0.6 that

$$
\begin{array}{lll}
x = x' + a, & & x' = x - a, \\
(2) \qquad y = y' + b, & \text{or} & y' = y - b, \\
z = z' + c, & & z' = z - c.
\end{array}
$$

Definition. *The change of coordinate axes from OX, OY, OZ to OX', OY', OZ' is called a translation of axes. Equations (2) are called the equations of the translation or of the transformation of coordinates effected by this translation.*

Theorem 0.1. *The coordinates of a vector are invariant under translation of axes.*

Proof. Let $\overrightarrow{AB}$ be any vector with coordinates $[x, y, z]$ relative to a given coordinate system with origin O. Then $P(x, y, z)$ is the terminal point of the vector $\overrightarrow{OP}$ equal to $\overrightarrow{AB}$. From Figure 0.7 it is clear that

$$
x = r \cos \alpha,
$$
$$
y = r \cos \beta,
$$
$$
z = r \cos \gamma,
$$

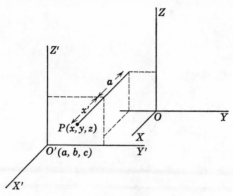

Figure 0.6

where r is the length of $\overrightarrow{OP}$ (or of $\overrightarrow{AB}$) and α, β, γ are the angles $\overrightarrow{OP}$ makes with OX, OY, and OZ.

Now translate to a new coordinate system with origin O'. (Figure 0.8). The vector $\overrightarrow{O'P'}$ equal to $\overrightarrow{AB}$ clearly makes the same angles α, β, γ with $O'X'$, $O'Y'$, and $O'Z'$ because the directions of $O'P'$, $O'X'$, $O'Y'$, $O'Z'$ are the same as the directions of OP, OX, OY, and OZ. Let (x', y', z') be the coordinates of P' in the new coordinate system. Since $\overrightarrow{O'P'}$ has the same length r as $\overrightarrow{OP}$, we have

$$x = r \cos \alpha = x'$$

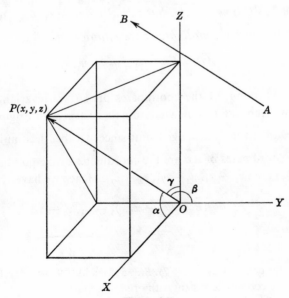

Figure 0.7

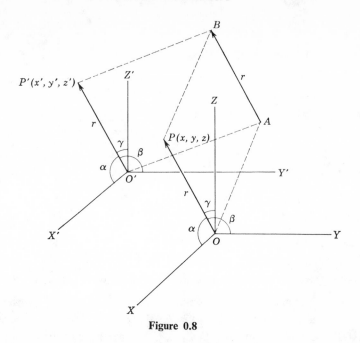

Figure 0.8

and, similarly, $y = y'$ and $z = z'$. Thus, $\overrightarrow{AB}$ has the same coordinates relative to the two sets of coordinate axes.

Theorem 0.2. *If a vector $\overrightarrow{AB}$ has initial point $A(x_1, y_1, z_1)$ and terminal point $B(x_2, y_2, z_2)$, then the coordinate representation of $\overrightarrow{AB}$ is*

$$\overrightarrow{AB} = [x_2 - x_1, y_2 - y_1, z_2 - z_1].$$

Proof. By Theorem 0.1 the coordinates of $\overrightarrow{AB}$ are invariant under translation. Hence, if we translate to a new coordinate system with origin at $A(x_1, y_1, z_1)$ the coordinates of $\overrightarrow{AB}$ are unchanged. But since $\overrightarrow{AB}$ then has initial point at the new origin, the coordinates of $\overrightarrow{AB}$ are the coordinates of the point B in the new system which by (2) are $(x_2 - x_1, y_2 - y_1, z_2 - z_1)$. Hence we have

$$\overrightarrow{AB} = [x_2 - x_1, y_2 - y_1, z_2 - z_1].$$

Exercise 0.1

1. Given the points $A(2, -1, -7)$, $B(3, 5, 2)$, and $C(9, 1, -4)$ relative to a rectangular isometric coordinate system with origin O, find:
   (a) The lengths of the segments OA, OB, OC, AB, BC, CA.

(b) The coordinates of the vectors $\overrightarrow{OA}$, $\overrightarrow{BO}$, $\overrightarrow{AB}$, $\overrightarrow{BA}$, $\overrightarrow{BC}$, $\overrightarrow{CA}$.

(c) Coordinates of points P, Q, R such that $\overrightarrow{OP} = \overrightarrow{AB}$, $\overrightarrow{OQ} = \overrightarrow{BC}$, $\overrightarrow{OR} = \overrightarrow{CA}$.

(d) Coordinates of a point M such that $\overrightarrow{AM} = \overrightarrow{BC}$.

2. Given the points $A(-1, 2, 4)$, $B(5, -1, 2)$, and $C(2, 3, 5)$, find three points P, Q, R such that each of them is the fourth vertex of a parallelogram of which A, B, and C are three vertices.

3. Write the equations of a translation of axes to a new coordinate system with origin at $(2, -3, 4)$, and find the coordinates in the new system of each of these points:

(a) $A(5, 3, -1)$.   (b) $B(2, 0, 3)$.

(c) $C(2, -3, 4)$.   (d) $D(0, 0, 0)$.

4. Show that the point $P(x, y, z)$ lies on a sphere of radius 3 and center at $(3, -1, 2)$ if and only if

$$(x - 3)^2 + (y + 1)^2 + (z - 2)^2 = 9.$$

This equation is called an equation of the sphere.

5. Show that an equation of the sphere with center at $(0, 0, 0)$ and radius r is

$$x^2 + y^2 + z^2 = r^2.$$

6. If coordinate axes are translated to a new origin $(3, -1, 2)$, show that in the new coordinate system the sphere of Problem 4 has equation

$$x'^2 + y'^2 + z'^2 = 9.$$

7. Show that the point $P(x, y, z)$ is equidistant from the two points $A(3, -2, 1)$ and $B(1, 2, 4)$ if and only if $4x - 8y - 6z + 7 = 0$. Where do the points P whose coordinates satisfy this equation lie?

8. Find a necessary and sufficient condition on the coordinates of P in Problem 7 in order that the distance from P to A is twice the distance from P to B.

9. Show that the points P of Problem 8 lie on a sphere.

0.5 Addition of Vectors

To define the sum of two vectors $\overrightarrow{OP}$ and $\overrightarrow{OQ}$ we construct a vector $\overrightarrow{PR}$ with initial point P such that $\overrightarrow{PR} = \overrightarrow{OQ}$. This uniquely determines the terminal point R, and we define $\overrightarrow{OP} + \overrightarrow{OQ}$ to be $\overrightarrow{OR}$ (Figure 0.9). In case $\overrightarrow{OP}$ and $\overrightarrow{OQ}$ do not lie in the same straight line, OR is the diagonal of the parallelogram of which OP and OQ are adjacent edges. This rule for adding vectors is called the parallelogram law. The original reason for it was the experimental fact that if the vectors $\overrightarrow{OP}$ and $\overrightarrow{OQ}$ represent forces or velocities, then $\overrightarrow{OP} + \overrightarrow{OQ}$ represents the resultant force or velocity.

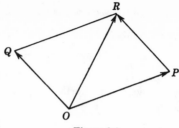

Figure 0.9

Theorem 0.3. *If, relative to a given coordinate system with origin O,*

$$\overrightarrow{OP} = [x_1, y_1, z_1] \qquad and \qquad \overrightarrow{OQ} = [x_2, y_2, z_2],$$

then

(3) $$\overrightarrow{OP} + \overrightarrow{OQ} = [x_1 + x_2, y_1 + y_2, z_1 + z_2].$$

Proof. Let $\overrightarrow{OR} = \overrightarrow{OP} + \overrightarrow{OQ} = [x, y, z]$. Then, by Theorem 0.2, we have

$$\overrightarrow{PR} = [x - x_1, y - y_1, z - z_1].$$

But since $\overrightarrow{PR} = \overrightarrow{OQ} = [x_2, y_2, z_2]$, we have $x - x_1 = x_2, y - y_1 = y_2, z - z_1 = z_2$, and hence,

$$\overrightarrow{OR} = \overrightarrow{OP} + \overrightarrow{OQ} = [x_1 + x_2, y_1 + y_2, z_1 + z_2].$$

Corollary 1. *Addition of vectors is commutative and associative. That is, if X, Y, Z are any three vectors, then*

$$X + Y = Y + X$$

and

$$X + (Y + Z) = (X + Y) + Z.$$

Although this corollary can be proved geometrically, it follows immediately from the theorem which reduces addition of vectors to addition of corresponding coordinates.

A vector $\overrightarrow{AA}$ for which the initial and terminal points coincide is called a zero vector. By convention any two zero vectors are equal. This unique zero vector is denoted by O. It is clear from Theorem 0.2 that the coordinates of O are $[0, 0, 0]$ in any coordinate system. Corollary 2 follows immediately from Theorem 0.3.

Corollary 2. *If V is any vector and O is the zero vector, then $V + O = V$.*

From our definition of addition of vectors it follows that $\overrightarrow{AB} + \overrightarrow{BA} = O$ and, therefore, we define the negative of $\overrightarrow{AB}$, denoted by $-\overrightarrow{AB}$, to be the vector

$\overrightarrow{BA}$ having the same length as $\overrightarrow{AB}$ but the opposite direction. It follows that if

$$\overrightarrow{AB} = V = [x_1, y_1, z_1],$$

then

$$\overrightarrow{BA} = -V = [-x_1, -y_1, -z_1].$$

Now if $W = [x_2, y_2, z_2]$ we define subtraction by

$$W - V = W + (-V) = [x_2 - x_1, y_2 - y_1, z_2 - z_1].$$

Comparison with Theorem 0.2 shows that if O is the origin and $A(x_1, y_1, z_1)$, $B(x_2, y_2, z_2)$ are any two points, then

$$\overrightarrow{AB} = \overrightarrow{OB} - \overrightarrow{OA}.$$

0.6 Multiplication by Scalars

Let k be any real number and let V be any vector. If $k > 0$ the notation kV will be used for the vector having the same direction as V whose length is k times the length of V. If $k < 0$ we define kV to mean $|k|(-V)$, that is, the vector whose length is the length of V multiplied by $|k|$ and whose direction is opposite to that of V. Clearly, also, in this case $kV = -(|k| V)$. Finally, if $k = 0$ we define $0V$ to be the zero vector O. Traditionally, a real number, used as a multiplier in this way, has been called a *scalar*, and the vector kV is called a *scalar multiple* of V.

Assuming $V \neq O$ and $k \neq 0$, V and kV have either the same or opposite directions and are therefore parallel. Conversely, if a vector W is parallel to V ($V \neq O$), then W is equal to a vector with the same initial point as V which must lie in the same straight line as V. Hence, $W = kV$ for same scalar k. We state this formally.

Theorem 0.4. *If $\overrightarrow{AB}$ and $\overrightarrow{CD}$ are nonzero vectors, then $\overrightarrow{CD}$ is a scalar multiple of $\overrightarrow{AB}$ if and only if the line CD is parallel to the line AB. (Of course, these two lines may coincide. We adopt the convention that every line is parallel to itself.)*

Theorem 0.5. *If a vector V has coordinates $[x, y, z]$ and k is any scalar, then kV has coordinates $[kx, ky, kz]$.*

Proof. Let $V = \overrightarrow{OP}$ and $kV = \overrightarrow{OP'}$ where P has coordinates (x, y, z) and P' has coordinates (x', y', z'). Figure 0.10 illustrates the case $k > 0$.

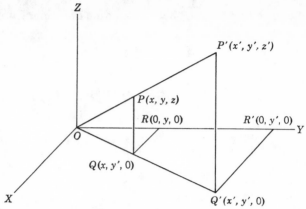

Figure 0.10

By similar triangles we have

$$\frac{y'}{y} = \frac{OR'}{OR} = \frac{OQ'}{OQ} = \frac{OP'}{OP} = k$$

and, therefore, $y' = ky$. Similarly, by projecting onto the x- and z-axes, $x' = kx$, $z' = kz$, and

$$\overrightarrow{OP'} = [x', y', z'] = [kx, ky, kz].$$

A similar proof holds when $k < 0$. When $k = 0$ the theorem is obvious.

The following theorem states rules for multiplication by scalars which follow at once from Theorem 0.5 or from the geometrical definition of kV.

Theorem 0.6. *If V, V_1, V_2 are any vectors and k, k_1, k_2 are any scalars, then:*
(a) $k(V_1 + V_2) = kV_1 + kV_2$.
(b) $(k_1 + k_2)V = k_1V + k_2V$.
(c) $k_1(k_2V) = (k_1k_2)V$.
(d) $1V = V, (-1)V = -V$ and $0V = O$.

We close this section with some examples of the use of vectors to answer geometric questions.

Example 1. Given points $A(1, -2, 3)$, $B(5, 2, 5)$, $C(-4, 2, 9)$, and $D(-2, 4, 10)$, show that AB is parallel to CD but that AC is not parallel to BD.

Solution. By Theorem 0.2,

$$\overrightarrow{AB} = [5 - 1, 2 + 2, 5 - 3] = [4, 4, 2],$$

$$\overrightarrow{CD} = [2, 2, 1].$$

Hence, $\overrightarrow{AB} = 2(\overrightarrow{CD})$ and AB is parallel to CD by Theorem 0.4. On the other hand, $\overrightarrow{AC} = [-5, 4, 6]$ and $\overrightarrow{BD} = [-7, 2, 5]$ are not parallel, since $\overrightarrow{AC}$ is not a scalar multiple of $\overrightarrow{BD}$.

Example 2. Let A, B, C be the points given in Example 1.
  (a) Find a point P such that $ABPC$ is a parallelogram.
  (b) Find a point Q such that $ABCQ$ is a parallelogram.

Solution.
  (a) Let P have coordinates (x, y, z). Since $ABPC$ is a parallelogram if and only if $\overrightarrow{AB} = \overrightarrow{CP}$, we must have

$$\overrightarrow{AB} = [4, 4, 2] = [x + 4, y - 2, z - 9] = \overrightarrow{CP}$$

or $x = 0$, $y = 6$, $z = 11$, and P is the point $(0, 6, 11)$.
  (b) If $ABCQ$ is a parallelogram, $\overrightarrow{AB} = \overrightarrow{QC}$, and hence, if Q is the point (x, y, z),

$$[4, 4, 2] = [-4 - x, 2 - y, 9 - z]$$

or $x = -8$, $y = -2$, $z = 7$, and Q is the point $(-8, -2, 7)$.

Example 3. Given points $A(5, -1, 4)$ and $B(3, 5, 2)$, find the coordinates of M, the midpoint of the segment AB.

Solution. The point $M(x, y, z)$ is the midpoint of AB if $\overrightarrow{AM} = \frac{1}{2}\overrightarrow{AB}$, or

$$[x - 5, y + 1, z - 4] = \tfrac{1}{2}[3 - 5, 5 + 1, 2 - 4] = [-1, 3, -1].$$

Hence, $x = 4$, $y = 2$, $z = 3$, and M is the point $(4, 2, 3)$.

Clearly, by replacing the scalar $\frac{1}{2}$ by $\frac{1}{3}$ and $\frac{2}{3}$ in this solution one could find the points of trisection of the segment AB.

Example 4. Given the points A and B of Example 3, show that a point $P(x, y, z)$ is on the line AB if and only if there is a real number t such that

$$x = 5 - 2t,$$

(4)
$$y = -1 + 6t,$$

$$z = 4 - 2t.$$

Solution. The point $P(x, y, z)$ is on the line through A and B if and only if $\overrightarrow{AP}$ is a scalar multiple of $\overrightarrow{AB}$ and, therefore, if and only if there exists a scalar t such that $\overrightarrow{AP} = t\overrightarrow{AB}$ or

$$[x - 5, y + 1, z - 4] = t[-2, 6, -2],$$

which is equivalent to (4).

Example 5. Find the point in which the line AB of Example 4 cuts the xy-plane.

Solution. If $P(x, y, z)$ is on the line AB, then x, y, z have the form (4). If P is on the xy-plane, $z = 0$ and therefore $4 - 2t = 0$ or $t = 2$. The required point, found by substituting $t = 2$ in (4), is therefore $(1, 11, 0)$.

Exercise 0.2

1. Given the three points $A(5, -3, 1)$, $B(-2, 4, 3)$, and $C(3, 1, -4)$, find:
   (a) The lengths of the sides of the triangle ABC.
   (b) The coordinates of the vectors $\overrightarrow{AB}$, $\overrightarrow{BC}$, $\overrightarrow{CA}$.
   (c) The coordinates of $\overrightarrow{AB} + \overrightarrow{BC}$ and of $\overrightarrow{AB} + \overrightarrow{BC} + \overrightarrow{CA}$.
   (d) The midpoints M of BC, N of AB, and Q of AC.
   (e) The coordinates of the centroid of the triangle ABC, that is, the point P on AM such that $AP = 2(PM)$. Show that P also trisects BQ and CN.

2. Show that the following sets of four points are the vertices of parallelograms.
   (a) $(0, 0, 0)$, $(5, 2, 7)$, $(-2, 2, 3)$, $(3, 4, 10)$.
   (b) $(4, 1, 2)$, $(-2, 3, 1)$, $(5, 1, 4)$, $(-1, 3, 3)$.

3. Given the points $A(x_1, y_1, z_1)$ and $B(x_2, y_2, z_2)$, show that the midpoint of AB has coordinates

$$\left(\frac{x_1 + x_2}{2}, \frac{y_1 + y_2}{2}, \frac{z_1 + z_2}{2} \right).$$

4. Given the points $A(4, -1, 6)$ and $B(2, 5, -4)$, find the point C such that B is the midpoint of AC, and the point D such that A is the midpoint of DB.

5. Given three points $A_i(x_i, y_i, z_i)$ where $i = 1, 2, 3$, show that the centroid of the triangle $A_1A_2A_3$ is the point

$$\left(\frac{x_1 + x_2 + x_3}{3}, \frac{y_1 + y_2 + y_3}{3}, \frac{z_1 + z_2 + z_3}{3} \right).$$

6. Find the points in which the line through $A(2, 1, 4)$ and $B(-3, 5, 2)$ cuts each of the coordinate planes (see Examples 4 and 5).

7. Given $A(x_1, y_1, z_1)$ and $B(x_2, y_2, z_2)$, show that if $P(x, y, z)$ is on the line through

A and B, then there exist numbers s and t such that $s + t = 1$ and

$$x = sx_1 + tx_2,$$
$$y = sy_1 + ty_2,$$
$$z = sz_1 + tz_2.$$

Also show that P is between A and B if and only if s and t are both positive.

8. Given the points $A(2, 5, -1)$, $B(-2, 1, 4)$, and $C(0, -6, 4)$, find:

   (a) A point P such that $\overrightarrow{AP} = 3\overrightarrow{AB}$.

   (b) A point P such that $\overrightarrow{CP} = 2\overrightarrow{AB}$.

   (c) A point P such that $\overrightarrow{AP} = -5\overrightarrow{BP}$.

9. In a triangle ABC, P is the midpoint of AB and Q the midpoint of AC. Use vector methods to show that PQ is parallel to BC and $PQ = \frac{1}{2}(BC)$.

10. Use vector methods to prove that the midpoints of the sides of a quadrilateral are the vertices of a parallelogram.

0.7 Inner Products

In this section we shall obtain a formula for the angle between two vectors in terms of their coordinates. The angle between the vectors $\overrightarrow{AB}$ and $\overrightarrow{CD}$ is, of course, the angle between $\overrightarrow{OP}$ and $\overrightarrow{OQ}$ where $\overrightarrow{OP} = \overrightarrow{AB}$ and $\overrightarrow{OQ} = \overrightarrow{CD}$. The angle θ between $\overrightarrow{OP}$ and $\overrightarrow{OQ}$ satisfies the inequalities $0 \leq \theta \leq \pi$. If $\theta = 0$ the vectors have the same direction and if $\theta = \pi$ they have opposite directions. A useful tool in discussing the angle between two vectors is the inner product of vectors defined as follows.

Definition. *The inner product of two vectors is the product of their lengths times the cosine of the angle between the vectors.*

When a vector is denoted by a single letter V its length is denoted by the notation $\|V\|$. The inner product of the vectors V_1 and V_2 is denoted by $V_1 \cdot V_2$ and is also sometimes called the *dot product*. Using this notation, our definition states that

(5) $$V_1 \cdot V_2 = \|V_1\| \, \|V_2\| \cos \theta,$$

where θ is the angle between V_1 and V_2.

We draw attention to the fact that the inner product $V_1 \cdot V_2$ is a scalar, not a vector.

Theorem 0.7. *If, relative to any rectangular isometric coordinate system,*

$$V_1 = [x_1, y_1, z_1] \quad \text{and} \quad V_2 = [x_2, y_2, z_2],$$

then

(6) $$V_1 \cdot V_2 = x_1 x_2 + y_1 y_2 + z_1 z_2.$$

Proof. Let $V_1 = \overrightarrow{OP_1}$ and $V_2 = \overrightarrow{OP_2}$. We apply both the distance formula and the law of cosines to find $(P_1P_2)^2$. If we let $r_1 = \|V_1\|$ and $r_2 = \|V_2\|$, we have

$$(P_1P_2)^2 = (x_1 - x_2)^2 + (y_1 - y_2)^2 + (z_1 - z_2)^2,$$
$$(P_1P_2)^2 = r_1^2 + r_2^2 - 2r_1r_2 \cos \theta.$$

(See Figure 0.11.) Because $r_1^2 = x_1^2 + y_1^2 + z_1^2$ and $r_2^2 = x_2^2 + y_2^2 + z_2^2$, when we equate these two expressions for $(P_1P_2)^2$ we get

$$-2r_1r_2 \cos \theta = -2x_1x_2 - 2y_1y_2 - 2z_1z_2$$

or

$$V_1 \cdot V_2 = r_1r_2 \cos \theta = x_1x_2 + y_1y_2 + z_1z_2.$$

Corollary 1. *The angle θ between the vectors V_1, V_2 is given by the formula*

$$\cos \theta = \frac{V_1 \cdot V_2}{\|V_1\| \|V_2\|} = \frac{x_1x_2 + y_1y_2 + z_1z_2}{\sqrt{x_1^2 + y_1^2 + z_1^2} \sqrt{x_2^2 + y_2^2 + z_2^2}}.$$

Corollary 2. *If V_1, V_2 have coordinates $[x_1, y_1, z_1]$, $[x_2, y_2, z_2]$ and $[x_1', y_1', z_1']$, $[x_2', y_2', z_2']$ relative to two different (rectangular isometric) coordinate systems (with the same origin and unit), then*

$$x_1'x_2' + y_1'y_2' + z_1'z_2' = x_1x_2 + y_1y_2 + z_1z_2.$$

Proof. This follows because by (6) both expressions are equal to $\|V_1\| \|V_2\| \cos \theta$ which depends only on the vectors V_1 and V_2 and not on the coordinate system. However (6) holds only if the coordinate system is rectangular and isometric and this must therefore be assumed.

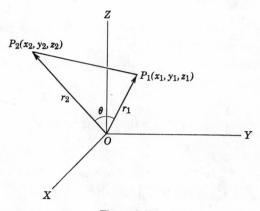

Figure 0.11

Theorem 0.8. *If V, V_1, V_2 are any three vectors and k is any scalar, then:*

(a) $V \cdot V \geq O$ and $V \cdot V = O$ if and only if $V = O$.

(b) $V_1 \cdot V_2 = V_2 \cdot V_1$.

(c) $V \cdot (V_1 + V_2) = V \cdot V_1 + V \cdot V_2$.

(d) $V_1 \cdot (kV_2) = k(V_1 \cdot V_2)$

Proof. These properties of the inner product follow easily from (6). If $V = [x, y, z]$, then $V \cdot V = x^2 + y^2 + z^2$ from which (a) follows. Part (b) is obvious and (c) and (d) almost so. For example, if $V_i = [x_i, y_i, z_i]$,

$$V \cdot (V_1 + V_2) = x(x_1 + x_2) + y(y_1 + y_2) + z(z_1 + z_2)$$

$$= (xx_1 + yy_1 + zz_1) + (xx_2 + yy_2 + zz_2)$$

$$= V \cdot V_1 + V \cdot V_2.$$

Theorem 0.8 makes it possible to do computations with inner products such as the following.

Example 1.

$$(2V_1 - V_2) \cdot (3V_1 + 5V_2) = (2V_1 - V_2) \cdot 3V_1 + (2V_1 - V_2) \cdot 5V_2$$

$$= (3V_1) \cdot (2V_1 - V_2) + (5V_2) \cdot (2V_1 - V_2)$$

$$= 6(V_1 \cdot V_1) - 3(V_1 \cdot V_2) + 10(V_2 \cdot V_1) - 5(V_2 \cdot V_2)$$

$$= 6(V_1 \cdot V_1) + 7(V_1 \cdot V_2) - 5(V_2 \cdot V_2).$$

We note that if $V = [x, y, z]$, then

$$\|V\| = \sqrt{x^2 + y^2 + z^2} = \sqrt{V \cdot V},$$

and hence $\|V\| = 0$ implies $V = O$. Therefore, if V_1 and V_2 are nonzero vectors, then by (5) we see that $V_1 \cdot V_2 = 0$ if and only if $\cos \theta = 0$ or $\theta = 90°$. Two non-zero vectors are said to be *orthogonal* or perpendicular to each other if the angle between them is a right angle. We adopt the convention that the zero vector is orthogonal to every vector and may then state the following.

Theorem 0.9. *The vectors V_1 and V_2 are orthogonal if and only if $V_1 \cdot V_2 = O$.*

Example 2. Find a nonzero vector that is orthogonal to both

$$V_1 = [1, -1, 2] \quad \text{and} \quad V_2 = [3, 2, -1].$$

Solution. Let $V = [x, y, z]$ be the required vector. By (6) and Theorem 0.9 w
must have

(7) $$V_1 \cdot V = x - y + 2z = 0,$$
(8) $$V_2 \cdot V = 3x + 2y - z = 0.$$

To solve for x, y, z we multiply (7) by 3 and subtract from (8), getting

(9) $$5y - 7z = 0.$$

An obvious solution of (9) is $y = 7$, $z = 5$, and substituting these in (7) give
$x = -3$. Hence, $V = [-3, 7, 5]$ is orthogonal to both V_1 and V_2. This answer
not unique. Any nonzero scalar multiple of V will serve as well.

Example 3. Find two vectors V_1, V_2 orthogonal to each other and both of whic
are orthogonal to $V = [2, 4, -3]$.

Solution. We first find $V_1 = [x, y, z]$ orthogonal to V by solving

$$2x + 4y - 3z = 0.$$

Two of the unknowns can be chosen arbitrarily and the third computed. An obv
ous solution is $V_1 = [1, 1, 2]$. We must now choose $V_2 = [x, y, z]$ orthogonal t
both V and V_1 so x, y, z must satisfy

$$x + y + 2z = 0,$$
$$2x + 4y - 3z = 0.$$

A solution, found as in Example 2, is $V_2 = [-11, 7, 2]$.

Definition. *A vector whose length is 1 is called a* unit vector.

If $V = [x, y, z]$ and k is any scalar,

$$\|kV\| = (k^2x^2 + k^2y^2 + k^2z^2)^{1/2} = |k|\, (x^2 + y^2 + z^2)^{1/2} = |k|\, \|V\|.$$

Hence, if $V \neq O$ and we choose k so that $|k| = \dfrac{1}{\|V\|}$, then kV is a unit vecto.
There are exactly two values of k, namely, $k = \pm \dfrac{1}{\|V\|}$ such that kV is a unit vecto

Example 4. Find a unit vector which is a scalar multiple of $V = [2, -1, 3]$.

Solution. Since $\|V\| = \sqrt{4 + 1 + 9} = \sqrt{14}$, it follows that

$$\frac{1}{\sqrt{14}} V = \left[\frac{2}{\sqrt{14}}, -\frac{1}{\sqrt{14}}, \frac{3}{\sqrt{14}} \right]$$

is a unit vector. The second solution is $-\dfrac{1}{\sqrt{14}} V$.

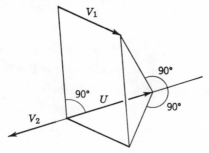

Figure 0.12

The (orthogonal) projection U of a vector V_1 on a vector V_2 is a vector whose initial and terminal points are the projections of the initial and terminal points of V_1 onto the line of V_2 (Figure 0.12). The length of the projection U is clearly

$$\|V_1\| \,|\cos \theta| = \|V_1\| \frac{|V_1 \cdot V_2|}{\|V_1\| \,\|V_2\|} = \frac{|V_1 \cdot V_2|}{\|V_2\|},$$

where θ is the angle between V_1 and V_2. The direction of U is the same as that of V_2 if $\theta < 90°$ but opposite to that of V_2 if $\theta > 90°$. If $\theta < 90°$, $V_1 \cdot V_2 > 0$ and if $\theta > 90°$, $V_1 \cdot V_2 < 0$. Hence the projection U is found by multiplying the unit vector in the direction of V_2 by the scalar $\dfrac{V_1 \cdot V_2}{\|V_2\|}$, and we have

(10) $$U = \left(\frac{V_1 \cdot V_2}{\|V_2\|^2}\right) V_2.$$

Example 5. Find the projection of the vector $[2, -1, 4]$ on $[3, 3, -1]$.

Solution. By (10) the projection is the vector

$$\left(\frac{6 - 3 - 4}{9 + 9 + 1}\right)[3, 3, -1] = \left[-\frac{3}{19}, -\frac{3}{19}, \frac{1}{19}\right].$$

Exercise 0.3

1. Find unit vectors having the same direction as:
   (a) $[2, 6, -3]$.   (b) $[2, 2, 1]$.   (c) $[4, 1, -3]$.

2. Find unit vectors having the opposite direction to each of the three vectors in Problem 1.

3. Find three unit vectors each of which is orthogonal to $[5, 2, -3]$.

4. Find two mutually orthogonal vectors each of which is orthogonal to $[7, 3, -2]$.

5. Find two unit vectors orthogonal to both the vectors $[1, 5, -2]$ and $[2, 3, 1]$.

6. Given the points $A(1, 3, 5)$, $B(-2, 1, 3)$, $C(4, 1, 2)$, find the cosine of the angle ABC. (*Hint.* This is the angle between the vectors $\overrightarrow{BA}$ and $\overrightarrow{BC}$.)

7. Find the cosines of the angles BAC and ACB in Problem 6.

8. Find a vector that is perpendicular to the plane of the triangle ABC in Problem 6. (*Hint.* A vector is perpendicular to a plane if it is perpendicular to two nonparallel vectors in the plane.)

9. Find the projection of:
   (a) The vector $[2, 1, -6]$ on the vector $[1, 1, 1]$.
   (b) The vector $[4, 1, 1]$ on the vector $[4, -2, 4]$.
   (c) The vector $[7, 2, 4]$ on each of the three vectors $[1, 0, 0]$, $[0, 1, 0]$, and $[0, 0, 1]$.

10. If V_1 and V_2 are any two vectors with the same initial point and k_1, k_2 are any two scalars, prove that $k_1V_1 + k_2V_2$ lies in the same plane as V_1 and V_2.

11. If V_1, V_2 are nonzero vectors with the same initial point which do not lie in the same straight line, prove that every vector V in the same plane as V_1 and V_2 can be written in the form $V = k_1V_1 + k_2V_2$ and that the scalars k_1, k_2 are uniquely determined by V.

12. If a vector is orthogonal to each of the vectors V_1 and V_2, show that it is orthogonal to $k_1V_1 + k_2V_2$ for any scalars k_1, k_2. Prove this by using Theorem 0.8 and interpret the result geometrically.

13. If V_1, V_2, V_3 are any three nonzero vectors with common initial point O which do not lie in one plane, prove that every vector $V = \overrightarrow{OP}$ can be expressed in the form

$$V = k_1V_1 + k_2V_2 + k_3V_3$$

where k_1, k_2, and k_3 are scalars.

14. Prove that if the diagonals of a parallelogram are perpendicular to each other, the sides of the parallelogram are equal. (*Hint.* Represent adjacent sides of the parallelogram by vectors V_1 and V_2 with common initial point. The two diagonals are then $V_1 + V_2$ and $V_1 - V_2$.)

0.8 Equation of a Plane

If V_1, V_2 are two vectors with common initial point O which do not lie in the same straight line, then every vector V in the plane of V_1 and V_2 can be written in the form

$$V = k_1V_1 + k_2V_2.$$

This is because V is equal to a vector, with initial point O, which is the diagonal of a parallelogram with adjacent edges lying in the lines of V_1 and V_2 (see Figure 0.13). Hence, V is the sum of scalar multiples of V_1 and V_2.

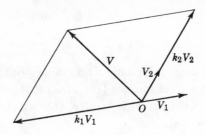

Figure 0.13

Now if a vector N is orthogonal to both V_1 and V_2 we have by Theorem 0.8

$$N \cdot V = N \cdot (k_1 V_1 + k_2 V_2) = k_1 (N \cdot V_1) + k_2 (N \cdot V_2) = 0,$$

and hence N is orthogonal to V. If we are given any two vectors V_1 and V_2 we can always find a vector N orthogonal to both of them by the method of Example 2, Section 0.7. (This fact will be proved in general in Theorem 1.1 of Chapter 1.) Hence we can always find a nonzero vector that is orthogonal to every vector in a given plane. Such a vector is said to be orthogonal (or perpendicular) to the given plane.

Definition. *A nonzero vector, or a line, that is orthogonal to a plane is called a* normal *to the plane.*

Now suppose that a plane has a normal vector $N = [a, b, c]$ and passes through the point $P_0(x_0, y_0, z_0)$. The point $P(x, y, z)$ then lies in the plane if and only if the vector

$$\overrightarrow{P_0 P} = [x - x_0, y - y_0, z - z_0]$$

is orthogonal to N and, therefore, if and only if

(11) $$a(x - x_0) + b(y - y_0) + c(z - z_0) = 0.$$

Therefore (11) is an equation of the plane. It can also be written in the form

$$ax + by + cz = ax_0 + by_0 + cz_0.$$

Conversely, suppose

(12) $$ax + by + cz = d$$

is any linear equation in x, y, z so that a, b, c are not all zero. We can certainly choose values x_1, y_1, z_1 for x, y, z which satisfy (12) so that

(13) $$ax_1 + by_1 + cz_1 = d.$$

Then x, y, z satisfy (12) if and only if they satisfy the equation

(14) $$a(x - x_1) + b(y - y_1) + c(z - z_1) = 0$$

obtained by subtracting (13) from (12). In other words (12) and (14) have the same graph. Since we know the graph of (14) is the plane through (x_1, y_1, z_1) with normal $[a, b, c]$, we have proved that the graph of (12) is a plane with normal $[a, b, c]$. We state our results formally.

Theorem 0.10. (a) *An equation of the plane through the point (x_0, y_0, z_0) with normal $[a, b, c]$ is*

$$a(x - x_0) + b(y - y_0) + c(z - z_0) = 0$$

(b) *The graph of the first-degree equation*

$$ax + by + cz = d$$

is a plane perpendicular to the vector $[a, b, c]$.

Example 1. Find an equation of the plane through the point $A(2, -1, 6)$ which is perpendicular to the line joining A to the point $B(7, 2, 3)$.

Solution. Since $\overrightarrow{AB} = [5, 3, -3]$ is normal to the plane the required equation is

$$5(x - 2) + 3(y + 1) - 3(z - 6) = 0$$

or

$$5x + 3y - 3z + 11 = 0$$

Example 2. Find an equation of the plane through the three points $A(1, 1, 2)$, $B(5, 0, 3)$, and $C(4, 1, -2)$.

Solution. The vectors $\overrightarrow{AB} = [4, -1, 1]$ and $\overrightarrow{AC} = [3, 0, -4]$ both lie in the plane. Hence the normal $[a, b, c]$ is orthogonal to both these vectors and

$$4a - b + c = 0,$$
$$3a - 4c = 0.$$

The second equation has solution $a = 4$, $c = 3$ and the first then yields $b = 19$. Therefore an equation of the plane is

$$4(x - 1) + 19(y - 1) + 3(z - 2) = 0$$

or

$$4x + 19y + 3z - 29 = 0.$$

Two planes are parallel if and only if their normal vectors are parallel. Thus the planes

(15)
$$a_1 x + b_1 y + c_1 z = d_1,$$
$$a_2 x + b_2 y + c_2 z = d_2$$

are parallel if and only if there exists a scalar k such that $[a_2, b_2, c_2] = k[a_1, b_1, c_1]$ or $a_2 = ka_1, b_2 = kb_1, c_2 = kc_1$. Since k is necessarily nonzero, equations of parallel planes can always be found that differ only in the constant term. For example, the second equation in (15) is equivalent to

$$a_1 x + b_1 y + c_1 z = \frac{d_2}{k}.$$

Similarly, two planes are perpendicular if and only if their normals are perpendicular to each other. Thus the planes (15) are perpendicular if and only if

$$a_1 a_2 + b_1 b_2 + c_1 c_2 = 0.$$

Example 3. Find an equation of a plane parallel to the plane $x - 2y + 6z = 4$ and containing the point $(4, 1, 2)$.

Solution. Since the normal to the plane is $[1, -2, 6]$, its equation is

$$x - 2y + 6z = 4 - 2(1) + 6(2)$$

or

$$x - 2y + 6z = 14.$$

Example 4. Find the point of intersection of the three planes

$$x + 2y - z = 6,$$
$$2x - y + z = 4,$$
$$x + 3y - 6z = 2.$$

Solution. It is only necessary to find values of x, y, z which satisfy all three equations. By multiplying the first equation by 2, subtracting from the second, and then subtracting the first equation from the third, we get

$$-5y + 3z = -8,$$
$$y - 5z = -4.$$

Eliminating y we get $-22z = -28$ or $z = \frac{14}{11}$, and hence $y = \frac{70}{11} - 4 = \frac{26}{11}$ and, finally, from the first equation, $x = 6 - \frac{52}{11} + \frac{14}{11} = \frac{28}{11}$. Therefore the point of intersection is $\left(\frac{28}{11}, \frac{26}{11}, \frac{14}{11}\right)$.

0.9 Equations of a Line

A straight line is determined by a point on the line and a vector parallel to the line. Suppose a line l is parallel to the vector $V = [a, b, c]$ and passes through the point $P_0(x_0, y_0, z_0)$. Then a point $P(x, y, z)$ other than P_0 is on the line if and only if the vector $\overrightarrow{P_0P} = [x - x_0, y - y_0, z - z_0]$ is parallel to V. By Theorem 0.4 these vectors are parallel if and only if $\overrightarrow{P_0P} = tV$ for some scalar t, or

$$[x - x_0, y - y_0, z - z_0] = [at, bt, ct].$$

Hence, $P(x, y, z)$ is on the line l if and only if

$$x = x_0 + at,$$

(16) $$y = y_0 + bt,$$

$$z = z_0 + ct.$$

Equations (16) are called parametric equations of the line l, and t is called a parameter. Every real value of t, when substituted in (16), yields a point (x, y, z) on the line l. Conversely, every point of l has the form $(x_0 + at, y_0 + bt, z_0 + ct)$ for some value of t. Note that the parametric equations of a line are not unique. Using a different point P_1 instead of P_0 or a scalar multiple of V instead of V would give a different set of equations for the line l.

Example 1. Find parametric equations of the line through $A(9, 2, 4)$ and $B(6, -1, 7)$ and find the point in which this line cuts the plane $x + y + 3z = 6$.

Solution. The line is parallel to the vector $\overrightarrow{AB} = [-3, -3, 3]$ and hence also to the vector $[1, 1, -1]$ which is a scalar multiple of $\overrightarrow{AB}$. Therefore parametric equations are

$$x = 9 + t,$$

(17) $$y = 2 + t,$$

$$z = 4 - t.$$

If the point $(9 + t, 2 + t, 4 - t)$ is on the plane $x + y + 3z = 6$, we have

$$9 + t + 2 + t + 12 - 3t = 6$$

whence $t = 17$ and, by substituting $t = 17$ in (17), the point common to the line and the plane is found to be $(26, 19, -13)$.

Example 2. Prove that the line (17) is parallel to the plane $x + 2y + 3z = 10$.

Solution. If the point $(9 + t, 2 + t, 4 - t)$ is on the plane, then

$$9 + t + 2(2 + t) + 3(4 - t) = 10,$$

which simplifies to $25 = 10$, a contradiction. Hence, no point of the line (17) lies in the given plane; the line and plane do not intersect and, therefore, are parallel.

A line can be specified not only by two points on it or by one point and a parallel vector, but also as the intersection of two nonparallel planes. Thus the line of intersection of the two planes

(18) $$x - 2y + 6z = 4,$$

(19) $$2x + y - 3z = 2$$

consists of all points (x, y, z) whose coordinates satisfy *both* equations (18) and (19). These two equations are called *equations of the line*. Note the plural. The graph of each individual equation is a plane, but the simultaneous solutions of the two equations, give the coordinates of the points which lie on the line of inter-section of the planes. Equations of a line in this sense are by no means unique. Any pair of planes whose intersection is the given line would supply a pair of equations for the line.

Example 3. Find a vector parallel to the line

$$x - 2y + 6z = 4,$$

$$2x + y - 3z = 2.$$

Solution. Let $V = [a, b, c]$ be a vector lying in the given line. Then V lies in both planes (18) and (19) and, therefore, is orthogonal to the normal to each plane. Hence,

$$a - 2b + 6c = 0,$$

$$2a + b - 3c = 0.$$

We eliminate a by multiplying the first equation by 2 and subtracting from the second to get

$$5b - 15c = 0,$$

a solution of which is $b = 3$, $c = 1$. Substitution now gives $a = 0$. Therefore the line is parallel to $[0, 3, 1]$. Parametric equations for the line could be found by choosing an arbitrary point (x_0, y_0, z_0) whose coordinates satisfy (18) and (19) and then using (16).

If $V_1 = [a_1, b_1, c_1]$ and $V_2 = [a_2, b_2, c_2]$ are any two vectors, it is easy to check that the vector

(20) $$[b_1c_2 - b_2c_1, c_1a_2 - c_2a_1, a_1b_2 - a_2b_1]$$

is orthogonal to both V_1 and V_2. The vector (20) is called the *cross product* of V_1 and V_2 and is denoted by $V_1 \times V_2$. The expression $b_1c_2 - b_2c_1$ is called a *determinant* of order 2 and is denoted by

$$\begin{vmatrix} b_1 & b_2 \\ c_1 & c_2 \end{vmatrix} = b_1c_2 - b_2c_1.$$

Determinants are discussed systematically in Chapter 3. They are introduced here simply as a device for writing the vector (20). Note that if

$$V_1 = [a_1, b_1, c_1],$$

$$V_2 = [a_2, b_2, c_2],$$

then the coordinates of $V_1 \times V_2$ are

$$\begin{vmatrix} b_1 & c_1 \\ b_2 & c_2 \end{vmatrix}, \quad -\begin{vmatrix} a_1 & c_1 \\ a_2 & c_2 \end{vmatrix}, \quad \begin{vmatrix} a_1 & b_1 \\ a_2 & b_2 \end{vmatrix}.$$

Example 4. Find a nonzero vector that is orthogonal to both $V_1 = [2, 3, 5]$ and $V_2 = [4, -1, 2]$.

Solution. Write the coordinates of the two vectors in this way:

$$\begin{matrix} 2 & 3 & 5 \\ 4 & -1 & 2. \end{matrix}$$

The coordinates of $V_1 \times V_2$ are

$$\begin{vmatrix} 3 & 5 \\ -1 & 2 \end{vmatrix} = 11, \quad -\begin{vmatrix} 2 & 5 \\ 4 & 2 \end{vmatrix} = 16, \quad \begin{vmatrix} 2 & 3 \\ 4 & -1 \end{vmatrix} = -14$$

Hence the vector $V_1 \times V_2 = [11, 16, -14]$ is orthogonal to both V_1 and V_2.

Exercise 0.4

1. Find equations for the planes satisfying the following conditions:
   (a) Passing through the point $(2, 9, -1)$ and having normal vector $[3, -1, 5]$.
   (b) Passing through the point $(1, 5, 2)$ and perpendicular to the line joining $A(1, 4, -3)$ and $B(-2, 5, 1)$.
   (c) Passing through the point $(2, 3, 4)$ and perpendicular to the y-axis.

(d) Passing through the three points $A(1, -2, 4)$, $B(5, 1, 6)$, and $C(6, 3, 2)$.

(e) Passing through the three points $(1, 0, 0)$, $(0, 1, 0)$, and $(0, 0, 1)$.

(f) Containing the x-axis and the point $(2, 1, 5)$.

(g) Containing the y-axis and the point $(4, -1, 2)$.

(h) Containing the point $(1, 5, 7)$ and the line

$$x = 2 + t,$$
$$y = 3 - 2t,$$
$$z = 5t.$$

(i) Containing the point $(3, 1, -2)$ and perpendicular to the line

$$x - 2y + 6z = 2,$$
$$3x + y - 2z = 4.$$

(j) Containing the point $(2, 0, 5)$ and the line of intersection of the two planes

$$x + y - z = 6,$$
$$2x + 5y - 3z = 1.$$

2. Find parametric equations of:

(a) The line through $(1, 5, 0)$ parallel to $[2, -7, 4]$.

(b) The line through the two points $(1, 0, 9)$ and $(-2, 4, 4)$.

(c) The line through the point $(1, 7, 6)$ parallel to the line through $(1, 2, 0)$ and $(5, -1, 3)$.

(d) The line through the point $(4, 1, -2)$ and normal to the plane $2x - y + 3z = 6$.

(e) The line through the point $(0, 5, 6)$ and parallel to the line

$$x = 2 + 4t,$$
$$y = -2t,$$
$$z = 3 + 5t.$$

(f) The line through the point $(7, -1, 4)$ and parallel to the line

$$x - y + 2z = 1,$$
$$2x + 3y - z = 6.$$

(g) The line through the point $(4, 2, -1)$ and parallel to each of the planes $2x + y + z = 0$ and $x - 5y + 2z = 9$.

3. Find the points in which the line through $A(2, 1, 7)$ and $B(3, -2, 4)$ cuts:

(a) The plane $2x + y + z = 18$.　　　　(b) Each of the coordinate planes.

4. Find the point of intersection of the three planes

$$x - 2y + z = 4,$$
$$3x + y - 2z = 1,$$
$$4x - y + 3z = 6.$$

5. Show that the three planes

$$2x - 3y - z = 1,$$
$$x + 2y + 3z = 0,$$
$$5x - y + 4z = 7,$$

intersect in a line. (*Hint*. Show that the line of intersection of the first two planes lies in the third plane.)

6. Show that the three planes

$$x + y - 2z = 2,$$
$$2x - 2y + 3z = 1,$$
$$3x - y + z = 7$$

do not intersect. (*Hint*. Either show that the three equations have no common solution or that the line of intersection of the first two planes is parallel to but not contained in the third plane.)

7. Find a vector parallel to the lines:

(a) $2x - y + 4z = 1,$ (b) $x - 6y + 5z = 1,$
    $x + 5y + z = 6.$ $2x + y + z = 4.$

8. Prove that:
   (a) The plane $ay + bz = c$ is parallel to the x-axis.
   (b) The plane $ax + bz = c$ is parallel to the y-axis.
   (c) The plane $ax + by = c$ is parallel to the z-axis.
   (d) The plane $ax = c$ is parallel to both the y- and z-axes.

9. Find the length of the perpendicular from the point $P(2, -1, 6)$ to the plane

$$2x + 6y - 3z = 28.$$

(*Hint*. Choose any point Q on the plane, for example $(14, 0, 0)$, and find the length of the projection of $\overrightarrow{PQ}$ onto a vector normal to the plane.

10. Find the perpendicular distance from $(3, 4, -2)$ to the plane $x + y + 2z = 10$.

11. Find a formula for the length of the perpendicular from the point $P(x_1, y_1, z_1)$ to the plane $ax + by + cz = d$.

12. If U, V, and W are any three nonzero vectors not all parallel to one plane, prove that:
    (a) $U \times V = -(V \times U)$.
    (b) $U \times (V + W) = (U \times V) + (U \times W)$.
    (c) For any scalar k, $V \times (kV) = 0$.
    (d) $U \times (V \times W)$ is in (or parallel to) the plane of V and W.
    (e) $(U \times V) \times W$ is in (or parallel to) the plane of U and V.

0.10 Surfaces in Three-Dimensional Space

It will be assumed that the reader is familiar with the relationship in two-dimensional analytic geometry between an equation and its graph. The graph of an equation $f(x, y) = 0$ is the set of all points, and only those points, whose coordinates (x, y) satisfy this equation. The graph may take various forms, some of which are illustrated by the following examples.

(a) The equation $x^2 + y^2 = -7$ is not satisfied by any pair of real numbers x and y. Therefore there are no points on its graph.

(b) The graph of the equation $(x - 1)^2 + (y - 2)^2 = 0$ is the single point $(1, 2)$.

(c) The graph of the equation

$$(x^2 + y^2)[x^2 + (y - 1)^2] = 0$$

consists of the two isolated points $(0, 0)$ and $(0, 1)$.

(d) The graph of the equation $x^2 + \sin^2 y = 0$ consists of an infinite number of isolated points $(0, n\pi)$, where n is any integer.

(e) The graph of the equation $(x - 1)^2 + (y + 2)^2 = 16$ is a circle with center at $(1, -2)$ and radius 4.

Although the first four of these examples illustrate interesting possibilities, the student is no doubt aware that they are not typical of the graphs studied in analytic geometry. Normally the graph of an equation $f(x, y) = 0$ is a *curve* as is the case in (e). The term curve here is to be interpreted broadly and includes a straight line as a special case. The equations in (a) through (d) may be given a somewhat different interpretation by considering "points" whose coordinates are complex numbers. From this point of view, by analogy with the equation $x^2 + y^2 = 7$, the equation $x^2 + y^2 = -7$ is said to represent an *imaginary circle*, since all the "points" whose coordinates satisfy the equation are "imaginary points" with one or more imaginary coordinates. Similarly the equation $(x - 1)^2 + (y - 2)^2 = 0$ in (b) may be written

$$y - 2 = \pm\sqrt{-1}\,(x - 1)$$

and is said to represent two imaginary lines intersecting in the real point $(1, 2)$. Although terminology of this kind is useful and revealing and may be used from time to time, the graph of an equation for most purposes consists only of the real points whose coordinates satisfy it.

A similar situation exists in space. The set of all points (x, y, z) whose coordinates satisfy an equation of the form

(21) $$F(x, y, z) = 0$$

constitutes the *graph* of this equation. This graph is normally, though subject to exceptions analogous to those in examples (a) through (d), a *surface*. This surface may be a plane, as we have seen is the case when equation (21) is linear in x, y, z, or it may be a curved surface. The term surface, like the term curve in plane geometry, will not be defined precisely. The following examples will illustrate its meaning.

Example 1. From equation (1) it is known that the square of the distance from the origin to the point (x, y, z) is $x^2 + y^2 + z^2$. Hence, all the points whose coordinates satisfy the equation

$$x^2 + y^2 + z^2 = 16$$

are 4 units distant from $(0, 0, 0)$ and, therefore, lie on the surface of a sphere with center at the origin and radius 4. Conversely the coordinates of every point on the sphere satisfy this equation. Similarly,

$$(x - 2)^2 + (y - 3)^2 + (z + 6)^2 = 20$$

is an equation of a sphere with center at $(2, 3, -6)$ and radius $\sqrt{20}$.

Example 2. What is the graph *in three-dimensional space* of the equation $9x^2 + 4y^2 = 36$?

It is known that *in the plane* this equation represents an ellipse with semiaxes 2 and 3. Hence, if we draw in the xy-plane the ellipse whose equation in the plane is $9x^2 + 4y^2 = 36$, the points $(x, y, 0)$ on this ellipse certainly satisfy the equation and form part of the three-dimensional graph. However, a point (x, y, k) satisfies the equation if and only if $(x, y, 0)$ does. Moreover the points (x, y, k), k arbitrary, are precisely the points of the line through $(x, y, 0)$ perpendicular to the xy-plane. The surface, a segment of which is shown in Figure 0.14, is therefore composed of all lines perpendicular to the xy-plane that intersect that plane in a point of the ellipse. Such a surface is called an *elliptic cylinder*.

Example 3. A point moves so that the line joining it to the origin makes a constant angle θ with the positive ray of the z-axis. Find an equation of the surface so generated.

Solution. Let $P(x, y, z)$ be any point on the surface. Draw a line PQ perpendicular to the z-axis to meet either OZ or OZ' in Q (see Figure 0.15). The coordinates of Q are then $(0, 0, z)$ and $PQ^2 = x^2 + y^2$. Since $OQ^2 = z^2$ and since P is on the surface

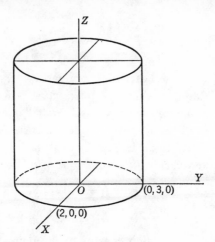

Figure 0.14

if and only if $QP^2 = OQ^2 \tan^2 \theta$, we have

$$x^2 + y^2 = z^2 \tan^2 \theta,$$

which is an equation of the surface. The surface is a cone, with vertex at the origin and axis of symmetry along the z-axis, which extends to infinity both above and below the xy-plane. It may be thought of as the surface generated by revolving the line OP about the z-axis.

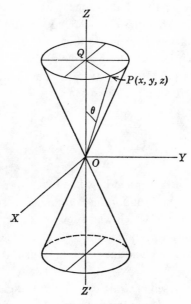

Figure 0.15

0.11 Curves in Space

Two surfaces that intersect each other ordinarily intersect in a curve, although there are exceptional cases in which the intersection is itself a surface or in which it reduces to one or more isolated points. We shall assume, unless otherwise stated, that the two surfaces considered intersect in a curve. This curve may lie in a plane, in which case it is called a *plane curve*, or it may not, in which case it is called a *space curve* or skew curve. If the equations of the two intersecting surfaces are

(22)
$$f_1(x, y, z) = 0,$$
$$f_2(x, y, z) = 0,$$

then the curve of intersection consists of the points (x, y, z), whose coordinates satisfy *both* equations (22). A curve in space, then, is defined by two equations in x, y, z, and the curve consists of those points whose coordinates are simultaneous solutions of these equations. The equations of a curve are not uniquely determined by the curve. For example, if k_1 and k_2 are real numbers different from zero, either of equations (22) can be replaced by

$$k_1 f_1(x, y, z) + k_2 f_2(x, y, z) = 0,$$

and the resulting pair of equations will have the same simultaneous solutions as (22) and therefore represent the same curve. If equations (22) are linear, our discussion reduces to that of Section 0.9 in which the equations of a line are defined as the equations of any two planes that intersect in the given line. It should be mentioned also that it is usually possible to define a curve in space by parametric equations analogous to (16), in which the coordinates of an arbitrary point on the curve are given as functions of a parameter.

0.12 Cylinders

Many surfaces can be generated by a line that moves according to some prescribed condition. Of particular importance among the surfaces so generated are the cylinders, an example of which was discussed in Section 0.10.

A *cylinder* is a surface generated by a straight line that moves so that it remains perpendicular to a fixed plane and its point of intersection with this plane generates a curve.

A cylinder is composed of straight lines perpendicular to a fixed plane and cutting this plane in a curve. These lines are called *generating lines* or *generators* of the cylinder. All generating lines of a cylinder are parallel to each other, since they are perpendicular to a fixed plane. All plane sections perpendicular to the generating lines are congruent. Suppose that the generating lines of a cylinder are parallel to the z-axis and the equations of the curve in which the cylinder cuts the xy-plane are

$$f(x, y) = 0,$$

$$z = 0.$$

Then, for the reason described in Example 2, Section 0.10, the equation of this cylinder is $f(x, y) = 0$. In general, *if one of the three variables x, y, and z does not occur in the equation of a surface, then the surface is a cylinder with generators parallel to the axis of the missing variable.* For example, the equation $4x^2 + z^2 = 16$ is an elliptic cylinder with generators parallel to the y-axis and cutting the plane $y = k$ in the ellipse

$$4x^2 + z^2 = 16,$$

$$y = k.$$

Also see Example 2, Section 0.10 and Figure 0.14.

Now consider any space curve C defined by equations (22). If through every point of this curve a line is drawn perpendicular to the xy-plane, these lines are generators of a cylinder that cuts the xy-plane in another curve C', called the (orthogonal) projection of C on the xy-plane. The cylinder is called the projecting cylinder of C onto the xy-plane.

Example 1.   Find equations of the projections of the curve

(23)
$$x^2 + y^2 + z^2 = 25,$$
$$x + 2y - z = 0$$

onto the coordinate planes.

Solution.   The first equation is that of a sphere with center at the origin and radius 5. The second is an equation of a plane through the origin. The given curve is a circle with center at the origin and lying in the plane $x + 2y - z = 0$.

Solving the second equation of (23) for z and substituting in the first, we find

$$x^2 + y^2 + (x + 2y)^2 = 25$$

or

(24)
$$2x^2 + 4xy + 5y^2 = 25.$$

Equation (24) is satisfied by all simultaneous solutions of (23) and hence is the equation of a surface containing the curve (23). Since every simultaneous solution of

(25)
$$2x^2 + 4xy + 5y^2 = 25,$$
$$x + 2y - z = 0$$

also satisfies $x^2 + y^2 + z^2 = 25$, it follows that equations (25) define the same curve as equations (23). Since (24) does not contain z, it is the equation of a cylinder with generators parallel to the z-axis and is therefore the projecting cylinder of the curve (23) onto the xy-plane. This cylinder cuts the xy-plane in the curve

$$2x^2 + 4xy + 5y^2 = 25,$$
$$z = 0,$$

which is therefore the projection of the given curve onto the xy-plane.

By similar methods the projections on the yz- and xz-planes are found to be

$$5y^2 - 4yz + 2z^2 = 25,$$
$$x = 0$$

and

$$5x^2 - 2xz + 5z^2 = 100,$$
$$y = 0.$$

All three projections are ellipses.

This example illustrates the general principle that an equation of the projecting cylinder of a curve with generators parallel to the x-, y-, or z-axis can normally be found by eliminating x, y, or z, respectively, from the equations of the curve.

Example 2. Find equations of the projection on the xy-plane of the line

$$x - 2y + z = 6,$$
$$3x - y + 2x = 2.$$

Solution. The equation

$$2(x - 2y + z - 6) - (3x - y + 2z - 2) = 0,$$

which reduces to

$$x + 3y + 10 = 0,$$

is the equation of a plane through the given line parallel to the z-axis. Hence the projection in the xy-plane is the line

$$x + 3y + 10 = 0,$$
$$z = 0.$$

0.13 Cones

A surface is called a *cone* if it is generated by straight lines all of which intersect a given plane curve and pass through a fixed point not in the plane of this curve. See Example 3 and Figure 0.15 in Section 0.10.

The generating lines are called *generators* of the cone, and the fixed point through which all the generators pass is called the *vertex* of the cone.

Example. Find an equation of the cone with vertex at $V(0, 0, 4)$ that cuts the xy-plane in the ellipse

$$4x^2 + y^2 = 4,$$

$$z = 0.$$

Solution. Let $P(x, y, z)$ be any point on the cone other than the vertex and let $Q(a, b, 0)$ be the point in which the line PV cuts the xy-plane. Since P, Q, and V are collinear, $\overrightarrow{VQ} = k\overrightarrow{VP}$ or

(26) $$[a, b, -4] = k[x, y, z - 4].$$

Since Q is on the given ellipse, we have from (26)

$$4 = 4a^2 + b^2 = 4k^2x^2 + k^2y^2 = k^2(4x^2 + y^2)$$

and

$$-4 = k(z - 4)$$

and, therefore,

$$16 = 4k^2(4x^2 + y^2) = k^2(z - 4)^2$$

or

$$16x^2 + 4y^2 = (z - 4)^2,$$

which is an equation of the cone, since it is also satisfied by the coordinates of V.

0.14 Surfaces of Revolution

Let $x = f(y)$, $z = 0$ be equations of a curve in the xy-plane. If this curve is rotated about the y-axis, it generates a surface with circular cross sections in any plane perpendicular to the y-axis. Such a surface is called a *surface of revolution*. To find its equation we refer to Figure 0.16 and note that if $P(x, y, z)$ is on the surface, a plane through P perpendicular to the y-axis cuts the y-axis in $Q(0, y, 0)$ and cuts

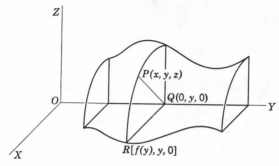

Figure 0.16

the given curve in $R(f(y), y, 0)$. Since $QP^2 = QR^2$, we have

$$x^2 + z^2 = [f(y)]^2,$$

which is an equation of the surface.

Example. Find an equation of the surface generated by rotating the parabola $y^2 = 4x$ about the x-axis.

Solution. Let $P(x, y, z)$ be any point on the surface. Then $Q(x, 0, 0)$ is the point at which the plane through P parallel to the yz-plane cuts the x-axis. The circular cross section of the surface in this plane contains the point $R(x, 2\sqrt{x}, 0)$ and has center Q. Since $QP^2 = QR^2$ we get

$$y^2 + z^2 = 4x,$$

which is an equation of the surface.

Exercise 0.5

1. Find equations of the following surfaces.
   (a) The cone generated by rotating the line $y = 2x$, $z = 0$ about the x-axis.
   (b) The cylinder with generators parallel to the y-axis that cuts the xz-plane in a circle with radius 3 and center at $(2, 0, -1)$.
   (c) The surface generated by revolving the ellipse $4x^2 + y^2 = 16$, $z = 0$ about the y-axis.
   (d) The surface generated by revolving the ellipse in (c), about the x-axis.
   (e) The cylinder with generators parallel to the x-axis that cuts the yz-plane in the parabola $z^2 = 8y$, $x = 0$.
   (f) The plane perpendicular to the xy-plane that cuts the xy-plane in the line $2x - y = 6$, $z = 0$.

2. Find equations of the projections onto the xy-plane of the curves of intersection of the following surfaces.
   (a) $y^2 + z^2 = 4$, $x + y + z = 2$.
   (b) $4x^2 + y^2 = 4z^2$, $x^2 + y^2 = 4z$.
   (c) $x^2 + y^2 + z^2 = 4$, $x + y + z = 2$.
   (d) $2x - y + z = 1$, $x + 5y + 2z = 6$.

3. Find an equation of the circular cone with vertex at the origin and axis along the vector $V = [6, -2, 3]$ if the angle between V and the generators is equal to $30°$.

   (*Hint*. The point $P(x, y, z)$ is on the cone if and only if the angle between V and $\overrightarrow{OP}$ is either $30°$ or $150°$.)

0.15 The Quadric Surfaces

The simplest curved surfaces are the *quadric surfaces*, whose equations are of the second degree in x, y, and z. In addition to the quadric cylinders and cones, examples of which were given in Examples 2 and 3, Section 0.10, there are five nondegenerate real quadric surfaces: the ellipsoid, two hyperboloids, and two paraboloids. We shall describe these surfaces by deriving their principal features from certain standard second-degree equations. A general classification of all surfaces that can be represented by a second-degree equation in x, y, and z will be found in Chapter 7. The method used here will be to sketch the surface from its equation by finding the curves in which it cuts the coordinate planes and other planes parallel to them. The curve of intersection of a surface and a plane is sometimes called the *trace* of the surface in the plane.

The Ellipsoid

Consider the surface whose equation is

$$\frac{x^2}{a^2} + \frac{y^2}{b^2} + \frac{z^2}{c^2} = 1.$$

In this equation and in what follows, a, b, and c are assumed to be positive. Directly from its equation we can draw several conclusions about this surface. It is symmetric with respect to each of the three coordinate planes, since if (x, y, z) is on the surface, so also are $(-x, y, z)$, $(x, -y, z)$, and $(x, y, -z)$. If (x, y, z) is a point on the surface, then $|x| \leq a$, $|y| \leq b$, and $|z| \leq c$. Therefore the surface is bounded, lying inside the box bounded by the planes $x = \pm a$, $y = \pm b$, and $z = \pm c$ and touching the faces of this box at the points $(\pm a, 0, 0)$, $(0, \pm b, 0)$, and $(0, 0, \pm c)$. The trace of the surface in the xy-plane is the ellipse

$$\frac{x^2}{a^2} + \frac{y^2}{b^2} = 1,$$
$$z = 0.$$

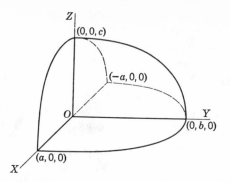

Figure 0.17. Ellipsoid.

If $|k| < c$, the trace in the plane $z = k$ is the ellipse

$$\frac{x^2}{a^2\left(1 - \frac{k^2}{c^2}\right)} + \frac{y^2}{b^2\left(1 - \frac{k^2}{c^2}\right)} = 1,$$

$$z = k,$$

whose semiaxes decrease as $|k|$ increases. Similarly the traces in the planes $x = k$, where $|k| < a$, and $y = k$, where $|k| < b$, are also ellipses. It is now possible to sketch the surface, a segment of which is shown in Figure 0.17. This surface is called an *ellipsoid*. If two of the numbers a, b, and c are equal, the ellipsoid is an *ellipsoid of revolution*, the surface generated by revolving an ellipse about its major or minor axis. If $a = b = c$, the surface is a sphere with center at the origin.

The Hyperboloid of One Sheet

Consider the surface whose equation is

(27)
$$\frac{x^2}{a^2} + \frac{y^2}{b^2} - \frac{z^2}{c^2} = 1.$$

Its trace in the xy-plane is the ellipse

$$\frac{x^2}{a^2} + \frac{y^2}{b^2} = 1,$$

$$z = 0$$

and, in the plane $z = k$, the ellipse

$$\frac{x^2}{a^2\left(1 + \frac{k^2}{c^2}\right)} + \frac{y^2}{b^2\left(1 + \frac{k^2}{c^2}\right)} = 1,$$

$$z = k.$$

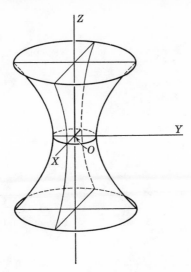

Figure 0.18. Hyperboloid of one sheet.

Therefore the semiaxes of the ellipse in which the surface cuts the plane $z = k$ increase as $|k|$ increases.

The trace in the xz-plane is the hyperbola

$$\frac{x^2}{a^2} - \frac{z^2}{c^2} = 1,$$

$$y = 0,$$

and the trace in the yz-plane is the hyperbola

$$\frac{y^2}{b^2} - \frac{z^2}{c^2} = 1,$$

$$x = 0.$$

A sketch of the surface, which is called a *hyperboloid of one sheet*, is shown in Figure 0.18. If $a = b$, the trace in any plane $z = k$ is a circle and the surface is a hyperboloid of revolution generated by revolving a hyperbola about its conjugate axis.

The Hyperboloid of Two Sheets

The surface whose equation is

$$-\frac{x^2}{a^2} + \frac{y^2}{b^2} - \frac{z^2}{c^2} = 1$$

does not cut the xz-plane nor, indeed, any plane $y = k$, where $|k| < b$. However,

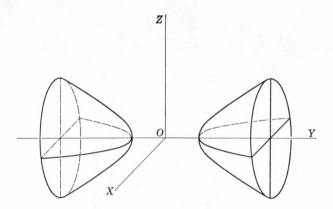

Figure 0.19. Hyperboloid of two sheets.

if $|k| > b$, its trace in the plane $y = k$ is the ellipse

$$\frac{x^2}{a^2} + \frac{z^2}{c^2} = \frac{k^2}{b^2} - 1,$$

$$y = k.$$

In the xy- and yz-planes the traces are, respectively, the hyperbolas

$$\frac{x^2}{a^2} - \frac{y^2}{b^2} = -1, \qquad z = 0$$

and

$$\frac{y^2}{b^2} - \frac{z^2}{c^2} = 1, \qquad x = 0.$$

The surface is symmetrical with respect to each of the coordinate planes and consists of two parts or *sheets*, one in region $y \geq b$ and one in the region $y \leq -b$. A sketch of the surface, a *hyperboloid of two sheets*, is shown in Figure 0.19. If $a = b$, the surface has circular cross sections in planes perpendicular to the y-axis. It is then a hyperboloid of revolution generated by revolving a hyperbola about its transverse axis.

The Elliptic Paraboloid

The equation

$$\frac{x^2}{a^2} + \frac{y^2}{b^2} = cz$$

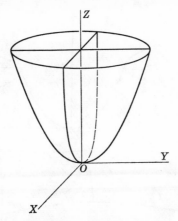

Figure 0.20. Elliptic paraboloid.

represents an *elliptic paraboloid* a sketch of which for the case $c > 0$ is shown in Figure 0.20. If $k > 0$, its trace in the plane $z = k$ is the ellipse

$$\frac{x^2}{cka^2} + \frac{y^2}{ckb^2} = 1,$$
$$z = k.$$

The surface does not extend below the xy-plane. Its traces in the xz- and yz-planes are the parabolas

$$x^2 = a^2cz, \qquad y = 0$$

and

$$y^2 = b^2cz, \qquad x = 0.$$

The surface is symmetric with respect to the xz- and yz-planes. If $a = b$, it is a paraboloid of revolution.

The Hyperbolic Paraboloid

The surface whose equation is

$$\frac{y^2}{b^2} - \frac{x^2}{a^2} = cz$$

is called a *hyperbolic paraboloid*. Its trace in the xz-plane is the parabola $x^2 = -a^2cz$, $y = 0$, which opens downward if $c > 0$ and has the z-axis as axis of symmetry. Its trace in the yz-plane is the parabola $y^2 = b^2cz$, $x = 0$, which opens upward if $c > 0$ and has the z-axis as axis of symmetry. Its trace in the plane $z = k$, $k \neq 0$, is the hyperbola

$$\frac{x^2}{a^2ck} - \frac{y^2}{b^2ck} = -1,$$
$$z = k.$$

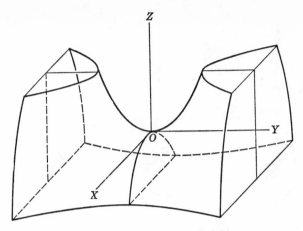

Figure 0.21. Hyperbolic paraboloid.

If $k > 0$, the transverse axis of the hyperbola is parallel to the y-axis, but, if $k < 0$, its transverse axis is parallel to the x-axis. The trace in the plane $z = 0$ is the two straight lines

$$\frac{x^2}{a^2} - \frac{y^2}{b^2} = 0,$$

$$z = 0.$$

A sketch of this surface for the case $c > 0$ is shown in Figure 0.21.

Example. Identify the graph of the equation

$$2x^2 + y^2 - 3z^2 + 6x - 4y -- 12z = 26.$$

Solution. By completing squares the equation can be written

$$2(x^2 + 3x + \tfrac{9}{4}) + (y^2 - 4y + 4) - 3(z^2 + 4z + 4) = 26 + \tfrac{9}{2} + 4 - 12$$

or

$$4(x + \tfrac{3}{2})^2 + 2(y - 2)^2 - 6(z + 2)^2 = 45.$$

We now translate to a new coordinate system with origin at $0'(-\tfrac{3}{2}, 2, -2)$ via the equations of transformation

$$x = x' - \tfrac{3}{2},$$

$$y = y' + 2,$$

$$z = z' - 2.$$

Dividing by 45, the new equation of the surface becomes

$$\frac{4x'^2}{45} + \frac{2y'^2}{45} - \frac{2z'^2}{15} = 1.$$

Comparison with the standard equation (27) shows that the surface is a hyperboloid of one sheet with $a^2 = \frac{45}{4}$, $b^2 = \frac{45}{2}$, $c^2 = \frac{15}{2}$.

Exercise 0.6

1. Identify each of the following surfaces:

   (a) $x^2 - y^2 - z^2 = 1$.
   (b) $x^2 - y^2 - z^2 = 0$.
   (c) $x^2 - y^2 - z^2 = -1$.
   (d) $y^2 - 4z = 0$.
   (e) $3x^2 + 4y^2 = 24$.
   (f) $y^2 - 4z^2 = 0$.

   (g) $y^2 - x^2 = 4z$.
   (h) $3x^2 + 9y^2 + 4z^2 = 18$.
   (i) $4x = y^2 + 16z^2$.
   (j) $x^2 + y^2 + 4z^2 = 4$.
   (k) $x^2 + y^2 + z^2 = 4$.

2. By suitable translation of axes, reduce each of the following equations to a form similar to one of the standard forms discussed in Section 0.15. Identify the surfaces that they represent.

   (a) $x^2 + y^2 + z^2 - 2x + y + 6z = 20$.
   (b) $x^2 + y^2 - 5x + 2y - 8z + 2 = 0$.
   (c) $2x^2 - y^2 + 4z^2 + 6x - 2y + z = 12$.
   (d) $4x^2 + y^2 + 9z^2 + 32x - 6y + 45z = 0$.
   (e) $x^2 - 4z^2 + 12z + 16y = 20$.

3. Describe the graphs in three-dimensional space that are defined by the following equations.

   (a) $x^2 + y^2 + z^2 = -4$.
   (b) $(x - 1)^2 + (y + 2)^2 + (z - 5)^2 = 0$.
   (c) $x^2 + (y - 1)^2 = 0$.
   (d) $[x^2 + y^2 + (z - 1)^2][(x - 1)^2 + (z - 2)^2] = 0$.
   (e) $(x + y + z - 2)^2 + (x^2 + 4y^2 - 4)^2 = 0$.
   (f) $x^2 + y^2 + \sin^2 z = 0$.
   (g) $x^2 + \sin^2 z = 0$.
   (h) $x^2 = y$.
   (i) $x = y$.
   (j) $x^2 = y^2$.

| VECTOR SPACES

1.1 Sets

The word *set* is used in mathematics for a collection of objects. The nature of the individual objects that are the members of the set is not necessarily specified. They may be points, numbers, functions, or other mathematical entities. The only requirement is that there be an unambiguous answer to the question whether a given object is a member of a given set. The members of a set are called *elements* of the set. We shall use script capital letters for sets. The fact that a is an element of the set $\mathscr{A}$ is expressed by the notation $a \in \mathscr{A}$. If every element of a set $\mathscr{A}$ is also an element of the set $\mathscr{B}$ we say $\mathscr{A}$ is a *subset* of $\mathscr{B}$ and write $\mathscr{A} \subset \mathscr{B}$. If $\mathscr{A} \subset \mathscr{B}$ and also $\mathscr{B} \subset \mathscr{A}$, then $\mathscr{A}$ and $\mathscr{B}$ consist of the same elements and we write $\mathscr{A} = \mathscr{B}$. If $\mathscr{A} \subset \mathscr{B}$ but $\mathscr{A} \neq \mathscr{B}$ we say $\mathscr{A}$ is a *proper* subset of $\mathscr{B}$.

We use the standard notation $\mathscr{A} \cap \mathscr{B}$ for the set of all elements that belong to both $\mathscr{A}$ and $\mathscr{B}$ and $\mathscr{A} \cup \mathscr{B}$ for the set of all elements that belong to either $\mathscr{A}$ or $\mathscr{B}$ (or both). The set $\mathscr{A} \cap \mathscr{B}$ is called the *intersection* of $\mathscr{A}$ and $\mathscr{B}$ and $\mathscr{A} \cup \mathscr{B}$ is called the *union* of $\mathscr{A}$ and $\mathscr{B}$. We define an *empty set* ϕ which contains no elements. By convention the empty set ϕ is a subset of every set. It then follows that there is only one empty set, for if there were two, ϕ_1 and ϕ_2, we would have $\phi_1 \subset \phi_2$ and $\phi_2 \subset \phi_1$ and, hence, $\phi_1 = \phi_2$. If $\mathscr{A}$ and $\mathscr{B}$ have no elements in common, then $\mathscr{A} \cap \mathscr{B} = \phi$ and $\mathscr{A}$ and $\mathscr{B}$ are said to be *disjoint*. There is no difficulty in extending the definition of intersection and union to any collection $\{\mathscr{S}_\alpha\}$ of sets where α varies over some set $\mathscr{I}$ of indices. The intersection $\bigcap_{\alpha \in \mathscr{I}} \mathscr{S}_\alpha$ is the set of all elements x such that $x \in \mathscr{S}_\alpha$ for all $\alpha \in \mathscr{I}$ and the union $\bigcup_{\alpha \in \mathscr{I}} \mathscr{S}_\alpha$ is the set of all x such that $x \in \mathscr{S}_\alpha$ for at least one $\alpha \in \mathscr{I}$.

All sets considered are usually assumed to be subsets of some *universal set* $\mathscr{U}$. A set can then be defined by some property which characterizes those elements of $\mathscr{U}$ that belong to it. For example, let $\mathscr{R}$ be the set of all real numbers. If $x \in \mathscr{R}$ the property $0 < x < 2$ determines the subset of $\mathscr{R}$ consisting of those real numbers between 0 and 2. The notation

$$\{x \in \mathscr{R} \mid 0 < x < 2\},$$

which is read "the set of all x in $\mathscr{R}$ such that $0 < x < 2$," is used to denote this set. In general, if P is some property that elements of a set $\mathscr{S}$ may or may not have, the notation

$$\{x \in \mathscr{S} \mid x \text{ has property } P\}$$

is used for the set of all x in $\mathscr{S}$ which have property P. For example, we can write

$$\mathscr{A} \cap \mathscr{B} = \{x \mid x \in \mathscr{A} \text{ and } x \in \mathscr{B}\},$$
$$\mathscr{A} \cup \mathscr{B} = \{x \mid x \in \mathscr{A} \text{ or } x \in \mathscr{B}\}.$$

If $\mathscr{E}_2$ is the set of all points (x, y) in the Cartesian plane, then

$$\{(x, y) \in \mathscr{E}_2 \mid x^2 + y^2 = 1\}$$

is the set of all points (x, y) on the circle whose equation is $x^2 + y^2 = 1$.

1.2 Mappings

Let $\mathscr{S}$ and $\mathscr{T}$ be sets. A *mapping* of $\mathscr{S}$ into $\mathscr{T}$ is a correspondence that assigns to each element s of $\mathscr{S}$ a unique element $f(s)$ in $\mathscr{T}$. The element $f(s)$ is called the *image* of s under f. The set $\mathscr{S}$ is called the *domain* of f (dom f), the set $\mathscr{T}$ is called the *range* of f and the set $\{f(s) \mid s \in \mathscr{S}\}$ is called the image of f, abbreviated im f. Clearly im $f \subset \mathscr{T}$. If im $f = \mathscr{T}$, then f is called a mapping of $\mathscr{S}$ *onto* $\mathscr{T}$ because every element of $\mathscr{T}$ is then the image of at least one element of $\mathscr{S}$. A mapping f with domain $\mathscr{S}$ and range $\mathscr{T}$ is usually denoted by the notation $f \colon \mathscr{S} \to \mathscr{T}$. The terms mapping and function are synonymous. The former is more commonly used in algebra and the latter in analysis, but the meaning is the same.

Example 1. Let $\mathscr{R}$ be the set of all real numbers; let $f \colon \mathscr{R} \to \mathscr{R}$ be the mapping that assigns to each real number x the real number $x^2 + 1$. Thus, $f(x) = x^2 + 1$, dom $f = \mathscr{R}$, range $f = \mathscr{R}$, and im $f = \{x \in \mathscr{R} \mid x \geq 1\}$.

Example 2. Let $\mathscr{S}$ be the set of points on the sphere $x^2 + y^2 + z^2 = 1$ and let $\mathscr{T}$ be the set of points on the sphere $x^2 + y^2 + z^2 = 4$. Let $f: \mathscr{S} \to \mathscr{T}$ be the mapping that maps the point (x, y, z) of $\mathscr{S}$ on the point $(2x, 2y, 2z)$ of $\mathscr{T}$. Thus, if P is a point of $\mathscr{S}$ the image $f(P)$ of P is found by joining the origin O to P and extending OP to cut the sphere $\mathscr{T}$ in the point $f(P)$. Here, dom $f = \mathscr{S}$ and im $f = \mathscr{T}$.

Definition. If $f_1: \mathscr{S} \to \mathscr{T}$ and $f_2: \mathscr{S} \to \mathscr{T}$ are two mappings with the same domain and range, we say f_1 and f_2 are equal and write $f_1 = f_2$ if and only if $f_1(s) = f_2(s)$ for every element s of $\mathscr{S}$. Note that equal mappings have the same domain and range.

Definition. If $\mathscr{S}, \mathscr{T}, \mathscr{U}$ are sets and $f: \mathscr{S} \to \mathscr{T}$, $g: \mathscr{T} \to \mathscr{U}$ are mappings, the composite mapping (or the composition or product of g and f) is the mapping $gf: \mathscr{S} \to \mathscr{U}$ defined by

(1) $\qquad\qquad gf(s) = g[f(s)] \qquad$ for all $s \in \mathscr{S}$.

Note that gf is defined if and only if dom g = range f.

Associative Law. If $f: \mathscr{S} \to \mathscr{T}$, $g: \mathscr{T} \to \mathscr{U}$, and $h: \mathscr{U} \to \mathscr{V}$ are mappings then the products (compositions) $(hg)f$ and $h(gf)$ are both defined and

(2) $\qquad\qquad\qquad h(gf) = (hg)f.$

Proof. Both $h(gf)$ and $(hg)f$ have domain $\mathscr{S}$ and range $\mathscr{V}$. If $s \in \mathscr{S}$ it is easy to verify that

(3) $\qquad\qquad h(gf)(s) = (hg)f(s) = h\{g[f(s)]\}$

and (2) follows from (3) by the definition of equal mappings. Equation (2) is called the associative law for composition of mappings.

Example 3. Choose rectangular coordinate axes in three-dimensional space and let $\mathscr{S}$ be the set of all points $P(x, y, z)$ in space. Let $\mathscr{P}$ be the set of all points in the xy plane and let $\mathscr{R}$ be the set of all real numbers. Let $f: \mathscr{S} \to \mathscr{P}$ be the mapping that maps each point of $\mathscr{S}$ on its projection in the xy-plane, that is, $f[(x, y, z)] = (x, y, 0)$. Now let $g: \mathscr{P} \to \mathscr{R}$ be the mapping that maps each point of P on its distance from the origin, that is, $g[(x, y, 0)] = \sqrt{x^2 + y^2}$. The composite $gf: \mathscr{S} \to \mathscr{R}$ is defined by $gf(P) = \sqrt{x^2 + y^2}$ where P is the point (x, y, z). Its domain is $\mathscr{S}$, its range $\mathscr{R}$, and im $gf = \{t \in \mathscr{R} \mid t \geq 0\}$.

A case of special interest arises when the range of a mapping f is the same set $\mathscr{S}$ as the domain of f. We then write $f: \mathscr{S} \to \mathscr{S}$ and call f a mapping of $\mathscr{S}$ into itself. If f and g are mappings of $\mathscr{S}$ into itself, then the composites fg and gf are both defined and are mappings of $\mathscr{S}$ into itself. They are not, in general, equal.

Example 4. Let $\mathscr{R}$ be the set of real numbers. Let $f: \mathscr{R} \to \mathscr{R}$ be the mapping defined by $f(x) = 2x + 1$ and let $g: \mathscr{R} \to \mathscr{R}$ be the mapping defined by $g(x) = x^2$. Then

$$(4) \qquad gf(x) = g[f(x)] = (2x + 1)^2 = 4x^2 + 4x + 1$$

whereas

$$(5) \qquad fg(x) = f[g(x)] = 2x^2 + 1.$$

It is clear from (4) and (5) that gf and fg are different mappings so that $gf \neq fg$ and composition of mappings is not commutative.

Definition. *The mapping $f: \mathscr{S} \to \mathscr{S}$ defined by $f(s) = s$ for all s in $\mathscr{S}$ is called the* identity mapping *on $\mathscr{S}$ and will be denoted by I or by $I_{\mathscr{S}}$ when it is necessary to specify the set $\mathscr{S}$ which is the domain range and image of this mapping. If $g: \mathscr{S} \to \mathscr{T}$ is any mapping, then clearly*

$$(6) \qquad gI_{\mathscr{S}} = I_{\mathscr{T}}g = g.$$

In particular, if $\mathscr{T} = \mathscr{S}$, then $I_{\mathscr{S}}g = gI_{\mathscr{S}} = g$ for every mapping $g: \mathscr{S} \to \mathscr{S}$.

Definition. *A mapping $f: \mathscr{S} \to \mathscr{T}$ is said to be one-to-one if $f(s_1) = f(s_2)$ implies $s_1 = s_2$.*

If $f: \mathscr{S} \to \mathscr{T}$ is a one-to-one mapping of $\mathscr{S}$ onto $\mathscr{T}$, then every element t of $\mathscr{T}$ is the image of exactly one element s of $\mathscr{S}$. Thus the equation $f(s) = t$ not only defines t uniquely when s is given but also defines s uniquely for each t in $\mathscr{T}$. We then write $s = f^{-1}(t)$ and $f^{-1}: \mathscr{T} \to \mathscr{S}$ is a mapping with domain $\mathscr{T}$ and range $\mathscr{S}$ which satisfies $f^{-1}[f(s)] = s$ and $f[f^{-1}(t)] = t$. Hence we have $f^{-1}f = I_{\mathscr{S}}$ and $ff^{-1} = I_{\mathscr{T}}$.

Definition. *If $f: \mathscr{S} \to \mathscr{T}$ is a one-to-one mapping of $\mathscr{S}$ onto $\mathscr{T}$ the mapping $f^{-1}: \mathscr{T} \to \mathscr{S}$, defined by $f^{-1}(t) = s$ if and only if $s = f(t)$, is called the* inverse *of f.*

Note that f^{-1} exists if and only if $f: \mathscr{S} \to \mathscr{T}$ is both one-to-one and onto. A mapping that satisfies these conditions is said to be *invertible*. In particular, if $f: \mathscr{S} \to \mathscr{S}$ is an invertible mapping of $\mathscr{S}$ onto itself, then we have $ff^{-1} = f^{-1}f = I$, where, of course, $I = I_{\mathscr{S}}$.

Example 5. Let $\mathscr{R}$ be the set of real numbers and let $f: \mathscr{R} \to \mathscr{R}$ be defined by $f(x) = x^3 + 4$. In this case f^{-1} exists because every real number has exactly one real cube root. Thus, if $f(x_1) = f(x_2)$ we have $x_1^3 + 4 = x_2^3 + 4$ which implies $x_1 = x_2$, so that f is one-to-one. Also, if y is an arbitrary real number there exists a real number x such that $x^3 + 4 = y$, namely, $x = \sqrt[3]{y - 4}$. Thus, f maps $\mathscr{R}$ onto $\mathscr{R}$ and f^{-1} exists.

Example 6. Consider the mapping $f: \mathscr{R} \to \mathscr{R}$ defined by $f(x) = \cos x$ where $\mathscr{R}$ is the set of all real numbers. Because $f(x + 2\pi) = f(x)$, the mapping is not one-to-one and f^{-1} does not exist. However, if we let $\mathscr{S} = \{x \in \mathscr{R} \mid 0 \le x \le \pi\}$ and let $g: \mathscr{S} \to \mathscr{R}$ be the mapping defined by $g(x) = \cos x$, then $g(x_1) = g(x_2)$, $x_1, x_2 \in \mathscr{S}$ does imply $x_1 = x_2$, so that g is one-to-one. But since im $g \ne \mathscr{R}$, g is not onto $\mathscr{R}$ and g^{-1} fails to exist. Now let $\mathscr{T} = \text{im } g = \{x \in \mathscr{R} \mid -1 \le x \le 1\}$. The mapping $h: \mathscr{S} \to \mathscr{T}$ defined by $h(x) = \cos x$ is now one-to-one and onto, so that $h^{-1}: \mathscr{T} \to \mathscr{S}$ exists.

In many circumstances it is unnecessary to distinguish between mappings, such as g and h in Example 6. These mappings have the same domain $\mathscr{S}$, the same image $\mathscr{T}$ and, moreover, $g(s) = h(s)$ for all s in $\mathscr{S}$. They differ only in that range $g \ne$ range h and, in a sense, this is an artificial distinction. There are advantages, however, in regarding these as different mappings, and the modern tendency is to define two mappings f and g to be equal if *and only if* dom $f =$ dom g, range $f =$ range g, and $f(x) = g(x)$ for all x in their common domain. It then follows, of course, that im $f =$ im g. Each mapping $f: \mathscr{S} \to \mathscr{T}$ induces a mapping $f^*: \mathscr{S} \to \mathscr{U}$ where $\mathscr{U} = \text{im } f$ and $f^*(s) = f(s)$ for all $s \in \mathscr{S}$. Clearly, f^* is onto $\mathscr{U}$ and hence, if f is one-to-one, f^* is invertible.

Theorem 1.0. *Let $f: \mathscr{S} \to \mathscr{T}$ be a mapping of $\mathscr{S}$ into $\mathscr{T}$. If there exists a mapping $g: \mathscr{T} \to \mathscr{S}$ such that*

(7)
$$fg = I_{\mathscr{T}}$$

and

(8)
$$gf = I_{\mathscr{S}},$$

then f is invertible and $f^{-1} = g$.

Proof. If t is any element of $\mathscr{T}$, then by (7)

$$f[g(t)] = fg(t) = I_{\mathscr{T}}t = t.$$

Hence, $t \in \text{im } f$, im $f = \mathscr{T}$, and f is onto.

Now if $f(s_1) = f(s_2)$, then $g[f(s_1)] = g[f(s_2)]$ and by (8), $I_{\mathscr{S}}s_1 = I_{\mathscr{S}}s_2$ or $s_1 = s_2$. Hence, f is one-to-one, as well as onto, and is therefore invertible. Finally, for every $t \in \mathscr{T}$ we have by (6) and (7)

$$f^{-1}(t) = f^{-1}I_{\mathscr{T}}t = f^{-1}fg(t) = I_{\mathscr{S}}g(t) = g(t)$$

and, hence, $f^{-1} = g$.

Corollary 1. *If $f: \mathscr{S} \to \mathscr{T}$ is invertible, f^{-1} is invertible and $(f^{-1})^{-1} = f$.*

Proof. Since $f^{-1}f = I_{\mathscr{S}}$ and $ff^{-1} = I_{\mathscr{T}}$ the theorem implies that f^{-1} is invertible and $(f^{-1})^{-1} = f$.

Corollary 2. *If $f: \mathscr{S} \to \mathscr{S}$ is a mapping of $\mathscr{S}$ into itself and if there exists a mapping $g: \mathscr{S} \to \mathscr{S}$ such that $fg = gf = I$, then f is invertible and $g = f^{-1}$.*

Exercise 1.1

Notation. Let $\mathscr{R}$ be the set of all real numbers and $\mathscr{X}$ the set of all integers. Denote by $\mathscr{R}^+$ the set of all positive real numbers, by $\mathscr{X}^+$ the set of positive integers, and by $\mathscr{R}_0$ the set $\{0\} \cup \mathscr{R}^+$. Let $\mathscr{E}_2$ be the set of all points (x, y) in a plane and $\mathscr{E}_3$ the set of all points (x, y, z) in 3-dimensional space.

1. What is the domain, range, and image of each of the following mappings?
   (a) $f_1: \mathscr{R} \to \mathscr{X}$, where $f_1(x) = [x] =$ the greatest integer $\leq x$.
   (b) $f_2: \mathscr{X} \to \mathscr{X}$, where $f_2(x) = x + 1$.
   (c) $f_3: \mathscr{X}^+ \to \mathscr{X}^+$, where $f_3(x) = x + 1$.
   (d) $f_4: \mathscr{X} \to \mathscr{X}$, where $f_4(x) = x^2$.
   (e) $f_5: \mathscr{R}_0 \to \mathscr{R}_0$, where $f_5(x) = x^2$.
   (f) $f_6: \mathscr{R}_0 \to \mathscr{R}_0$ where $f_6(x) = \sqrt{x}$.
   (g) $f_7: \mathscr{R} \to \mathscr{R}_0$ where $f_7(x) = x^2$.
   (h) $f_8: \mathscr{E}_2 \to \mathscr{E}_2$ where $f_8[(x, y)] = (x, 2x - y)$.
   (i) $f_9: \mathscr{E}_3 \to \mathscr{E}_2$ where $f_9[(x, y, z)] = (x + y, x - y)$.
   (j) $f_{10}: \mathscr{E}_3 \to \mathscr{R}$ where $f_{10}[(x, y, z)] = x + 2y - 3z$.

2. Which of the mappings in Problem 1 are:
   (a) One-to-one.   (b) Onto.   (c) Invertible.
   (d) Define the inverse mappings for those in (c).

3. Let
$$\mathscr{A} = \{x \in \mathscr{R} \mid -\pi/2 < x < \pi/2\}, \qquad \mathscr{B} = \{x \in \mathscr{R} \mid -\pi/2 \leq x \leq \pi/2\},$$
$$\mathscr{C} = \{x \in \mathscr{R} \mid 0 \leq x \leq \pi\}, \qquad \mathscr{D} = \{x \in \mathscr{R} \mid -1 \leq x \leq 1\},$$
$$\mathscr{F} = \{x \in \mathscr{R} \mid 0 \leq x \leq 1\}.$$

Give the domain, range, and image of the following mappings and determine which
of them are invertible.

(a) $f_1: \mathscr{A} \to \mathscr{R}$ where $f_1(x) = \tan x$.

(b) $f_2: \mathscr{B} \to \mathscr{R}$ where $f_2(x) = \sin x$.

(c) $f_3: \mathscr{B} \to \mathscr{D}$ where $f_3(x) = \sin x$.

(d) $f_4: \mathscr{C} \to \mathscr{D}$ where $f_4(x) = \sin x$.

(e) $f_5: \mathscr{C} \to \mathscr{F}$ where $f_5(x) = \sin x$.

4. With reference to Problem 1, define the composite mappings $f_2 f_1, f_2 f_4, f_4 f_2, f_6 f_7$ $f_8 f_9, f_5 f_6$, and $f_6 f_5$.

5. If $f: \mathscr{S} \to \mathscr{T}, g: \mathscr{T} \to \mathscr{U}$ are mappings, prove that:

(a) If gf is onto, g is onto.

(b) If gf is one-to-one, f is one-to-one.

6. Define a mapping from $\mathscr{Z}^+$ into the set of all even positive integers which is:

(a) One-to-one and onto.

(b) One-to-one but not onto.

(c) Onto but not one-to-one.

(d) Neither onto nor one-to-one.

1.3 Fields

In the discussion of the algebra of vectors in Chapter 0 the scalars play an important
role. For most elementary applications of vectors the scalars are assumed to be
real numbers. However, in many mathematical situations and in some applications
to physics it is desirable to allow the scalars to be complex numbers. The algebra
of vectors can easily be formulated in such a way that the nature of the scalars is
unimportant provided they have certain basic algebraic properties. These properties
can be summarized in the requirement that the scalars must form a *field* in the sense
of the definition which we now give.

Let $\mathscr{F}$ be a set of elements on which two operations are defined as follows. The
first operation, called addition, associates with each ordered pair x, y of elements
in $\mathscr{F}$, a unique element of $\mathscr{F}$ denoted by $x + y$ and called the *sum* of x and y. The
second operation, called multiplication, associates with each ordered pair x, y a
unique element of $\mathscr{F}$ denoted by xy and called the *product* of x and y. If these two
operations satisfy the following postulates, $\mathscr{F}$ is called a *field*.

F1. *Commutative laws.* For all x, y in $\mathscr{F}$:

(a) $x + y = y + x$.   (b) $xy = yx$.

F2. *Associative laws.* For all x, y, z in $\mathscr{F}$:

(a) $(x + y) + z = x + (y + z)$.   (b) $x(yz) = (xy)z$.

F3. *Distributive law.* For all x, y, z in $\mathscr{F}$:

$$x(y + z) = xy + xz.$$

F4. *Identity elements.* There exist in $\mathscr{F}$ elements denoted by 0 and 1 such that $0 \neq 1$ and for each x in $\mathscr{F}$:

    (a) $x + 0 = x$.   (b) $1x = x$.

These are called the zero element and the unity element, respectively.

F5. *Inverse elements.*

    (a) For each x in $\mathscr{F}$ there exists an element denoted by $-x$, called the negative or additive inverse of x, such that $x + (-x) = 0$.

    (b) For each $x \neq 0$ in $\mathscr{F}$ there exists an element x^{-1} called the multiplicative inverse of x, such that $xx^{-1} = 1$. The element x^{-1} will also be written $1/x$ when convenient.

We shall refer to these five rules as the *field postulates*. The student will recognize them as basic properties of real numbers which are used more or less unconsciously in all computations. It is immediately clear that the set $\mathscr{R}$ of all real numbers, with operations addition and multiplication, is a field. So, too, is the set $\mathscr{C}$ of all complex numbers, under addition and multiplication of complex numbers.

If $\mathscr{F}$ is a field and if a subset $\mathscr{F}_1$ of $\mathscr{F}$ is also a field under the same operations as those of $\mathscr{F}$, then $\mathscr{F}_1$ is called a *subfield* of $\mathscr{F}$. Clearly the real field $\mathscr{R}$ is a subfield of the complex field $\mathscr{C}$. The set $\mathscr{Q}$ of all rational numbers, with the usual operations, is a subfield both of $\mathscr{R}$ and of $\mathscr{C}$. To prove that a subset $\mathscr{F}_1$ of a field $\mathscr{F}$ is a subfield it is only necessary to show that 0 and 1 belong to $\mathscr{F}_1$ and that if $x \in \mathscr{F}_1$ and $y \in \mathscr{F}_1$,

$$x + y \in \mathscr{F}_1, \quad -x \in \mathscr{F}_1, \quad xy \in \mathscr{F}_1 \quad \text{and} \quad (\text{if } x \neq 0) \ x^{-1} \in \mathscr{F}_1.$$

The other postulates automatically hold in $\mathscr{F}_1$ because they hold in $\mathscr{F}$.

In order to show the existence of fields which are not subfields of the complex numbers, we give two other examples although no further mention of these will be made. The set of all rational functions of a real variable x, with the usual addition and multiplication[1] of functions, is a field. (A rational function is a quotient of two polynomials with nonzero denominator.) Finally, we give an example of a

---

[1] The product of two functions f and g in this context is the function that maps x onto $f(x)g(x)$, not the composite $x \rightarrow f[g(x)]$.

field with a finite number of elements. The two elements 0, 1 constitute a field when addition and multiplication are defined by the following tables.

Addition				Multiplication		
	0	1			0	1
0	0	1		0	0	0
1	1	0		1	0	1

It can be shown that there exist fields with p^n elements where p is any prime and n any positive integer.

We now list a few simple consequences of the field postulates which as properties, for example, of the real numbers are as familiar as the postulates themselves.

1. The zero element of a field is unique.

Proof. If 0 and 0′ both have the property of postulate F4(a),

$$0' = 0' + 0 = 0 + 0' = 0.$$

2. The unity element of a field is unique.

Proof. The proof is similar to that of (1) and is left as an exercise.

3. For all x in F, $x0 = 0$.

Proof. $x0 + x0 = x(0 + 0)$ by Postulate 3

$\qquad\qquad = x0 \qquad$ by Postulate 4a.

Add $-(x0)$ to each side of this equation and

$$(x0 + x0) + [-(x0)] = x0 + [-(x0)] = 0.$$

But

$$(x0 + x0) + [-(x0)] = x0 + [x0 + \{-(x0)\}]$$
$$= x0 + 0$$
$$= x0,$$

and hence $x0 = 0$ as required.

4. If $xy = 0$ either $x = 0$ or $y = 0$.

Proof. If $x \neq 0$ it has an inverse x^{-1} and

$$0 = x^{-1}0 = x^{-1}(xy) = (x^{-1}x)y = 1y = y.$$

Hence, either

$$x = 0 \qquad \text{or} \qquad y = 0.$$

Exercise 1.2

1. Which of the following subsets of the set of complex numbers are fields under ordinary addition and multiplication? Give reasons for your answer in each case.
   (a) The set of all positive integers.
   (b) The set of all integers (positive, negative, and zero).
   (c) The set consisting of the two numbers 0 and 1.
   (d) The set of all rational numbers.
   (e) The set of all complex numbers $a + bi$ where a and b are integers.
   (f) The set of all complex numbers $a + bi$ where a and b are rational.
   (g) The set of all numbers of the form $a + b\sqrt{2}$ where a and b are rational.

2. If m is a fixed positive integer, prove that the set of all numbers of the form $a + b\sqrt{m}$, where a and b are rational, is a subfield of the complex numbers.

3. If in Problem 2, $\sqrt{m}$ is replaced by $\sqrt[3]{m}$, is the resulting set of numbers a field? Why?

4. Prove that if a, b, c are elements of a field such that $a \neq 0$ and $ab = ac$, then $b = c$.

5. Prove that every subfield of the complex numbers contains the field $\mathscr{Q}$ of all rational numbers.

1.4 Vector Spaces

Roughly speaking, the essential ingredients of vector algebra developed in Sections 0.5 and 0.6 are:

1. Any two vectors X, Y can be added to give a vector $X + Y$.
2. Any vector X can be multiplied by any scalar k to give a vector kX.
3. These operations, vector addition and multiplication by scalars, satisfy certain basic rules which imply in particular that if $X_1, X_2, \ldots, X_r$ are vectors and $c_1, c_2, \ldots, c_r$ are arbitrary scalars, the "linear combination"

$$c_1 X_1 + c_2 X_2 + \cdots + c_r X_r$$

is a uniquely determined vector.

Algebraic systems of this type occur frequently in mathematics and have many important applications. Therefore it is advantageous to give them a name and a precise definition. To this end we now define an important type of algebraic system called a vector space.

Let $\mathscr{V}$ be a nonempty set of elements which we shall call *vectors* and let $\mathscr{F}$ be any field. The elements of $\mathscr{F}$ will be called *scalars*. Suppose further that an operation, vector addition, is defined in $\mathscr{V}$ whereby to any ordered pair X, Y of elements of

$\mathscr{V}$ a unique sum $X + Y$ is defined which is also an element of $\mathscr{V}$. Moreover, for any vector X in $\mathscr{V}$ and any scalar k in $\mathscr{F}$, there is defined a unique scalar product kX which is an element of $\mathscr{V}$. If these two operations satisfy the following eight postulates, $\mathscr{V}$ is called a *vector space over $\mathscr{F}$*.

Postulates for a Vector Space

V1. If $X, Y \in \mathscr{V}$, then $X + Y = Y + X$.

V2. If $X, Y, Z \in \mathscr{V}$, then $(X + Y) + Z = X + (Y + Z)$.

V3. There is a vector O in $\mathscr{V}$ such that $X + O = X$ for all $X \in \mathscr{V}$.

V4. If $X \in \mathscr{V}$ there is a vector $-X$ in $\mathscr{V}$ such that $X + (-X) = O$. The vector $-X$ is called the negative or additive inverse of X.

V5. If $X, Y \in \mathscr{V}$ and $k \in \mathscr{F}$, then $k(X + Y) = kX + kY$.

V6. If $X \in \mathscr{V}$ and $k_1, k_2 \in \mathscr{F}$, then $(k_1 + k_2)X = k_1X + k_2X$.

V7. If $X \in \mathscr{V}$ and $k_1, k_2 \in \mathscr{F}$, then $k_1(k_2X) = (k_1k_2)X$

V8. If $X \in \mathscr{V}$, then $1X = X$.

Throughout this book the field of scalars will be either the field of all real numbers or the field of all complex numbers although many of the results obtained hold for an arbitrary field of scalars. When it is of no importance whether the scalars are real or complex numbers, we shall denote the field of scalars by $\mathscr{F}$. When it is necessary to use special properties of the real or complex numbers, we denote the real field by $\mathscr{R}$ and the complex field by $\mathscr{C}$. From this point on, $\mathscr{F}$ is to denote either $\mathscr{R}$ or $\mathscr{C}$ at the reader's choice. The nature of the "vectors," that is, elements of $\mathscr{V}$ is unspecified. They can be interpreted in a variety of different ways to suit the particular area of mathematics or the specific application that is of interest at the moment. For example, $\mathscr{V}$ may be taken as the set of all directed line segments with fixed initial point O, together with a zero vector, addition and scalar multiplication being defined geometrically as in Sections 0.5 and 0.6. Or $\mathscr{V}$ may be taken to be the set of all ordered triples $[x_1, x_2, x_3]$ of real numbers with addition and scalar multiplication defined coordinatewise as in Example 1 below. Many other interpretations of $\mathscr{V}$ are possible, several of which are described in the examples which follow.

Example 1. For a fixed positive integer n, let $V_n(F)$ be the set of all ordered "n-tuples"

$$[x_1, x_2, \ldots, x_n],$$

where $x_1, x_2, \ldots, x_n$ are arbitrary elements of $\mathscr{F}$. If $X = [x_1, x_2, \ldots, x_n]$ and $Y = [y_1, y_2, \ldots, y_n]$ we define $X + Y$ and kX, where $k \in \mathscr{F}$, by the equations

$$X + Y = [x_1 + y_1, x_2 + y_2, \ldots, x_n + y_n],$$
$$kX = [kx_1, kx_2, \ldots, kx_n].$$

It is easy to verify that the postulates V1 through V8 are satisfied, with $O = [0, 0, \ldots, 0]$ and $-X = [-x_1, -x_2, \ldots, -x_n]$. Hence, $\mathcal{V}_n(\mathcal{F})$ is a vector space over $\mathcal{F}$. It is clear from Theorems 0.3 and 0.5 and their two-dimensional equivalents that $\mathcal{V}_2(\mathcal{R})$ can be identified with the sets of all vectors in a plane and $\mathcal{V}_3(\mathcal{R})$ with the set of all vectors in three-dimensional space.

Example 2. Let $\mathcal{V}$ be the set of all directed line segments in space with initial point at a fixed point O together with a "zero vector" O. Define addition of vectors by the parallelogram law (Section 0.5) and multiplication by real scalars as in Section 0.6. Then $\mathcal{V}$ is a vector space over the real field $\mathcal{R}$.

Example 3. Let $\mathcal{P}$ be the set of all polynomials

$$a_0 + a_1 x + a_2 x^2 + \cdots a_m x^m$$

with coefficients in $\mathcal{F}$. These polynomials may be added and multiplied by scalars in the usual way and form a vector space over $\mathcal{F}$. Note that multiplication of one polynomial by another has no place in the vector space algebra which is concerned only with addition and scalar multiplication.

Example 4. Let $\mathcal{P}_n$ be the set of all polynomials in $\mathcal{P}$ of degree $\leq n$ for some fixed integer n. Then $\mathcal{P}_n$ is also a vector space over $\mathcal{F}$. Note that in this example and in Example 3, the zero vector is the polynomial O whose coefficients are all zero.

Example 5. The set of all continuous real-valued functions of x defined on the closed interval $a \leq x \leq b$ is a vector space over the real field. Here functions are added and multiplied by real numbers in the usual way.

Example 6. Consider the linear differential equation

(9)
$$\frac{d^n y}{dx^n} + a_1 \frac{d^{n-1}y}{dx^{n-1}} + \cdots + a_n y = 0,$$

in which $a_1, a_2, \ldots, a_n$ are real constants. Because

$$\frac{d}{dx}(y_1 + y_2) = \frac{dy_1}{dx} + \frac{dy_2}{dx} \quad \text{and} \quad \frac{d}{dx}(ky) = k\frac{dy}{dx} \text{ if } k \in \mathcal{R},$$

it is easy to verify that the sum of two solutions of (9) is again a solution as is any multiple of a solution by a real number. Since the solutions are continuous functions of x, the postulates are easily verified. Moreover, the set $\mathcal{V}$ of all real solutions of (9) is not empty because the zero function is a solution. Hence, $\mathcal{V}$ is a vector space over $\mathcal{R}$.

We require a few algebraic properties of vectors that are simple consequences of the postulates.

1. The zero vector O is unique. The proof is the same as that of the uniqueness of the zero element of a field.

2. If $A, B \in \mathscr{V}$ there is a unique vector X such that $A + X = B$.

Proof. Clearly, $X = -A + B$ satisfies this equation because $A + (-A + B) = [A + (-A)] + B = O + B = B + O = B$. Conversely, if $A + X = B$, adding $-A$ to each side of this equation gives $X = -A + B$. In particular, if we put $B = O$ we see that the additive inverse $-A$ is uniquely determined by A.

We agree to write $X - Y$ for the vector $X + (-Y)$ and the usual rules for subtraction hold. In particular, $Y - X = -(X - Y)$ and $k(X - Y) = kX - kY$.

3. If $X \in \mathscr{V}$, then $0X = O$.

Proof.
$$\begin{aligned} X + 0X &= 1X + 0X &&\text{by V8,} \\ &= (1 + 0)X &&\text{by V6,} \\ &= 1X = X &&\text{by V8.} \end{aligned}$$

Hence, $0X = O$ by (2) and V3.

4. If $X \in \mathscr{V}$, then $(-1)X = -X$.

Proof. $X + (-1)X = 1X + (-1)X = [1 + (-1)]X = 0X = O$.

Hence, $(-1)X = -X$ by (2) and V4.

5. If $X_1, X_2, \ldots, X_n$ are in $\mathscr{V}$ and $c_1, c_2, \ldots, c_n$ are in $\mathscr{F}$, then

$$X = c_1X_1 + c_2X_2 + \cdots + c_nX_n$$

is a vector of $\mathscr{V}$ uniquely determined by the X_i and c_i. Such a vector X, which is a sum of scalar multiples of $X_1, X_2, \ldots, X_n$, is called a *linear combination* of $X_1, X_2, \ldots, X_n$.

1.5 Subspaces

Let $\mathscr{V}$ be a vector space over $\mathscr{F}$. A subset $\mathscr{S}$ of $\mathscr{V}$ is called a *subspace* of $\mathscr{V}$ if it is a vector space over $\mathscr{F}$ relative to the addition and scalar multiplication which it inherits from $\mathscr{V}$. This means that if X, $Y \in \mathscr{S}$ their sum as vectors of the subspace $\mathscr{S}$ must be the same as their sum as vectors of $\mathscr{V}$ and similarly for scalar products. Also note that the field of scalars for the subspace $\mathscr{S}$ must be the same as for the parent space $\mathscr{V}$.

In order to prove that a nonempty subset $\mathscr{S}$ of $\mathscr{V}$ is a subspace, it is only necessary to prove two things:

(a) If X, $Y \in \mathscr{S}$, then $X + Y \in \mathscr{S}$.
(b) If $X \in \mathscr{S}$ and $k \in \mathscr{F}$, then $kX \in \mathscr{S}$.

For if we put $k = 0$, (b) tells us that $O \in \mathscr{S}$, and putting $k = -1$ we see that if $X \in \mathscr{S}$, then $-X = (-1)X \in \mathscr{S}$. The other vector space postulates hold in $\mathscr{S}$ because they hold in $\mathscr{V}$.

Note that according to our definition, $\mathscr{V}$ itself is a subspace of $\mathscr{V}$. A subspace $\mathscr{S}$ of $\mathscr{V}$ such that $\mathscr{S} \neq \mathscr{V}$ will be called a *proper* subspace of $\mathscr{V}$.

Example 1. Let $\mathscr{V}$ be a vector space over $\mathscr{F}$ and let $X_1, X_2, \ldots, X_n$ be any n vectors of $\mathscr{V}$. The set $\mathscr{S}$ of all linear combinations of $X_1, X_2, \ldots, X_n$ is a subspace of $\mathscr{V}$, since if $Y, Z \in \mathscr{S}$ we have

$$Y = a_1 X_1 + a_2 X_2 + \cdots + a_n X_n,$$

$$Z = b_1 X_1 + b_2 X_2 + \cdots + b_n X_n,$$

and hence

$$Y + Z = (a_1 + b_1)X_1 + \cdots + (a_n + b_n)X_n$$

and

$$kY = (ka_1)X_1 + (ka_2)X_2 + \cdots + (ka_n)X_n.$$

Thus, $Y + Z \in \mathscr{S}$ and $kY \in \mathscr{S}$, and hence $\mathscr{S}$ is a subspace of $\mathscr{V}$.

Definition. *The subspace $\mathscr{S}$ of $\mathscr{T}$ consisting of all linear combinations of the vectors $X_1, X_2, \ldots, X_n$ is called the subspace* spanned *by (or generated by) the vectors $X_1, X_2, \ldots, X_n$. If $\mathscr{S} = \mathscr{V}$, then $X_1, X_2, \ldots, X_n$ are said to span (or generate) $\mathscr{V}$.*

Example 2. The subset of $\mathscr{V}$ consisting of the zero vector only is a subspace of $\mathscr{V}$, called the zero space, because $O + O = O$ and $kO = O$ for any scalar k. Note that this is a special case of Example 1, since the zero space is the space spanned by the zero vector. The zero space will be denoted by the same symbol O as the zero vector.

If $\mathscr{S}$ and $\mathscr{T}$ are subspaces of $\mathscr{V}$, denote by $\mathscr{S} + \mathscr{T}$ the set of all vectors of the form $S + T$ where $S \in \mathscr{S}$ and $T \in \mathscr{T}$. It is easy to verify that $\mathscr{S} + \mathscr{T}$ is a subspace of $\mathscr{V}$. For if $S_1 + T_1$ and $S_2 + T_2$ are two elements of $\mathscr{S} + \mathscr{T}$, we have

$$(S_1 + T_1) + (S_2 + T_2) = (S_1 + S_2) + (T_1 + T_2) \in \mathscr{S} + \mathscr{T}$$

and

$$k(S_1 + T_1) = kS_1 + kT_1 \in \mathscr{S} + \mathscr{T}$$

because $S_1, S_2 \in \mathscr{S}$ and $T_1, T_2 \in \mathscr{T}$. The space $\mathscr{S} + \mathscr{T}$ is called the *sum* or the *join* of $\mathscr{S}$ and $\mathscr{T}$. Similarly, if $\mathscr{S}_1, \mathscr{S}_2, \ldots, \mathscr{S}_r$ are subspaces of $\mathscr{V}$ we can define the sum $\mathscr{S}_1 + \mathscr{S}_2 + \cdots + \mathscr{S}_r$ to be the set $\mathscr{S}$ of all vectors $S_1 + S_2 + \cdots + S_r$ where $S_i \in \mathscr{S}_i$. It is easy to see as in the case $r = 2$ that $\mathscr{S}$ is a subspace of $\mathscr{V}$.

Example 3. Let $c_1, c_2, \ldots, c_n$ be fixed scalars. A vector $X = [x_1, x_2, \ldots, x_n]$ in $\mathscr{V}_n(\mathscr{F})$ whose coordinates satisfy the equation

(10) $$c_1 x_1 + c_2 x_2 + \cdots + c_n x_n = 0$$

is called a *solution vector* of equation (10). Let $\mathscr{S}$ be the set of all solution vectors of (10). Note that $\mathscr{S}$ is not empty because $O \in \mathscr{S}$. If, too, $Y \in \mathscr{S}$ where $Y = [y_1, y_2, \ldots, y_n]$ we have

(11) $$c_1 y_1 + c_2 y_2 + \cdots + c_n y_n = 0.$$

By adding (10) and (11) we see that

$$c_1(x_1 + y_1) + c_2(x_2 + y_2) + \cdots + c_n(x_n + y_n) = 0,$$

so that $X + Y \in \mathscr{S}$. Also $kx \in \mathscr{S}$ because

$$c_1(kx_1) + c_2(kx_2) + \cdots + c_n(kx_n) = k(c_1 x_1 + c_2 x_2 + \cdots + c_n x_n) = 0.$$

Hence, $\mathscr{S}$ is a subspace of $\mathscr{V}_n(\mathscr{F})$. It is called the *solution space* of equation (10).

Note that if $n = 3$ and $\mathscr{F} = \mathscr{R}$, a linear equation of the form

(12) $$c_1 x_1 + c_2 x_2 + c_3 x_3 = 0$$

represents a plane through the origin and the corresponding solution space $\mathscr{S}$ is the set of all vectors in $\mathscr{V}_3(\mathscr{R})$ that lie in this plane. In fact, equation (12) can be written

$$C \cdot X = 0$$

where $C = [c_1, c_2, c_3]$ and $X = [x_1, x_2, x_2]$. Therefore it states that $\mathscr{S}$ consists of all vectors X that are perpendicular to the fixed vector C, that is, of all vectors in the plane through the origin which is perpendicular to C.

Let $\mathscr{S}$ and $\mathscr{T}$ be subspaces of a vector space $\mathscr{V}$ over $\mathscr{F}$. The set $\mathscr{S} \cap \mathscr{T}$ is then a subspace of $\mathscr{V}$. For if $X, Y \in \mathscr{S} \cap \mathscr{T}$, then $X + Y \in \mathscr{S}$ because $X \in \mathscr{S}$ and $Y \in \mathscr{S}$. Similarly, $X + Y \in \mathscr{T}$ and hence $X + Y \in \mathscr{S} \cap \mathscr{T}$. Also, if $X \in \mathscr{S} \cap \mathscr{T}$, $kX \in \mathscr{S}$ and $kX \in \mathscr{T}$ and hence $kX \in \mathscr{S} \cap \mathscr{T}$. Finally, $O \in \mathscr{S} \cap \mathscr{T}$ and hence

$\mathscr{S} \cap \mathscr{T}$ is nonempty. The subspace $\mathscr{S} \cap \mathscr{T}$ is called the *intersection* of $\mathscr{S}$ and $\mathscr{T}$. The same proof will show that the intersection of any set of subspaces of $\mathscr{V}$ is a subspace of $\mathscr{V}$.

Example 4. Let $\mathscr{S}$ be the set of all vectors $X = [x_1, x_2, \ldots, x_n]$ in $\mathscr{V}_n(\mathscr{F})$ whose coordinates $x_1, x_2, \ldots, x_n$ satisfy each of the equations in the system

(13)
$$
\begin{aligned}
a_{11}x_1 + a_{12}x_2 + \cdots + a_{1n}x_n &= 0, \\
a_{21}x_1 + a_{22}x_2 + \cdots + a_{2n}x_n &= 0, \\
&\cdots\cdots\cdots\cdots\cdots\cdots\cdots\cdots \\
a_{r1}x_1 + a_{r2}x_2 + \cdots + a_{rn}x_n &= 0,
\end{aligned}
$$

where the coefficients a_{ij} are fixed elements of $\mathscr{F}$. It is easy to show, just as in Example 2, that $\mathscr{S}$ is a subspace of $\mathscr{V}_n(\mathscr{F})$. In fact, if $\mathscr{S}_1$ is the solution space of the first equation, $\mathscr{S}_2$ that of the second ... and $\mathscr{S}_r$ that of the rth equation, then the space $\mathscr{S}$, called the *solution space* of the system (13), is precisely the intersection $\mathscr{S}_1 \cap \mathscr{S}_2 \cap \cdots \cap \mathscr{S}_r$ of the solution spaces of the individual equations.

Now suppose we take $n = 3$, $r = 2$, and $\mathscr{F} = \mathscr{R}$, and consider the two equations

(14)
$$
\begin{aligned}
a_1x_1 + a_2x_2 + a_3x_3 &= 0, \\
b_1x_1 + b_2x_2 + b_3x_3 &= 0,
\end{aligned}
$$

with solution spaces $\mathscr{S}_1$ and $\mathscr{S}_2$, respectively. It was seen in Example 2 that the spaces $\mathscr{S}_1$ and $\mathscr{S}_2$ are planes through the origin. The solution space of (14), provided these two planes are distinct, is the set of all vectors that lie in their line of intersection.

A system of linear equations of the form (13) is called a *homogeneous* system because all the constant terms are zero. We have seen in Example 4 that the set of all solution vectors of (13) is a subspace of $\mathscr{V}_n(\mathscr{F})$, called the solution space of (13). We wish to show that if $n > r$ (that is, the number of unknowns is greater than the number of equations), the solution space contains nonzero vectors. Before formally stating and proving this result, we need a few preliminaries.

Definition. *Two systems of linear equations in $x_1, x_2, \ldots, x_n$ are said to be* equivalent *if they have the same solutions.*

Example 5. The system

(15)
$$
\begin{aligned}
x + 3y + 5z &= 0, \\
2x - y - 2z &= 0
\end{aligned}
$$

is equivalent to the system

(16)
$$7x - z = 0,$$
$$7y + 12z = 0.$$

To prove this, we denote the left-hand sides of (15) by L_1 and L_2. Then equations (16) are

$$L_1 + 3L_2 = 0,$$

$$2L_1 - L_2 = 0.$$

That is to say (16) have been obtained from (15) by eliminating first y and then x in the usual way. Now if $L_1 = 0$ and $L_2 = 0$, certainly $L_1 + 3L_2 = 0$ and $2L_1 - L_2 = 0$. Hence, every solution of (15) is a solution of (16). Conversely, if $L_1 + 3L_2 = 0$ and $2L_1 - L_2 = 0$, then $L_2 = 2L_1$ and

$$L_1 + 3L_2 = L_1 + 6L_1 = 7L_1 = 0.$$

Hence, $L_1 = 0$ and $L_2 = 2L_1 = 0$. Thus every solution of (16) is a solution of (15) and, therefore, the two systems have precisely the same solutions and are equivalent. Geometrically this equivalence means that the line of intersection of the two planes represented by equations (15) is the same as the line of intersection of the two planes represented by equations (16). Equivalent systems will be studied in more detail in Chapter 2.

Theorem 1.1 *If $n > r$ every system of r homogeneous linear equations in n unknowns, with coefficients in $\mathscr{F}$, has a nonzero solution, that is, a nonzero solution vector in $\mathscr{V}_n(\mathscr{F})$.*

Proof. The proof will be by mathematical induction on r, the number of equations. Consider first the case $r = 1$ or one equation in n unknowns, $n > 1$. Suppose the equation is

$$a_1x_1 + a_2x_2 + \cdots + a_nx_n = 0.$$

If all the coefficients are zero arbitrary values of $x_1, \ldots, x_n$, provide a solution. If not all the coefficients are zero we can assume, by renaming the unknowns if necessary, that $a_1 \neq 0$. By assigning arbitrary nonzero values to $x_2, x_3, \ldots, x_n$ and choosing

$$x_1 = \frac{-a_2x_2 - a_3x_3 - \cdots - a_nx_n}{a_1},$$

we get a nonzero solution for our equation.

Now assume the theorem has been proved for $r - 1$ equations in more than $r - 1$ unknowns and consider the system (13). Again, if all the coefficients a_{ij} are zero, arbitrary values of the unknowns provide a solution and, therefore, a nonzero solution exists. We assume that at least one coefficient is not zero, and by reordering the equations and renaming the unknowns if necessary, we can assume without loss of generality that $a_{11} \neq 0$. We now use the first equation of (13) to eliminate x_1 from the remaining equations. Thus, if we denote by $L_1, L_2, \ldots, L_r$ the left-hand sides of equations (13), we get a new system of the form

(17)
$$
\begin{aligned}
L_1 &= a_{11}x_1 + a_{12}x_2 + \cdots + a_{1n}x_n = 0, \\
L_2 - \frac{a_{21}}{a_{11}} L_1 &= \qquad b_{22}x_2 + \cdots + b_{2n}x_n = 0, \\
&\cdots \cdots \cdots \cdots \cdots \cdots \cdots \cdots \cdots \\
L_r - \frac{a_{r1}}{a_{11}} L_1 &= \qquad b_{r2}x_2 + \cdots + b_{rn}x_n = 0.
\end{aligned}
$$

Clearly (13) and (17) are equivalent systems, for if $x_1, x_2, \ldots, x_n$ satisfy (13), then $L_1 = L_2 = \cdots = L_r = 0$ and hence equations (17) are also satisfied. Conversely, if $x_1, x_2, \ldots, x_n$ is a solution of (17), then $L_1 = 0, L_2 - \frac{a_{21}}{a_{11}} L_1 = 0, \ldots,$ $L_r - \frac{a_{r1}}{a_{11}} L_1 = 0$ and, therefore, $L_1 = L_2 = \cdots = L_r = 0$ and $x_1, x_2, \ldots, x_n$ is a solution of (13).

Now because the last $r - 1$ equations of (17) do not contain x_1 they comprise a system of $r - 1$ equations in the $n - 1$ unknowns $x_2, x_3, \ldots, x_n$. Since $r < n$ we have $r - 1 < n - 1$, and by the induction assumption the last $r - 1$ equations of (17) have a nonzero solution in $\mathscr{F}$ for $x_2, x_3, \ldots, x_n$. We then put

$$
x_1 = \frac{a_{12}x_2 - \cdots - a_{1n}x_n}{a_{11}},
$$

and $x_1, x_2, \ldots, x_n$ is then a nonzero solution of (1) and therefore of (13). This completes the induction, and the theorem is proved.

Example 6. Find a nonzero solution of the system

(18)
$$
\begin{aligned}
x - 2y + z - t &= 0, \\
2x + 4y - 3z &= 0, \\
3x + 2y + 2z - t &= 0.
\end{aligned}
$$

Solution. As in the proof of Theorem 1.1 we multiply the first equation by 2 and by 3 and subtract the resulting equations from the second and third equations of

(18), respectively. This gives the equivalent system

$$x - 2y + z - t = 0,$$

(19)
$$8y - 5z + 2t = 0,$$

$$8y - z + 2t = 0.$$

Subtracting the second equation of (19) from the third gives an equivalent system

$$x - 2y + z - t = 0,$$

(20)
$$8y - 5z + 2t = 0,$$

$$4z = 0.$$

Now the third equation of (20) yields $z = 0$, the second $t = -4y$ and the first $x = -2y$. Thus if we put $y = 1$ we get $x = -2$, $t = -4$, $z = 0$ or a nonzero solution vector $[-2, 1, 0, -4]$. Any scalar multiple of this is also a solution vector. Systematic solution of systems of equations by this method will be discussed in Chapter 2.

Exercise 1.3

1. Show that the field $\mathscr{F}$ may be considered as a vector space over $\mathscr{F}$ if scalar multiplication is identified with field multiplication.

2. Show that the complex field $\mathscr{C}$ is a vector space over the real field $\mathscr{R}$.

3. Let $\mathscr{S}$ be a subspace of $\mathscr{V}_2(\mathscr{R})$ that contains two vectors X_1 and X_2 that do not lie in the same straight line through the origin. Prove that $\mathscr{S} = \mathscr{V}_2(\mathscr{R})$. (*Hint.* Show geometrically that every vector of $\mathscr{V}_2(\mathscr{R})$ is a linear combination of X_1 and X_2.)

4. Let $\mathscr{S}$ be a subspace of $\mathscr{V}_3(\mathscr{R})$ that contains three vectors X_1, X_2, X_3 which do not all lie in the same plane through the origin. Prove that $\mathscr{S} = \mathscr{V}_3(\mathscr{R})$.

5. Prove that if $\mathscr{S}$ is a subspace of $\mathscr{V}_3(\mathscr{R})$, either:
   (a) $\mathscr{S} = O$.
   (b) $\mathscr{S}$ is the set of all vectors that lie in some straight line through the origin.
   (c) $\mathscr{S}$ is the set of all vectors that lie in some plane through the origin.
   (d) $\mathscr{S} = \mathscr{V}_3(\mathscr{R})$.

6. Let $\mathscr{V}$ be any vector space over $\mathscr{F}$ and let $X_1, X_2, \ldots, X_r \in \mathscr{V}$. Let $\mathscr{S}$ be the set of all vectors of the form

$$c_1 X_1 + c_2 X_2 + \cdots + c_r X_r,$$

where $c_1, c_2, \ldots, c_r$ are integers. Is $\mathscr{S}$ a subspace of $\mathscr{V}$? Why?

7. Let $\mathscr{S}$ be the set of all vectors in $\mathscr{V}_4(\mathscr{F})$ of the form

$$[x, x + y, y, 2x + 3y],$$

where $x, y \in \mathscr{F}$. Is $\mathscr{S}$ a subspace of $\mathscr{V}_4(\mathscr{F})$?

8. Let $\mathscr{S}$ be the set of all vectors in $\mathscr{V}_4(\mathscr{F})$ of the form

$$[x, y, x + 1, 2x + y - 3],$$

where $x, y \in \mathscr{F}$. Is $\mathscr{S}$ a subspace of $\mathscr{V}_4(\mathscr{F})$?

9. Let $\mathscr{S}$ be the set of all vectors in $\mathscr{V}_3(\mathscr{F})$ of the form

$$[x, y, x^2 + y^2],$$

where $x, y \in \mathscr{F}$. Is $\mathscr{S}$ a subspace of $\mathscr{V}_3(\mathscr{F})$?

10. Under what circumstances is the set of all solution vectors of the equation

$$a_1 x_1 + a_2 x_2 + \cdots + a_n x_n = c$$

(where $a_1, \ldots, a_n, c \in \mathscr{F}$) a subspace of $\mathscr{V}_n(\mathscr{F})$?

11. Describe geometrically the solution space of the equation

$$x_1 - 2x_2 + x_3 = 0.$$

12. Describe geometrically the solution space of the equations

$$x_1 - 2x_2 + x_3 = 0,$$
$$2x_1 + x_2 - x_3 = 0.$$

13. Find a nonzero solution vector of the equations

$$x_1 + x_2 + 2x_3 = 0,$$
$$2x_1 + 6x_2 + 3x_3 = 0,$$

and prove that every solution vector of these equations is a scalar multiple of it.

14. Find two solution vectors of the equation

$$x + y + 3z = 0$$

such that one of them is not a scalar multiple of the other. Prove that every solution vector is a linear combination of these two.

15. Prove that the two systems of equations,

$$x + y - 3z = 0, \qquad \text{and} \qquad 3x - 2z = 0,$$
$$2x - y + z = 0 \qquad\qquad\qquad 7x - 2y = 0$$

are equivalent.

16. If equations (13) have real coefficients, prove that every complex solution vector of this system has the form

$$[s_1 + it_1, s_2 + it_2, \ldots, s_n + it_n],$$

where $[s_1, s_2, \ldots, s_n]$ and $[t_1, t_2, \ldots, t_n]$ are real solution vectors of the system.

17. By successive application of the method used in the proof of Theorem 1.1 (see Example 6), find nonzero solutions of the following systems of equations:

(a) $2x - y + z - t = 0,$
 $x + y - z + 4t = 0,$
 $3x - y + 2z - t = 0.$

(c) $(2 + 2i)x + (4 + 3i)y - 2z = 0,$
 $2x - 3iy + (1 + i)z = 0.$

(b) $x + y + z + 2t - u = 0,$
 $2x - y - z + t + 2u = 0,$
 $x + 3y - 2z + t + u = 0.$

(d) $x + 2y = 0,$
 $(2 - i)x + 3iz = 0,$
 $x - iy + (2 + i)w = 0.$

1.6 Linear Dependence

Let $\mathscr{V}$ be a vector space over $\mathscr{F}$ and let $X_1, X_2, \ldots, X_r$ be r vectors of $\mathscr{V}$. It may happen that these vectors satisfy an equation of the form

(21) $$c_1 X_1 + c_2 X_2 + \cdots + c_r X_r = O,$$

where $c_1, \ldots, c_r \in \mathscr{F}$. Such an equation will be called a *linear relation*. In fact, *any* r vectors satisfy one linear relation, namely, the one in which $c_1 = c_2 = \cdots = c_r = 0$. This will be called the trivial linear relation. Any linear relation in which the coefficients are not all 0 will be called nontrivial.

Definition. *A finite set* $\{X_1, X_2, \ldots, X_r\}$ *of vectors that satisfy a nontrivial linear relation of the form* (21) *is said to be* linearly dependent. *A finite set of vectors that satisfy no linear relation except the trivial one is said to be* linearly independent.

This definition can be extended to infinite sets by defining a set to be linearly dependent if and only if it contains a finite subset which is linearly dependent. However, we shall have no occasion to use this generalization, and when we speak of linearly dependent or independent sets, it will be assumed, unless otherwise stated, that the sets are finite.

Example 1. The set of three vectors $S = [1, 2, 0, 4]$, $T = [-1, 0, 5, 1]$, and $U = [1, 6, 10, 14]$ is linearly dependent because $3S + 2T - U = O$. The reader should check this equation.

The following theorem gives another way of looking at linear dependence.

Theorem 1.2. *A finite set of vectors in* $\mathscr{V}$ *is linearly dependent if and only if one vector of the set is in the subspace spanned by the other vectors in the set.*

Proof. Suppose the set of vectors $\{X_1, X_2, \ldots, X_n\}$ is linearly dependent, so that there is a nontrivial linear relation

$$(22) \qquad c_1X_1 + c_2X_2 + \cdots + c_nX_n = O.$$

At least one coefficient is nonzero. Suppose $c_1 \neq 0$. Then

$$X_1 = -\frac{c_2}{c_1}X_2 - \cdots - \frac{c_n}{c_1}X_n,$$

and X_1 is in the subspace spanned by $X_2, \ldots, X_n$. Similarly, any vector that has a nonzero coefficient in (22) is in the subspace spanned by the other vectors of the set.

Conversely, if X_1 is in the subspace spanned by $X_2, \ldots, X_n$, then $X_1 = b_2X_2 + \cdots + b_nX_n$ and we have a linear relation

$$X_1 - b_1X_2 - \cdots - b_nX_n = O,$$

which is nontrivial because the coefficient of X_1 is nonzero. Hence, $X_1, X_2, \ldots, X_n$ are linearly dependent as required.

Corollary 1. *Any set of vectors that contains a linearly dependent subset is itself linearly dependent.*

Proof. If, for example, X_1 is in the subspace spanned by $X_2, \ldots, X_r, r < n$, then it is certainly in the space spanned by $X_2, \ldots, X_n$.

Corollary 2. *Any set of vectors that contains the zero vector is linearly dependent.*

Proof. The set consisting of the zero vector only is linearly dependent because $cO = O$, for any nonzero scalar c, is a nontrivial linear relation. Therefore the result follows from Corollary 1.

We note that a set of two vectors $\{X_1, X_2\}$ is linearly dependent if and only if one of the vectors is a scalar multiple of the other.

Example 2. The vectors $X = [2, -1, 3]$, $Y = [4, -2, 6]$, and $Z = [1, 1, 1]$ constitute a linearly dependent set because

$$2X - Y + 0Z = O.$$

Note that in this example the set $\{X, Y, Z\}$ is linearly dependent because the subset $\{X, Y\}$ is. Since the set $\{X, Z\}$ is linearly independent, it follows that in any linear relation satisfied by X, Y, Z the coefficient of Z must be 0. This stems from the fact that although X is in the space spanned by Y and Z and Y is in the space spanned by X and Z, Z is *not* in the space spanned by X and Y.

Example 3.   Prove that the vectors $X = [2, 1, 4]$, $Y = [1, -1, 2]$, and $Z = [3, 1, -2]$ of $\mathscr{V}_3(\mathscr{F})$ constitute a linearly independent set.

Proof.   If there exists a linear relation of the form

$$xX + yY + zZ = 0,$$

then x, y, z must satisfy the equations

$$2x + y + 3z = 0,$$
$$x - y + z = 0,$$
$$4x + 2y - 2z = 0.$$

If we divide the third equation by 2 and add to the second, we get $3x = 0$ or $x = 0$. The first two equations then yield $z = 0$ and $y = 0$. Thus, $x = y = z = 0$ is the only solution and X, Y, Z satisfy no nontrivial linear relation.

Example 4.   In the vector space $\mathscr{P}$ over $\mathscr{F}$, consisting of all polynomials in x with coefficients in $\mathscr{F}$, the set $1, x, x^2, \ldots, x^n$ is linearly independent because $c_0 + c_1 x + \cdots + c_n x^n$ is the zero polynomial only if $c_0 = c_1 = \cdots = c_n = 0$.

Example 5.   Let $\mathscr{V}$ be the vector space over $\mathscr{R}$ of all real valued functions defined on an interval $[a, b]$. The zero vector of $\mathscr{V}$ is the zero function, that is, the function whose value is 0 at each point of $[a, b]$. Linear dependence of a set of n functions $f_1, f_2, \ldots, f_n$ in $\mathscr{V}$ therefore means that there exist real numbers $c_1, c_2, \ldots, c_n$, not all zero, such that $c_1 f_1 + c_2 f_2 + \cdots + c_n f_n$ is the zero function or, in other words, such that

$$c_1 f_1(x) + c_2 f_2(x) + \cdots + c_n f_n(x) = 0$$

for all values of x such that $a \leq x \leq b$.

Although linear dependence and independence are strictly set properties, it is often convenient to use the phrase "the vectors X_1, X_2, \ldots, X_n are linearly dependent (independent)" rather than the more correct form "the set $\{X_1, X_2, \ldots, X_n\}$ is linearly dependent (independent)." In the interests of simplicity in certain situations, we adopt the convention that these two statements have the same meaning.

Exercise 1.4

1. Find nontrivial linear relations satisfied by the following sets of vectors.
   (a) $[2, 1, 1]$, $[3, -4, 6]$, $[4, -9, 11]$ in $\mathscr{V}_3(\mathscr{R})$.
   (b) $[2, 1]$, $[-1, 3]$, $[4, 2]$ in $\mathscr{V}_2(\mathscr{R})$.
   (c) $[1, 0, 2, 4]$, $[0, 1, 9, 2]$, $[-5, 2, 8, -16]$ in $\mathscr{V}_4(\mathscr{R})$.
   (d) $[1, 4]$, $[3, -1]$, $[2, 5]$ in $\mathscr{V}_2(\mathscr{R})$.

2. Show that the three vectors $[1, 1, -1]$, $[2, -3, 5]$, and $[-2, 1, 4]$ of $\mathscr{V}_3(\mathscr{F})$ are linearly independent.

3. If three vectors X, Y, Z are linearly independent, prove that X, $Y + aX$, $Z + bY + cZ$ are also linearly independent, where a, b, c are arbitrary scalars.

4. Prove that any set of three vectors in $\mathscr{V}_2(\mathscr{F})$ is linearly dependent.

5. Prove that two vectors in $\mathscr{V}_2(\mathscr{R})$ are linearly dependent if and only if they lie in the same straight line through the origin.

6. Prove that three vectors in $\mathscr{V}_3(\mathscr{R})$ are linearly dependent if and only if they lie in one plane.

7. If $f(x)$ is a polynomial of degree n with real coefficients, prove that $f(x)$ and its first n derivatives are linearly independent vectors of the space $\mathscr{P}_n$ of Example 4, Section 1.4.

8. If $n > r$, prove that any set of n vectors in $\mathscr{V}_r(\mathscr{F})$ is linearly dependent. *Hint.* Use Theorem 1.1.

9. Prove that any $n + 2$ polynomials in x of degree less than or equal to n are linearly dependent vectors in $\mathscr{P}_n$.

10. If α, β, γ are any three distinct real constants, prove that $\sin(x + \alpha)$, $\sin(x + \beta)$, $\sin(x + \gamma)$ are linearly dependent vectors of the space in Example 5, Section 1.4.

11. Prove that the functions $\sin x$, $\sin 2x$, $\sin 3x, \ldots$, $\sin nx$ are linearly independent in the interval $[0, \pi]$. *Hint.* Assume that for all x in $[0, \pi]$,

$$c_1 \sin x + c_2 \sin 2x + \cdots + c_n \sin nx = 0.$$

Multiply this equation by $\sin mx$, integrate over the interval 0 to π, and hence deduce that $c_m = 0$ for $m = 1, 2, \ldots, n$.

1.7 Bases and Dimension of a Vector Space

A vector space $\mathscr{V}$ over $\mathscr{F}$ is said to be *finitely generated* or *finite dimensional* if there exists a finite set of vectors $X_1, X_2, \ldots, X_n$ which span $\mathscr{V}$, that is, such that every vector of $\mathscr{V}$ is a linear combination of $X_1, X_2, \ldots, X_n$.

Example 1. The space $\mathscr{V}_n(\mathscr{F})$ is finite dimensional. For if $X = [x_1, x_2, \ldots, x_n]$ is an arbitrary vector of $\mathscr{V}_n(\mathscr{F})$, we have

$$X = x_1 E_1 + x_2 E_2 + \cdots x_n E_n$$

where

$$E_1 = [1, 0, 0, \ldots, 0],$$
$$E_2 = [0, 1, 0, \ldots, 0],$$
$$\cdots \cdots \cdots \cdots \cdots$$
$$E_n = [0, 0, 0, \ldots, 1].$$

Thus, $\mathscr{V}$ is spanned by the finite set of vectors $E_1, E_2, \ldots, E_n$.

Example 2. The space $\mathscr{P}$ of all polynomials with coefficients in $\mathscr{R}$ is not finite dimensional. For if $p_1, p_2, \ldots, p_n$ is any finite set of polynomials, let m be the maximum degree of the polynomials p_i. Since every linear combination of $p_1, p_2, \ldots, p_n$ is a polynomial of degree at most m, it is clear that these polynomials cannot span $\mathscr{P}$ which contains polynomials of degree greater than m. On the other hand, the space of all polynomials over $\mathscr{F}$ of degree $\leq n$ is finite dimensional, since it is spanned by the polynomials $1, x, x^2, \ldots, x^n$.

For the most part we shall be concerned in this book with finite dimensional vector spaces.

Definition. *Let $\mathscr{V}$ be a vector space over $\mathscr{F}$. A set of vectors which span $\mathscr{V}$ is called a generating system of $\mathscr{V}$. A linearly independent generating system is called a basis of $\mathscr{V}$.*

Theorem 1.3. *Every nonzero finite dimensional[2] vector space has a finite basis.*

Proof. Since $\mathscr{V}$ is finite dimensional, let

$$\mathscr{G} = \{X_1, X_2, \ldots, X_r\}$$

be a finite generating system of $\mathscr{V}$. Since $\mathscr{V}$ is a nonzero space, $\mathscr{G}$ must contain nonzero vectors; and since $\mathscr{G}$ is finite, it contains a maximal linearly independent subset $\mathscr{B}$ which contains at least one nonzero vector. Suppose $\mathscr{B}$ contains m vectors and suppose the numbering is chosen so that

$$\mathscr{B} = \{X_1, X_2, \ldots, X_m\}.$$

---

[2] It is also true that every nonzero vector space $\mathscr{V}$ has a basis. If $\mathscr{V}$ is not finite dimensional, however, the basis cannot be finite. For proof of this more general theorem, see [1]. Numbers in square brackets refer to references listed at the back of the book.

Because $\mathscr{B}$ is a maximal linearly independent subset of $\mathscr{G}$, the sets

$$\{X_1, X_2, \ldots, X_m, X_{m+p}\}, \quad p = 1, 2, \ldots, r - m,$$

are all linearly dependent and there exists a nontrivial linear relation

(23) $$c_1 X_1 + c_2 X_2 + \cdots + c_m X_m + c_{m+p} X_{m+p} = O.$$

Moreover, $c_{m+p} \neq 0$ for otherwise (23) would be a nontrivial linear relation satisfied by $X_1, X_2, \ldots, X_m$, contrary to the linear independence of the set $\mathscr{B}$. Hence we can divide (23) by c_{m+p} and express X_{m+p} as a linear combination of $X_1, X_2, \ldots, X_m$. Thus, each of the vectors $X_{m+1}, X_{m+2}, \ldots, X_r$ is in the space spanned by $\mathscr{B}$ and, since $\mathscr{G}$ spans $\mathscr{V}$, it follows that $\mathscr{B}$ also spans $\mathscr{V}$. Thus, $\mathscr{B}$ is a linearly independent generating system and, therefore, a basis of $\mathscr{V}$.

Corollary. *Every finite generating system of a finite dimensional vector space $\mathscr{V}$ contains a subset which is a basis of $\mathscr{V}$.*

The proof of this corollary is contained in the proof of the theorem.

Theorem 1.4. *If a vector space $\mathscr{V}$ has a basis $\mathscr{B} = \{X_1, X_2, \ldots, X_n\}$ containing n vectors, then any set $\mathscr{B}'$ containing more than n vectors is linearly dependent.*

Proof. Let $\mathscr{B}' = \{Y_1, Y_2, \ldots, Y_m\}$ where $m > n$. Since $\mathscr{B}$ is a basis, each Y_i is a linear combination of $X_1, X_2, \ldots, X_n$. Thus,

(24)
$$\begin{aligned}
Y_1 &= a_{11} X_1 + a_{12} X_2 + \cdots + a_{1n} X_n, \\
Y_2 &= a_{21} X_1 + a_{22} X_2 + \cdots + a_{2n} X_n, \\
&\cdots\cdots\cdots\cdots\cdots\cdots\cdots\cdots\cdots\cdots \\
Y_m &= a_{m1} X_1 + a_{m2} X_2 + \cdots + a_{mn} X_n.
\end{aligned}$$

Since $m > n$, by Theorem 1.1 we can choose scalars $c_1, c_2, \ldots, c_m$, not all zero, such that

(25)
$$\begin{aligned}
a_{11} c_1 + a_{21} c_2 + \cdots + a_{m1} c_m &= 0, \\
a_{12} c_1 + a_{22} c_2 + \cdots + a_{m2} c_m &= 0, \\
&\cdots\cdots\cdots\cdots\cdots\cdots\cdots\cdots \\
a_{1n} c_1 + a_{2n} c_2 + \cdots + a_{mn} c_m &= 0.
\end{aligned}$$

Equations (24) and (25) imply

$$c_1 Y_1 + c_2 Y_2 + \cdots + c_m Y_m = O,$$

so that $\mathscr{B}$ is linearly dependent.

Corollary 1. *If $\mathscr{V}$ is a finite dimensional vector space any two bases of $\mathscr{V}$ contain the same number of vectors.*

Proof. If $\mathscr{B}$ and $\mathscr{B}'$ are both bases of $\mathscr{V}$ and therefore linearly independent, the theorem shows that $\mathscr{B}$ cannot contain more vectors than $\mathscr{B}'$ and similarly that $\mathscr{B}'$ cannot contain more vectors than $\mathscr{B}$.

Definition. *Let $\mathscr{V}$ be a finite dimensional vector space. The dimension of $\mathscr{V}$ is defined to be the number of vectors in any basis of $\mathscr{V}$. It will be denoted by* dim $\mathscr{V}$.

Note that the term "finite dimensional" for a finitely generated vector space is now justified since we have shown that a finitely generated space has a finite basis and hence has finite dimension.

Theorem 1.5. *The vector space $\mathscr{V}_n(\mathscr{F})$ has dimension n.*

Proof. It was shown in Example 1 that $\mathscr{V}_n(\mathscr{F})$ is spanned by the n vectors

$$E_1 = [1, 0, 0, \ldots, 0],$$
$$E_2 = [0, 1, 0, \ldots, 0],$$
$$\cdots\cdots\cdots\cdots\cdots\cdots$$
$$E_n = [0, 0, 0, \ldots, 1].$$

However, these n vectors are linearly independent because

$$c_1 E_1 + c_2 E_2 + \cdots + c_n E_n = [c_1, c_2, \ldots, c_n]$$

and this is the zero vector only if $c_1 = c_2 = \cdots = c_n = 0$. Thus, $\{E_1, E_2, \ldots, E_n\}$ is a basis for $\mathscr{V}_n(\mathscr{F})$ and $\mathscr{V}_n(\mathscr{F})$ has dimension n.

Note that since $\mathscr{V}_2(\mathscr{R})$ has dimension 2 and $\mathscr{V}_3(\mathscr{R})$ has dimension 3, our algebraic definition of dimension coincides with our intuitive geometric notion of dimension when the latter is applicable.

Definition. *The basis $\{E_1, E_2, \ldots, E_n\}$ of $\mathscr{V}_n(\mathscr{F})$ described in the proof of Theorem 1.5 will be called the standard basis of $\mathscr{V}_n(\mathscr{F})$.*

The following is a useful result.

Lemma 1.6. *If $\mathscr{B} = \{X_1, X_2, \ldots, X_r\}$ is a linearly independent set of vectors in $\mathscr{V}$ and if X_{r+1} is any vector in $\mathscr{V}$ which is not in the subspace spanned by $X_1, X_2, \ldots, X_r$, then $\mathscr{B}' = \{X_1, X_2, \ldots, X_{r+1}\}$ is a linearly independent set.*

Proof. If there is a nontrivial linear relation of the form

(26) $$c_1 X_1 + c_2 X_2 + \cdots + c_{r+1} X_{r+1},$$

then $c_{r+1} \neq 0$ because $\mathscr{B}$ is linearly independent. But then (26) shows that X_{r+1} is in the space spanned by $X_1, X_2, \ldots, X_r$ contrary to the hypothesis. Thus, no nontrivial linear relation (26) can exist and $\mathscr{B}'$ is linearly independent.

Theorem 1.7. *In a vector space $\mathscr{V}$ of dimension n any linearly independent set of n vectors is a basis of $\mathscr{V}$.*

Proof. Let $B = \{X_1, X_2, \ldots, X_n\}$ be a linearly independent set of n vectors and let X be an arbitrary vector of $\mathscr{V}$. Since $\mathscr{V}$ has dimension n and hence has a basis of n vectors, the set $\mathscr{B}' = \{X_1, X_2, \ldots, X_n, X\}$ is linearly dependent by Theorem 1.4. Hence by Lemma 1.6, X is in the space spanned by $X_1, X_2, \ldots, X_n$ since otherwise $\mathscr{B}'$ would be linearly independent. Thus $\mathscr{B}$ is a generating system of $\mathscr{V}$ and hence, being linearly independent, a basis of $\mathscr{V}$.

Theorem 1.8. *Let $\mathscr{V}$ be a vector space of dimension n. Every linearly independent set of vectors of $\mathscr{V}$ is a subset of a basis of $\mathscr{V}$.*

Proof. Let $\mathscr{B} = \{X_1, X_2, \ldots, X_r\}$ be a linearly independent set of vectors. If $r < n$, $\mathscr{B}$ itself is not a basis of $\mathscr{V}$. Hence it does not span $\mathscr{V}$ and there exists a vector X_{r+1} in $\mathscr{V}$ that is not in the space spanned by $X_1, X_2, \ldots, X_r$. By Lemma 1.6 the set $\mathscr{B}_1 = \{X_1, X_2, \ldots, X_{r+1}\}$ is linearly independent. If $\mathscr{B}_1$ is not a basis of $\mathscr{V}$ we repeat the process until we get a linearly independent set

$$\{X_1, X_2, \ldots, X_r, \ldots, X_n\},$$

necessarily containing n vectors, that spans $\mathscr{V}$ and is therefore a basis.

Theorem 1.9. *Every subspace $\mathscr{S}$ of a finite dimensional space $\mathscr{V}$ is finite dimensional.*

Proof. Suppose dim $\mathscr{V} = n$. Since by Theorem 1.4 the number of vectors of $\mathscr{V}$ in a linearly independent set must be less than or equal to n, the same is true for the subspace $\mathscr{S}$. It follows that there exists a maximal linearly independent set $\mathscr{B} = \{S_1, S_2, \ldots, S_m\}$ in $\mathscr{S}$ with $m \leq n$. But by the maximal property of $\mathscr{B}$ this set must span $\mathscr{S}$ since otherwise, by Lemma 1.6, $\mathscr{B}$ could be embedded in a larger linearly independent set of vectors of $\mathscr{S}$. Hence, $\mathscr{S}$ is finite dimensional. Moreover, since $\mathscr{B}$ is a linearly independent generating system and hence a basis of $\mathscr{S}$, dim $\mathscr{S} = m \leq n = $ dim $\mathscr{V}$.

Corollary. *If $\mathscr{S}$ is a subspace of a finite dimensional vector space $\mathscr{V}$ and $\mathscr{B}$ is a basis of $\mathscr{S}$, there exists a basis $\mathscr{B}'$ of $\mathscr{V}$ such that $\mathscr{B}$ is a subset of $\mathscr{B}'$.*

Proof. Since $\mathscr{S}$ is finite dimensional, this result follows from Theorem 1.8.

Theorem 1.10. *If $\mathscr{S}$ and $\mathscr{T}$ are subspaces of a finite dimensional space $\mathscr{V}$ then*

$$\dim \mathscr{S} + \dim \mathscr{T} = \dim (\mathscr{S} + \mathscr{T}) + \dim \mathscr{S} \cap \mathscr{T}.$$

Proof. Let $\dim \mathscr{S} = s$, $\dim \mathscr{T} = t$, $\dim (\mathscr{S} + \mathscr{T}) = j$ and $\dim \mathscr{S} \cap \mathscr{T} = m$. Let $X_1, X_2, \ldots, X_m$ be a basis of $\mathscr{S} \cap \mathscr{T}$. By Theorem 1.8 we can choose a basis of $\mathscr{S}$ of the form $X_1, X_2, \ldots, X_m, Y_1, \ldots, Y_{s-m}$ and a basis of $\mathscr{T}$ of the form $X_1, X_2, \ldots, X_m, Z_1, \ldots, Z_{t-m}$. It follows that every vector $S + T$ of $\mathscr{S} + \mathscr{T}$ is a linear combination of the vectors

(27) $\qquad X_1, X_2, \ldots, X_m, Y_1, \ldots, Y_{s-m}, Z_1, \ldots, Z_{t-m},$

and these, therefore, form a generating system of $\mathscr{S} + \mathscr{T}$. We shall show that they are linearly independent and therefore form a basis. Let

$$a_1 X_1 + \cdots + a_m X_m + b_1 Y_1 + \cdots + b_{s-m} Y_{s-m} + c_1 Z_1 + \cdots + c_{t-m} Z_{t-m} = O.$$

Transposing, we get

(28) $\quad a_1 X_1 + \cdots + a_m X_m + \cdots + b_{s-m} Y_{s-m} = -c_1 Z_1 - \cdots - c_{t-m} Z_{t-m}.$

Now the left-hand side of equation (28) is a vector of $\mathscr{S}$, and the right-hand side is a vector of $\mathscr{T}$. Their equality implies that they both belong to $\mathscr{S} \cap \mathscr{T}$ and, therefore

$$-c_1 Z_1 - \cdots - c_{t-m} Z_{t-m} = d_1 X_1 + \cdots + d_m X_m.$$

But $X_1, \ldots, X_m, Z_1, \ldots, Z_{t-m}$ are linearly independent, being a basis of $\mathscr{T}$, and hence $c_1 = c_2 = \cdots = c_{t-m} = 0$. Therefore, equation (28) becomes a linear relation in the linearly independent vectors $X_1, \ldots, X_m, Y_1, \ldots, Y_{s-m}$ and, consequently,

$$a_1 = a_2 = \cdots = a_m = b_1 = \cdots = b_{s-m} = 0.$$

Therefore the vectors (27) are linearly independent and form a basis of $\mathscr{S} + \mathscr{T}$. Their number, namely, $s + t - m$, is then equal to the dimension j of $\mathscr{S} + \mathscr{T}$ and $s + t = m + j$ as required.

Exercise 1.5

1. Given the vectors $X = [2, -1, 4, 0]$, $Y = [1, 1, 2, 3]$, and $Z = [4, -5, 8, -6]$:
   (a) Show that the set $\{X, Y, Z\}$ is linearly dependent by finding a nontrivial linear relation satisfied by X, Y, Z.

(b) Show that any two of the three vectors constitute a basis of the space $\mathscr{S}$ spanned by X, Y, Z.

(c) Does $\mathscr{S}$ contain a nonzero vector of the form $[a, b, 0, 0]$?

(d) Does $\mathscr{S}$ contain a nonzero vector of the form $[0, c, 0, d]$?

(e) If $\mathscr{T}$ is the subspace of $\mathscr{V}_4(\mathscr{R})$ consisting of all vectors of the form $[0, c, 0, d]$, $c, d \in \mathscr{R}$, find the dimensions of the subspaces $\mathscr{T}$, $\mathscr{S}$, $\mathscr{T} \cap \mathscr{S}$, and $\mathscr{T} + \mathscr{S}$.

2. If an n-dimensional vector space $\mathscr{V}$ has subspaces $\mathscr{S}$ and $\mathscr{T}$, with dimensions s and t, which have no nonzero vector in common, what restriction must s and t satisfy?

3. If $\mathscr{S}$ is the subspace of $\mathscr{V}_3(\mathscr{R})$ generated by the vectors $[1, 2, -1]$ and $[3, 0, 1]$ and $\mathscr{T}$ is the subspace generated by $[-1, 1, 0]$ and $[2, 1, 3]$, find a nonzero vector that belongs to $\mathscr{S} \cap \mathscr{T}$.

4. Prove that if $\mathscr{S}$ is a proper subspace of a finite dimensional vector space $\mathscr{T}$, the dimension of $\mathscr{S}$ is less than that of $\mathscr{T}$.

5. Use Theorem 1.10 to prove that two planes that pass through the origin must either coincide or have a line in common.

6. If $\mathscr{S}$ is a subspace of a finite dimensional vector space $\mathscr{V}$, prove that there exists a subspace $\mathscr{T}$ of $\mathscr{V}$ such that $\mathscr{S} + \mathscr{T} = \mathscr{V}$ and $\mathscr{S} \cap \mathscr{T} = O$. *Hint.* Use the corollary to Theorem 1.9.

7. If $\mathscr{S}$, $\mathscr{T}$, and $\mathscr{U}$ are subspaces of a vector space $\mathscr{V}$ and if $\mathscr{T} \subset \mathscr{S}$, prove that $\mathscr{S} \cap (\mathscr{T} + \mathscr{U}) = \mathscr{T} + (\mathscr{S} \cap \mathscr{U})$.

8. Investigate the geometric meaning of the result proved in Problem 7 when $\mathscr{S}$ and $\mathscr{U}$ are two-dimensional subspaces of $\mathscr{V}_3(\mathscr{R})$. Show by example that the hypothesis $\mathscr{T} \subset \mathscr{S}$ in Problem 7 is necessary for the truth of the conclusion.

9. Show that the following three vectors of $\mathscr{V}_4(\mathscr{R})$ are linearly independent whatever values may be assigned to a, b, and c: $X_1 = [a, 1, 0, 0]$, $X_2 = [b, 0, 1, 0]$, $X_3 = [c, 0, 0, 1]$.

10. Find three linearly independent vectors of the form X_1, X_2, X_3 in Problem 9 which belong to the solution space $\mathscr{S}$ of the equation

$$2x_1 - x_2 + 3x_3 - 5x_4 = 0.$$

Prove that these three vectors span $\mathscr{S}$. What is the dimension of $\mathscr{S}$?

11. Prove that if $a_1, a_2, \ldots, a_n$ are real numbers not all zero, the solution space of the equation

$$a_1 x_1 + a_2 x_2 + \cdots + a_n x_n = 0$$

is an $(n - 1)$-dimensional subspace of $\mathscr{V}_n(\mathscr{R})$.

1.8 Bases and Coordinates

Until now, the term *basis* of a vector space $\mathscr{V}$ has meant a linearly independent set of vectors that spans $\mathscr{V}$. We now find it necessary to confer a definite ordering on the vectors of a basis so that a single basis $\mathscr{B} = \{F_1, F_2, \ldots, F_n\}$, in the sense of the original definition, will give rise to $n!$ different ordered bases corresponding to the $n!$ different ways in which the vectors $F_1, F_2, \ldots, F_n$ can be ordered. We shall use the term basis in the sense of ordered basis. For example, the basis $F_1, F_2, F_3, \ldots$ F_n will be different from the basis $F_2, F_1, F_3, \ldots, F_n$. To emphasize that a basis is now an ordered set rather than simply a set, we shall omit the braces and speak of the basis $F_1, F_2, \ldots, F_n$ rather than the basis $\{F_1, F_2, \ldots, F_n\}$.

Let $\mathscr{V}$ be an n-dimensional vector space over $\mathscr{F}$ and let $F_1, F_2, \ldots, F_n$ be an basis of $\mathscr{V}$. Every vector V of $\mathscr{V}$ can be written in the form

(29) $$V = x_1F_1 + x_2F_2 + \cdots + x_nF_n, \qquad x_i \in \mathscr{F}.$$

We shall show that, because the basis vectors $F_1, F_2, \ldots, F_n$ are linearly independent, the scalar coefficients $x_1, x_2, \ldots, x_n$ are uniquely determined. If we also have

$$V = y_1F_1 + y_2F_2 + \cdots + y_nF_n,$$

we can subtract (29) to get

$$(y_1 - x_1)F_1 + (y_2 - x_2)F_2 + \cdots + (y_n - x_n)F_n = O$$

and the linear independence of the basis implies that $y_i = x_i$, $(i = 1, 2, \ldots, n)$. Because of its importance we state this result formally.

Theorem 1.11. *Let $F_1, F_2, \ldots, F_n$ be a basis of the n-dimensional vector space $\mathscr{V}$ over $\mathscr{F}$. Then every vector V of $\mathscr{V}$ can be written in one and only one way in the form*

$$V = x_1F_1 + x_2F_2 + \cdots + x_nF_n,$$

where $x_i \in \mathscr{F}$, $(i = 1, 2, \ldots, n)$.

The scalars $x_1, x_2, \ldots, x_n$ in (29) are called the *coordinates of V relative to the basis $F_1, F_2, \ldots, F_n$.* It follows from Theorem 1.11 that corresponding to each vector V of $\mathscr{V}$, there is a uniquely determined vector

$$X = \alpha V = [x_1, x_2, \ldots, x_n]$$

in $\mathscr{V}_n(\mathscr{F})$ which we shall call the *coordinate vector* of V relative to the basis $F_1, F_2, \ldots, F_n$. We emphasize that αV depends on the basis as well as on V.

Definition. *Let $\mathscr{V}$ and $\mathscr{W}$ be vector spaces over the same field $\mathscr{F}$. A one-to-one mapping $\theta : \mathscr{V} \to \mathscr{W}$ of $\mathscr{V}$ onto $\mathscr{W}$ is called an isomorphism if it satisfies, for all vectors V and V' of $\mathscr{V}$ and all scalars k, the relations*

(30)
$$\theta(kV) = k\theta(V),$$
$$\theta(V + V') = \theta(V) + \theta(V').$$

If there exists an isomorphism from $\mathscr{V}$ to $\mathscr{W}$ we say the space $\mathscr{V}$ is isomorphic to $\mathscr{W}$.

Since an isomorphism is a one-to-one mapping of $\mathscr{V}$ onto $\mathscr{W}$, there is an inverse mapping $\theta^{-1}:\mathscr{W} \to \mathscr{V}$ defined by $\theta^{-1}W = V$ if and only if $\theta V = W$. Because $\theta(kV) = k(\theta V) = kW$ it follows that $\theta^{-1}(kW) = kV = k(\theta^{-1}W)$. Similarly, $\theta^{-1}(W + W') = \theta^{-1}W + \theta^{-1}W'$ because $\theta(\theta^{-1}W + \theta^{-1}W') = \theta(\theta^{-1}W) + \theta(\theta^{-1}W') = W + W'$. Hence, θ^{-1} is an isomorphism of $\mathscr{W}$ onto $\mathscr{V}$. It follows that if $\mathscr{V}$ is isomorphic to $\mathscr{W}$, via the isomorphism θ, then $\mathscr{W}$ is isomorphic to $\mathscr{V}$ via the isomorphism θ^{-1}. We may therefore say simply that the two spaces $\mathscr{V}$ and $\mathscr{W}$ are isomorphic.

Theorem 1.12. *Let $\mathscr{V}$ be an n-dimensional vector space over the field $\mathscr{F}$ and let $F_1, F_2, \ldots, F_n$ be a basis of $\mathscr{V}$. The mapping $\alpha:\mathscr{V} \to \mathscr{V}_n(\mathscr{F})$ which maps each vector of $\mathscr{V}$ on its coordinate vector relative to the F-basis is an isomorphism of $\mathscr{V}$ onto $\mathscr{V}_n(\mathscr{F})$. Thus, every n-dimensional vector space over $\mathscr{F}$ is isomorphic to $\mathscr{V}_n(\mathscr{F})$.*

Proof. By Theorem 1.11 every vector V has a unique image $X = \alpha V$ in $\mathscr{V}_n(\mathscr{F})$ so the mapping α is well-defined. Since every vector in $\mathscr{V}_n(\mathscr{F})$ is obviously the coordinate vector of some vector in $\mathscr{V}$, α maps $\mathscr{V}$ *onto* $\mathscr{W}$. Clearly, equality of the coordinate vectors of V_1 and V_2 implies $V_1 = V_2$ so the mapping is one-to-one. Finally, it is trivial to verify that for any two vectors V, V' in $\mathscr{V}$ and any scalar k

$$\alpha(V + V') = \alpha V + \alpha V',$$

$$\alpha(kV) = k\alpha V,$$

so that α is an isomorphism.

If in (30) we put $k = 0$ we see that $\theta(O) = O$. Thus, an isomorphism maps the zero vector of $\mathscr{V}$ on the zero vector of $\mathscr{W}$. Moreover the fact that θ is a one-to-one mapping shows that $\theta V = O$ implies $V = O$. We use this fact in proving the following.

Theorem 1.13. *If $\theta:\mathscr{V} \to \mathscr{W}$ is an isomorphism between vector spaces $\mathscr{V}$ and $\mathscr{W}$, a set of vectors $\theta V_1, \theta V_2, \ldots, \theta V_r$ is linearly dependent in $\mathscr{W}$ if and only if the set $V_1, V_2, \ldots, V_r$ is linearly dependent in $\mathscr{V}$.*

Proof. If $V_1, V_2, \ldots, V_r$ are linearly dependent there exists a nontrivial linear relation

$$c_1V_1 + c_2V_2 + \cdots + c_rV_r = O.$$

Applying θ to each side of this equation and using (30), we get

$$c_1(\theta V_1) + c_2(\theta V_2) + \cdots + c_r(\theta V_r) = \theta(O) = O$$

and, since $c_1, c_2, \ldots, c_r$ are not all zero, the set

$$\{\theta V_1, \theta V_2, \ldots, \theta V_r\}$$

is linearly dependent. Conversely, if there is a nontrivial linear relation

$$b_1(\theta V_1) + b_2(\theta V_2) + \cdots + b_r(\theta V_r) = O$$

then (30) gives

$$\theta(b_1 V_1 + b_2 V_2 + \cdots + b_r V_r) = O$$

which, we have already remarked, implies

$$b_1 V_1 + b_2 V_2 + \cdots + b_r V_r = O$$

because θ is one-to-one. Hence, if the set $\{\theta V_1, \theta V_2, \ldots, \theta V_r\}$ is linearly dependent, so is the set $\{V_1, V_2, \ldots, V_r\}$.

Corollary 1. *If $\theta: \mathscr{V} \to \mathscr{W}$ is an isomorphism, a set of vectors $\{\theta V_1, \theta V_2, \ldots \theta V_r\}$ in $\mathscr{W}$ is linearly independent if and only if the set $\{V_1, V_2, \ldots, V_r\}$ is linearly independent in $\mathscr{V}$.*

Corollary 2. *If $\theta: \mathscr{V} \to \mathscr{W}$ is an isomorphism and $F_1, F_2, \ldots, F_n$ is a basis of $\mathscr{V}$, then $\theta F_1, \theta F_2, \ldots, \theta F_n$ is a basis of $\mathscr{W}$ and conversely.*

Proof. If $W \in \mathscr{W}$, then $W = \theta V$ where $V \in \mathscr{V}$. If $F_1, F_2, \ldots, F_n$ is a basis of $\mathscr{V}$, then

$$V = c_1 F_1 + c_2 F_2 + \cdots + c_n F_n$$

and

$$W = \theta V = c_1(\theta F_1) + c_2(\theta F_2) + \cdots + c_n(\theta F_n).$$

This shows that $\theta F_1, \theta F_2, \ldots, \theta F_n$ span $\mathscr{W}$. But these vectors are linearly independent by Corollary 1 and hence constitute a basis of $\mathscr{W}$.

Corollary 3. *If $\mathscr{V}$ is an n-dimensional vector space over $\mathscr{F}$ the vectors $V_1, V_2, \ldots, V_r$ in $\mathscr{V}$ are linearly dependent (independent) if and only if their coordinate vectors relative to any fixed basis are linearly dependent (independent) in $\mathscr{V}_n(\mathscr{F})$.*

We conclude this section with a brief discussion of the relationship between bases of $\mathscr{V}_2(\mathscr{R})$ and $\mathscr{V}_3(\mathscr{R})$ and coordinate systems in the sense of elementary analytic geometry. Choose a rectangular Cartesian coordinate system in the plane with equal units on the two coordinate axes. Let E_1, E_2 be vectors of unit length with initial points at the origin and terminal points at the points $(1, 0)$ and $(0, 1)$, respectively, so that

$$E_1 = [1, 0],$$

$$E_2 = [0, 1]$$

and E_1, E_2 is the standard basis of $\mathscr{V}_2(\mathscr{R})$. If $P(x_1, x_2)$ is an arbitrary point of the plane it is clear that

$$\overrightarrow{OP} = [x_1, x_2] = x_1E_1 + x_2E_2$$

and hence $[x_1, x_2]$ is the coordinate vector of $\overrightarrow{OP}$ relative to the standard basis of $\mathscr{V}_2(\mathscr{R})$ and also (x_1, x_2) are the coordinates of the point P in the chosen coordinate system. Now let

$$F_1 = a_1E_1 + a_2E_2,$$

$$F_2 = b_1E_1 + b_2E_2$$

be any two linearly independent vectors of $\mathscr{V}_2(\mathscr{R})$. Since F_1, F_2 is a basis of $\mathscr{V}_2(\mathscr{R})$, we have

$$\overrightarrow{OP} = x_1'F_1 + x_2'F_2.$$

This means that $\overrightarrow{OP}$ is the diagonal of a parallelogram of which adjacent sides are $x_1'F_1$ and $x_2'F_2$ (Figure 1.1). Thus, $[x_1', x_2']$ is the coordinate vector of $\overrightarrow{OP}$ relative to the F-basis and (x_1', x_2') are the coordinates of the point P in a coordinate system with x_1'-axis along F_1 and x_2'-axis along F_2. The terminal points of F_1, F_2 are the unit points on these coordinate axes, so that the length of F_1 is the unit on the

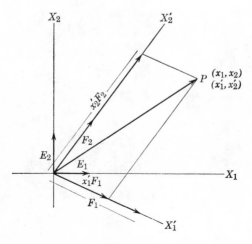

Figure 1.1

x_1'-axis and the length of F_2 is the unit on the x_2'-axis. Finally, since

$$\overrightarrow{OP} = x_1 E_1 + x_2 E_2 = x_1' F_1 + x_2' F_2$$
$$= x_1'(a_1 E_1 + a_2 E_2) + x_2'(b_1 E_1 + b_2 E_2)$$
$$= (a_1 x_1' + b_1 x_2')E_1 + (a_2 x_1' + b_2 x_2')E_2,$$

we have by Theorem 1.11

(31)
$$x_1 = a_1 x_1' + b_1 x_2',$$
$$x_2 = a_2 x_1' + b_2 x_2'.$$

These equations relate the E-coordinates of the point P (or of the vector $\overrightarrow{OP}$) with F-coordinates of the same point (or vector). In particular, if the basis vectors F_1, F_2 are simply the basis vectors E_1, E_2 rotated counterclockwise through an angle θ, then we have

$$F_1 = (\cos \theta)E_1 + (\sin \theta)E_2,$$
$$F_2 = (-\sin \theta)E_1 + (\cos \theta)E_2.$$

Equations (31), connecting the original coordinates of P with its coordinates relative to the rotated axes, now become

$$x_1 = x_1' \cos \theta - x_2' \sin \theta,$$
$$x_2 = x_1' \sin \theta + x_2' \cos \theta.$$

These are the equations of a rotation of axes in the plane.

Example 1. Find the coordinates of the vector $X = [2, 3]$ relative to the basis $F_1 = [-1, 5]$, $F_2 = [3, 1]$ of $\mathscr{V}_2(\mathscr{R})$.

Solution. Scalars x and y must be found such that $X = xF_1 + yF_2$. This requires that

$$-x + 3y = 2,$$
$$5x + y = 3.$$

Solving these two equations, we get $x = \frac{7}{16}$, $y = \frac{13}{16}$, which are the required coordinates.

Example 2. Find an equation of the circle $x_1^2 + x_2^2 = 9$ in the coordinate system based on the basis $F_1 = [2, 1]$, $F_2 = [3, 2]$ of $\mathscr{V}_2(\mathscr{R})$.

Solution. Let E_1, E_2 be the standard basis of $\mathscr{V}_2(\mathscr{R})$. Since

$$F_1 = 2E_1 + E_2,$$
$$F_2 = 3E_1 + 2E_2,$$

we have by (31)

(32)
$$x_1 = 2x_1' + 3x_2',$$
$$x_2 = x_1' + 2x_2',$$

where (x_1, x_2) and (x_1', x_2') are the coordinates of the same point in the E-coordinate system and the F-coordinate system, respectively. Hence, by substituting (32) in the equation $x_1^2 + x_2^2 = 9$ of the given circle, we get the equation of the circle in the new coordinate system, namely,

$$(2x_1' + 3x_2')^2 + (x_1' + 2x_2')^2 = 9$$

or

$$5x_1'^2 + 16x_1'x_2' + 13x_2'^2 = 9.$$

In a similar way we can set up a Cartesian coordinate system in space based on any linearly independent set of three vectors F_1, F_2, F_3, with common initial point O. We choose the origin at O, the x_1-axis along F_1, the x_2-axis along F_2, and the x_3-axis along F_3. The terminal points of F_1, F_2, F_3 are the unit points on the three coordinate axes. Now let $\overrightarrow{OP}$ be any vector. Through P pass a plane parallel to the plane of F_2 and F_3 to cut the x_1-axis in A_1, a plane parallel to the plane of F_1 and F_3 to cut the x_2-axis in A_2, and a plane parallel to the plane of F_1 and F_2 to cut the x_3-axis in A_3. The vectors $\overrightarrow{OA_1}$, $\overrightarrow{OA_2}$, and $\overrightarrow{OA_3}$ are scalar multiples of F_1, F_2, F_3 and

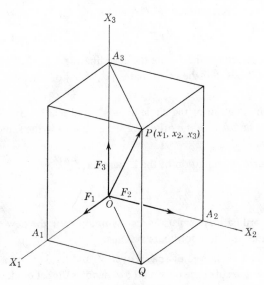

Figure 1.2

may be written $\overrightarrow{OA}_i = x_i F_i$ $(i = 1, 2, 3)$. Since OP is the diagonal of a parallele-piped with edges OA_1, OA_2, OA_3, we have (Figure 1.2)

$$\overrightarrow{OP} = \overrightarrow{OQ} + \overrightarrow{OA}_3$$
$$= \overrightarrow{OA}_1 + \overrightarrow{OA}_2 + \overrightarrow{OA}_3$$
$$= x_1 F_1 + x_2 F_2 + x_3 F_3.$$

Thus $[x_1, x_2, x_3]$ is the coordinate vector of $\overrightarrow{OP}$ relative to the F-basis and (x_1, x_2, x_3) are the coordinates of the point P in the Cartesian coordinate system based on these vectors. Note that we are here using the term Cartesian coordinate system in a very general sense. It is not assumed that the F-coordinate system is either rectangular or isometric.

Exercise 1.6

1. Find the coordinates of the vector $[2, 3]$ of $\mathscr{V}_2(\mathscr{R})$ relative to the basis $F_1 = [1, -1]$ and $F_2 = [3, 5]$.

2. Find the coordinates of the point $(3, 4)$ relative to coordinate axes that have equations $y = -2x$ and $y = 2x$ in rectangular coordinates, using the same unit in both coordinate systems.

3. Find the coordinates of the vector $[2, 1, -6]$ of $\mathscr{V}_3(\mathscr{R})$ relative to the basis $F_1 = [1, 1, 2]$, $F_2 = [3, -1, 0]$, $F_3 = [2, 0, -1]$.

4. If E_1, E_2, E_3 is a basis of $\mathscr{V}_3(\mathscr{R})$ and if F_1, F_2, F_3 is a second basis such that

$$F_i = a_i E_i + b_i E_2 + c_i E_3, \qquad (i = 1, 2, 3),$$

find the equations that relate the coordinates $[x, y, z]$ of an arbitrary vector relative to the E-basis and the coordinates $[x', y', z']$ of the same vector relative to the F-basis.

5. Find an equation of the circle $x^2 + y^2 = r^2$ relative to coordinate axes that make angles α and β with the original x-axis, and using the same unit in both coordinate systems.

6. Transform the equation of the hyperbola

$$\frac{x^2}{a^2} - \frac{y^2}{b^2} = 1$$

to a coordinate system using its asymptotes for the new coordinate axes, and the same unit in both coordinate systems.

7. Let $\mathscr{V}$ and $\mathscr{W}$ be vector spaces over $\mathscr{F}$ and let $\theta: \mathscr{V} \to \mathscr{W}$ be an isomorphism.
   (a) If $\mathscr{S}$ is a subspace of $\mathscr{V}$ and $\theta\mathscr{S}$ denotes the set of all vectors θV, $V \in \mathscr{S}$, prove that $\theta\mathscr{S}$ is a subspace of $\mathscr{W}$.

(b) Prove that dim $(\theta \mathscr{S}) = \dim \mathscr{S}$.

(c) If $\mathscr{T}$ is a subspace of $\mathscr{W}$ and $\theta^{-1}\mathscr{T}$ denotes the set of all vectors V of $\mathscr{V}$ such that $\theta V \in \mathscr{T}$, prove that $\theta^{-1}\mathscr{T}$ is a subspace of $\mathscr{V}$.

(d) Prove that dim $(\theta^{-1}\mathscr{T}) = \dim \mathscr{T}$.

8. Let $\mathscr{V}$ and $\mathscr{W}$ be vector spaces over $\mathscr{F}$ and let $\theta: \mathscr{V} \to \mathscr{W}$ be an isomorphism. If $\mathscr{S}$ and $\mathscr{T}$ are subspaces of $\mathscr{V}$, prove that $\theta\mathscr{S} \cap \theta\mathscr{T} = \theta(\mathscr{S} \cap \mathscr{T})$ and $\theta(\mathscr{S} + \mathscr{T}) = \theta\mathscr{S} + \theta\mathscr{T}$.

CHAPTER TWO

MATRICES AND SYSTEMS OF LINEAR EQUATIONS

2.1 Matrices

A matrix is a rectangular array of scalars of the form

$$(1) \qquad A = \begin{bmatrix} a_{11} & a_{12} & \cdots & a_{1n} \\ a_{21} & a_{22} & \cdots & a_{2n} \\ \cdots & \cdots & \cdots & \cdots \\ a_{m1} & a_{m2} & \cdots & a_{mn} \end{bmatrix}.$$

Each a_{ij} occurring in the matrix is an element of a fixed field $\mathscr{F}$ of scalars which, in accordance with our convention, is assumed to be either $\mathscr{R}$ or $\mathscr{C}$. The scalars a_{ij} are called the *elements* of the matrix. The matrix (1) has m *rows* and n *columns* and is therefore called an $m \times n$ matrix or a matrix of type $m \times n$. If $m = n$, it is called a *square* matrix of *order n*. If the number of rows and columns is clear from the context, the matrix (1) will often be written in the abbreviated form $[a_{ij}]$. Note that the element a_{ij} occurs in the ith row and jth column. Thus, the first index i is called the row index and the second index j the column index of the element a_{ij}.

If we denote the matrix (1) by A, the elements $a_{11}, a_{22}, a_{33}, \ldots$ are called the elements of the *main diagonal* of A. A square matrix in which all elements above and below the main diagonal are 0 is called a *diagonal matrix*. For example, A is a diagonal matrix if $m = n$ and $a_{ij} = 0$ if $i \neq j$ $(i, j = 1, 2, \ldots, n)$.

We can associate with A a second matrix obtained by reflecting the elements of A in the main diagonal; that is, by writing the rows of A as columns. This second matrix is called the transpose of A and is denoted by A^T. We have then

(2)
$$A^T = \begin{bmatrix} a_{11} & a_{21} & \cdots & a_{m1} \\ a_{12} & a_{22} & \cdots & a_{m2} \\ \cdot & \cdot \cdot \cdot \cdot \cdot \cdot & \cdot \\ a_{1n} & a_{2n} & \cdots & a_{mn} \end{bmatrix},$$

and A^T has n rows and m columns. A matrix $X = [x_1, x_2, \ldots, x_n]$ consisting of one row and n columns can be identified with a vector of $\mathscr{V}_n(\mathscr{F})$. Its transpose

$$X^T = \begin{bmatrix} x_1 \\ x_2 \\ \cdot \\ \cdot \\ \cdot \\ x_n \end{bmatrix}$$

is a matrix of n rows and one column, and will be called a *column vector* in contrast to X itself, which will be called a *row vector*. The rows of the matrix A will be called the row vectors of A, and the columns of A the column vectors of A.

Two matrices $A = [a_{ij}]$ and $B = [b_{ij}]$ are said to be *equal* if and only if they have the same number of rows and columns and $a_{ij} = b_{ij}$ for all i and j.

Matrices occur in many areas of mathematics and its applications. For example, a system of homogeneous linear equations such as (13) in Section 1.5 is completely determined by its matrix $[a_{ij}]$ of coefficients, where it is understood that a_{ij} is the coefficient of x_j in the ith equation of the system. In the next section we develop an algebra of matrices which not only will provide us with the basic equipment for the analysis of systems of linear equations, but will lead to many other useful and interesting applications.

2.2 The Algebra of Matrices

If $A = [a_{ij}]$ and $B = [b_{ij}]$ are two $m \times n$ matrices, their *sum* $A + B$ is defined to be the matrix $[a_{ij} + b_{ij}]$ obtained by adding corresponding elements of A and B. Note that the sum $A + B$ is defined only when A and B are of the same type, that is, have the same number of rows and columns. It is clear that addition of matrices is commutative and associative, that is, $A + B = B + A$ and $(A + B) + C = A + (B + C)$ for arbitrary $m \times n$ matrices A, B, and C.

A matrix of which all the elements are zero is called a zero matrix and will be denoted by O regardless of its type. It is clear that O acts as a zero element for matrix addition, that is, $A + O = A$ for every matrix A of the same type as O. Moreover, every matrix $A = [a_{ij}]$ has a negative, or additive inverse, $-A = [-a_{ij}]$, such that $A + (-A) = O$. Thus the set $\mathscr{M}$ of all $m \times n$ matrices, with our definition of addition, satisfies the first four postulates V1 through V4 for a vector space (Section 1.4). If we now define the product of a matrix $A = [a_{ij}]$ by a scalar k to be the matrix $[ka_{ij}]$, it is easy to verify that $\mathscr{M}$ satisfies the postulates V5 through V6 also. Hence, $\mathscr{M}$ is a vector space over $\mathscr{F}$. We state the result formally.

Theorem 2.1. *The set $\mathscr{M}$ of all $m \times n$ matrices with elements in the field $\mathscr{F}$ is a vector space over $\mathscr{F}$.*

We next define, in certain circumstances, the product of two matrices. Let $A = [a_{ij}]$ be a matrix of type $m \times n$ and let $B = [b_{ij}]$ be a matrix of type $n \times r$, so that the number of columns of A is equal to the number of rows of B. We define the product AB to be the $m \times r$ matrix $[c_{ij}]$ where

(3) $$c_{ij} = a_{i1}b_{1j} + a_{i2}b_{2j} + \cdots + a_{in}b_{nj}.$$

Thus, to obtain the element in the ith row and jth column of AB, we multiply the elements of the ith row of A by the corresponding elements of the jth column of B and add.

Example 1

$$\begin{bmatrix} 2 & 1 & 4 \\ 3 & -1 & 1 \\ 0 & 2 & -1 \end{bmatrix} \begin{bmatrix} 5 & 3 \\ 1 & 4 \\ 2 & -1 \end{bmatrix} = \begin{bmatrix} 19 & 6 \\ 16 & 4 \\ 0 & 9 \end{bmatrix}$$

because

$$(2)(5) + (1)(1) + (4)(2) = 19, \qquad (3)(3) + (-1)(4) + (1)(-1) = 4,$$

$$(2)(3) + (1)(4) + (4)(-1) = 6, \qquad (0)(5) + (2)(1) + (-1)(2) = 0,$$

$$(3)(5) + (-1)(1) + (1)(2) = 16, \qquad (0)(3) + (2)(4) + (-1)(-1) = 9.$$

Example 2. Let $A = \begin{bmatrix} 2 & 1 \\ -1 & 4 \end{bmatrix}$ and $B = \begin{bmatrix} 3 & 2 \\ 4 & 1 \end{bmatrix}$.

Then

$$AB = \begin{bmatrix} 2 & 1 \\ -1 & 4 \end{bmatrix} \begin{bmatrix} 3 & 2 \\ 4 & 1 \end{bmatrix} = \begin{bmatrix} 10 & 5 \\ 13 & 2 \end{bmatrix}$$

and

$$BA = \begin{bmatrix} 3 & 2 \\ 4 & 1 \end{bmatrix} \begin{bmatrix} 2 & 1 \\ -1 & 4 \end{bmatrix} = \begin{bmatrix} 4 & 11 \\ 7 & 8 \end{bmatrix}.$$

Note that in Example 2, $AB \neq BA$ and hence matrix multiplication is not commutative. Of course, individual factors *may* commute. For example, if

$$C = \begin{bmatrix} -1 & 2 \\ 3 & 5 \end{bmatrix} \quad \text{and} \quad D = \begin{bmatrix} 7 & 8 \\ 12 & 31 \end{bmatrix}$$

the student may check that $CD = DC$.

Note that our definition of the product AB applies only if the rows of A contain the same number of elements as the columns of B, that is, the number of columns in the left-hand factor A is equal to the number of rows in the right-hand factor B. Whenever a matrix product is indicated, it is assumed, even if not explicitly stated, that this condition is satisfied. Also note that if A is of type $m \times n$ and B of type $n \times r$, then AB is of type $m \times r$.

The student may well ask why we choose this apparently complicated definition of a matrix product rather than, for example, the more obvious procedure of multiplying corresponding elements in two $m \times n$ matrices $[p_{ij}]$ and $[q_{ij}]$ to get a "product" matrix $[p_{ij}q_{ij}]$. The answer is that we choose the definition that leads to the most interesting and fruitful mathematical development and the most useful applications. It will soon be made clear why this criterion dictates the definition embodied in equation (3). First, however, we require some basic properties of matrix products.

Theorem 2.2. (*The associative law for matrix multiplication*) *If A, B, C are three matrices of type $m \times n$, $n \times r$, and $r \times s$, respectively, $(AB)C = A(BC)$.*

Proof. Let $A = [a_{ij}]$, $B = [b_{ij}]$, and $C = [c_{ij}]$. Note that the products AB, $(AB)C$, BC, and $A(BC)$ all exist.

By (3) the element in the ith row and lth column of AB is

$$\sum_{k=1}^{n} a_{ik}b_{kl}$$

and the element in the ith row and jth column of $(AB)C$ is

(4)
$$\sum_{l=1}^{r} \sum_{k=1}^{n} a_{ik}b_{kl}c_{lj}.$$

On the other hand, the element in the kth row and jth column of BC is

$$\sum_{l=1}^{r} b_{kl}c_{lj}$$

and hence the element in the ith row and jth column of $A(BC)$ is

(5)
$$\sum_{k=1}^{n} \sum_{l=1}^{r} a_{ik} b_{kl} c_{lj}.$$

The two double sums (4) and (5) are identical except for the order of summation and hence $(AB)C = A(BC)$ as required.

Corollary. (*General associative law*) *If* $A_1, A_2, \ldots, A_s$ *is any finite ordered set of matrices such that each of the products* $A_i A_{i+1}, i = 1, 2, \ldots, s - 1$, *is defined, then the product*

$$A_1 A_2 \cdots A_s$$

is uniquely determined so that the insertion of brackets is unnecessary.

The associative law on three elements is well known to imply the general associative law. The proof is by induction and is omitted here. It may be found in [6].

Theorem 2.3. (*Distributive laws*) *If* A *and* B *are matrices of type* $m \times n$, C *a matrix of type* $r \times m$, *and* D *a matrix of type* $n \times s$, *then*

$$C(A + B) = CA + CB,$$
$$(A + B)D = AD + BD.$$

Proof. Note that all indicated products are, in fact, defined. Suppose that the elements in the ith row and jth column of A, B, and C are, respectively, a_{ij}, b_{ij}, and c_{ij}. Then the element in the ith row and jth column of $A + B$ is $a_{ij} + b_{ij}$ and that of $C(A + B)$ is

$$\sum_{k=1}^{m} c_{ik}(a_{kj} + b_{kj}).$$

This is equal to

$$\sum_{k=1}^{m} c_{ik} a_{kj} + \sum_{k=1}^{m} c_{ik} b_{kj},$$

which is the element in the ith row and jth column of $CA + CB$. This proves the first distributive law. The proof of the second is similar and is left as an exercise.

Let I_n be the $n \times n$ matrix

$$\begin{bmatrix} 1 & 0 & 0 & \cdots & 0 \\ 0 & 1 & 0 & \cdots & 0 \\ \cdots & \cdots & \cdots & \cdots & \cdots \\ 0 & 0 & 0 & \cdots & 1 \end{bmatrix}$$

in which the main diagonal elements are all equal to 1 and all other elements are 0. The matrix I_n acts as a multiplicative unit because $I_n A = AI_n = A$ for every $n \times n$ matrix A. It is called the *unit* or the *identity* matrix of order n. The subscript n will usually be omitted and an identity matrix will be denoted simply by I, its order being understood from the context.

It is clear also that if O is a zero matrix, then for any matrix A, AO and OA are both zero matrices if defined.

Theorem 2.4. *If A is an $m \times n$ matrix and B an $n \times r$ matrix, then*

$$(AB)^T = B^T A^T.$$

Proof. If $A = [a_{ij}]$ and $B = [b_{ij}]$, the element in the ith row and jth column of AB is

$$\sum_{k=1}^{n} a_{ik} b_{kj}$$

Now

$$B^T = \begin{bmatrix} b_{11} & b_{21} & \cdots & b_{n1} \\ b_{12} & b_{22} & \cdots & b_{n2} \\ \cdot & \cdot & \cdots & \cdot \\ b_{1r} & b_{2r} & \cdots & b_{nr} \end{bmatrix} \quad \text{and} \quad A^T = \begin{bmatrix} a_{11} & a_{21} & \cdots & a_{m1} \\ a_{12} & a_{22} & \cdots & a_{m2} \\ \cdot & \cdot & \cdots & \cdot \\ a_{1n} & a_{2n} & \cdots & a_{mn} \end{bmatrix}.$$

Therefore $B^T A^T$ is defined and the element in the ith row and jth column of this product is

$$\sum_{k=1}^{n} b_{ki} a_{jk} = \sum_{k=1}^{n} a_{jk} b_{ki}.$$

Since this is the element in the jth row and ith column of AB, it follows that $B^T A^T = (AB)^T$ as required.

Corollary. *If the product $A_1 A_2 \cdots A_s$ is defined, so is the product $A_s^T A_{s-1}^T \cdots A_1^T$*

and

$$(A_1 A_2 \cdots A_s)^T = A_s^T A_{s-1}^T \cdots A_1^T.$$

In other words, the transpose of a product is the product of the transposes in reverse order.

Proof. This follows by repeated application of the theorem.

The following is a useful result which is implicit in the definition of a matrix product. We state it formally for easy reference.

Theorem 2.5

(a) *The ith row in the matrix product AB is $R_i B$ where R_i is the ith row of A.*

(b) *The jth column in the matrix product AB is AC_j where C_j is the jth column of B.*

Proof. The statement (a) follows directly from the definition of AB and (b) follows either from the definition or from (a) and Theorem 2.4. The details are left to the student.

Exercise 2.1

1. Let

$$A = \begin{bmatrix} 1 & -2 & 5 \\ 3 & 1 & 0 \\ 2 & 2 & 3 \end{bmatrix}, \quad B = \begin{bmatrix} 2 & 1 \\ -1 & 4 \\ 6 & 2 \end{bmatrix}, \quad C = \begin{bmatrix} 4 & -1 \\ 2 & 3 \end{bmatrix}.$$

Which of the indicated products AB, BA, AC, BC, CB, CB^T, AA, and BB are defined?

2. Compute all the products listed in Problem 1 which are defined.

3. If $X = [3 \quad 1 \quad 2 \quad 4]$, compute XX^T and $X^T X$.

4. If $A = \begin{bmatrix} 1 & 2 \\ 2 & -1 \end{bmatrix}$ and $B = \begin{bmatrix} 2 & 1 \\ 0 & 4 \end{bmatrix}$, compute AB and BA.

5. If $A = \begin{bmatrix} 1 & -2 \\ 3 & -6 \end{bmatrix}$ and $B = \begin{bmatrix} 4 & 2 \\ 2 & 1 \end{bmatrix}$, compute AB and BA.

6. If P is any matrix, prove that the product PP^T is defined and that this product is equal to its own transpose.

7. Show that the matrix equation

$$\begin{bmatrix} 2 & -1 & 6 \\ 1 & 4 & 2 \\ 3 & 2 & -1 \end{bmatrix} \begin{bmatrix} x \\ y \\ z \end{bmatrix} = \begin{bmatrix} 2 \\ 1 \\ 7 \end{bmatrix}$$

is equivalent to three linear equations in x, y, and z.

8. Let A be any square matrix of order n.

(a) Write A^2 for the product AA and show that $A^2 A = A A^2$.

(b) Show that the definitions $A^2 = AA$ and $A^m = A A^{m-1}$ for $m = 3, 4, 5, \ldots$ define a unique matrix A^m for every positive integer m.

(c) Prove that for any two positive integers r and s,

$$A^r A^s = A^s A^r = A^{r+s}.$$

9. If $A = \begin{bmatrix} 2 & -1 \\ 3 & 4 \end{bmatrix}$, compute $3A^2 - 4A$.

10. Show that if A is any square matrix, the matrix $A^3 - 2A^2 + 7A$ commutes with A.

11. If A and B are square matrices of order n:
    (a) Is it always true that $(A + B)^2 = A^2 + 2AB + B^2$?
    (b) Is it always true that $(A + A^2)^2 = A^2 + 2A^3 + A^4$?
    (c) Is it always true that $(A + B)^2 = A^2 + AB + BA + B^2$?

12. If A and B are square matrices of order n in which all the elements above the main diagonal are zero, show that AB has the same property. If the word "above" is replaced by "below," is this still true?

13. Let

$$A = \begin{bmatrix} 2 & 1 & 1 & -2 \\ 1 & 3 & -1 & 2 \\ 2 & 0 & 1 & 4 \\ 3 & 1 & 0 & -1 \end{bmatrix}, \qquad B = \begin{bmatrix} 5 & 2 & 1 & 4 \\ 3 & -1 & 2 & 2 \\ 7 & 1 & 1 & -1 \\ 0 & 3 & 2 & 4 \end{bmatrix}.$$

    (a) Compute the product AB.
    (b) Verify [Theorem 2.5(a)] that the 2nd row of AB is R_2B where $R_2 = [1, 3, -1, 2]$ and, similarly, that the 4th row of AB is R_4B where R_4 is the 4th row of A.
    (c) Verify that the columns of AB are AC_1, AC_2, AC_3, and AC_4 where C_i is the ith column of B.

14. Let $\mathscr{M}$ be the vector space of all 2×2 matrices with elements in $\mathscr{F}$. Show that $\dim \mathscr{M} = 4$ by finding four matrices that form a basis for $\mathscr{M}$.

15. If $\mathscr{M}$ is the vector space of all $m \times n$ matrices with elements in $\mathscr{F}$, show that $\dim \mathscr{M} = mn$ and find a basis for $\mathscr{M}$.

16. Let $\mathscr{M}$ be the set of all $m \times n$ matrices $[a_{ij}]$ over $\mathscr{F}$ such that $\sum_{i=1}^{n} a_{ii} = 0$. Prove that $\mathscr{M}$ is a vector space over $\mathscr{F}$, and find $\dim \mathscr{M}$ and a basis for $\mathscr{M}$.

17. Multiply the matrices A and B of Problem 13 first on the left and then on the right by the diagonal matrix

$$\begin{bmatrix} 2 & 0 & 0 & 0 \\ 0 & 3 & 0 & 0 \\ 0 & 0 & 5 & 0 \\ 0 & 0 & 0 & -1 \end{bmatrix}.$$

18. State a general theorem about the effect of multiplying an $m \times n$ matrix:
    (a) On the left by an $m \times m$ diagonal matrix.
    (b) On the right by an $n \times n$ diagonal matrix.

2.3 Systems of Linear Equations

Consider the following system of m linear equations in n unknowns, $x_1, x_2, \ldots, x_n$:

(6)
$$
\begin{aligned}
a_{11}x_1 + a_{12}x_2 + \cdots + a_{1n}x_n &= c_1, \\
a_{21}x_1 + a_{22}x_2 + \cdots + a_{2n}x_n &= c_2, \\
\cdots \cdots \cdots \cdots \cdots \cdots \cdots \cdots \cdots \cdots \cdots \\
a_{m1}x_1 + a_{m2}x_2 + \cdots + a_{mn}x_n &= c_m.
\end{aligned}
$$

The system (6) will frequently be written in the abbreviated form

$$
\sum_{j=1}^{n} a_{ij}x_j = c_i \qquad (i = 1, 2, \ldots, m).
$$

If we let $A = [a_{ij}]$ be the matrix of coefficients and let

$$
X = \begin{bmatrix} x_1 \\ x_2 \\ \cdot \\ \cdot \\ \cdot \\ x_n \end{bmatrix} \quad \text{and} \quad C = \begin{bmatrix} c_1 \\ c_2 \\ \cdot \\ \cdot \\ \cdot \\ c_n \end{bmatrix},
$$

then (6) can also be written as the matrix equation

$$
AX = C
$$

because the matrix product AX is an $n \times 1$ matrix whose elements are precisely the left members of equations (6). The matrix A is called the coefficient matrix of the system (6) and the matrix

$$
B = \begin{bmatrix}
a_{11} & a_{12} & \cdots & a_{1n} & c_1 \\
a_{21} & a_{22} & \cdots & a_{2n} & c_2 \\
\cdots & \cdots & \cdots & \cdots & \cdots \\
a_{m1} & a_{m2} & \cdots & a_{mn} & c_m
\end{bmatrix}
$$

is called the *augmented matrix* of the system. Our immediate aim is to study the solutions of such a system by examining these matrices. In the process, we shall derive a systematic procedure for finding solutions and a number of important theoretical results. Before doing this, we shall illustrate the method by some examples.

Example 1. Solve the system

$$
\begin{aligned}
2x - y + 3z &= 4, \\
3x + 2y + z &= 6, \\
-5x + y - 2z &= 3.
\end{aligned}
$$

(7)

Solution.

Step 1. Multiply the first equation by $\frac{1}{2}$ so that the coefficient of x will be 1:

$$
\begin{aligned}
x - \tfrac{1}{2}y + \tfrac{3}{2}z &= 2, \\
3x + 2y + z &= 6, \\
-5x + y - 2z &= 3.
\end{aligned}
$$

(8)

Step 2. Multiply the first equation of (8) by -3 and add to the second, then multiply the first equation by 5 and add to the third to get

$$
\begin{aligned}
x - \tfrac{1}{2}y + \tfrac{3}{2}z &= 2, \\
\tfrac{7}{2}y - \tfrac{7}{2}z &= 0, \\
-\tfrac{3}{2}y + \tfrac{11}{2}z &= 13.
\end{aligned}
$$

(9)

Step 3. Multiply the second equation of (9) by $\frac{2}{7}$ so that the coefficient of y will be 1; multiply the new second equation by $\frac{3}{2}$ and add to the third to get

$$
\begin{aligned}
x - \tfrac{1}{2}y + \tfrac{3}{2}z &= 2, \\
y - z &= 0, \\
4z &= 13.
\end{aligned}
$$

(10)

Step 4. Multiply the third equation of (10) by $\frac{1}{4}$. Add the new third equation to the second. Multiply the new third equation by $-\frac{3}{2}$ and add it to the first. The result is

$$
\begin{aligned}
x - \tfrac{1}{2}y &= -\tfrac{23}{8}, \\
y &= \tfrac{13}{4}, \\
z &= \tfrac{13}{4}.
\end{aligned}
$$

(11)

Step 5. Multiply the second equation of (11) by $\frac{1}{2}$ and add it to the first; we get

$$
\begin{aligned}
x &= -\tfrac{5}{4}, \\
y &= \tfrac{13}{4}, \\
z &= \tfrac{13}{4}.
\end{aligned}
$$

(12)

By the argument used in the proof of Theorem 1.1, each of the systems (8), (9), (10), (11), and (12) is equivalent to the original system (7). Thus the solution of (7) is given by (12).

The alert student will be able to shorten the solution of Example 1 in several places. We have been more interested in describing a systematic procedure than in obtaining the solution of this particular example in the shortest possible way. The important features of the method described are:

1. It always yields the solutions of the system when these exist.
2. It demonstrates the nonexistence of solutions when solutions do not exist (see Example 2).
3. It requires only three basic operations on the system of equations. For this reason, it is easy to program an electronic computer to carry out the solution by this method. If the number of equations and unknowns is large, this is highly desirable.

The three basic operations referred to in (3) are:
 (a) Changing the order in which the equations are written.
 (b) Multiplying each term in an equation by a nonzero scalar.
 (c) Multiplying an equation by a non-zero scalar and then adding the new equation so formed to another equation of the system.
The operation (a) was not used in Example 1 but it is necessary if, for example, the coefficient of x is zero in the first equation.

We now note that the operations used to reduce the system (7) to system (12) can be considered as operations on the augmented matrix B of the system. From this point of view, the three operations (a), (b), and (c) become equivalent to the following operations on the rows of the matrix B:

 (α) Interchange of two rows
 (β) Multiplication of one row vector by a nonzero scalar
 (γ) Addition of a scalar multiple of one row vector to another.

Operations of the form (α), (β), and (γ) are called *elementary row transformations* of the matrix B. Our method of solution can be described as reduction of the augmented matrix of the system by elementary row transformations to a form from which the solutions of the system can be written down immediately. We now write the solution of Example 1 in matrix form. At each stage, we use R_1, R_2, R_3 for the first, second, and third row of the matrix on which we are operating and indicate at the left the operations performed.

We begin with the augmented matrix of (7), namely,

$$\begin{bmatrix} 2 & -1 & 3 & 4 \\ 3 & 2 & 1 & 6 \\ -5 & 1 & -2 & 3 \end{bmatrix}.$$

Successive elementary row transformations yield the following:

$$\begin{matrix} \frac{1}{2}R_1 \to \end{matrix} \begin{bmatrix} 1 & -\frac{1}{2} & \frac{3}{2} & 2 \\ 3 & 2 & 1 & 6 \\ -5 & 1 & -2 & 3 \end{bmatrix},$$

$$\begin{matrix} \\ R_2 - 3R_1 \to \\ R_3 + 5R_1 \to \end{matrix} \begin{bmatrix} 1 & -\frac{1}{2} & \frac{3}{2} & 2 \\ 0 & \frac{7}{2} & -\frac{7}{2} & 0 \\ 0 & -\frac{3}{2} & \frac{11}{2} & 13 \end{bmatrix},$$

$$\begin{matrix} R_1 \to \\ \frac{2}{7}R_2 \to \\ R_3 \to \end{matrix} \begin{bmatrix} 1 & -\frac{1}{2} & \frac{3}{2} & 2 \\ 0 & 1 & -1 & 0 \\ 0 & -\frac{3}{2} & \frac{11}{2} & 13 \end{bmatrix},$$

$$\begin{matrix} R_1 \to \\ R_2 \to \\ R_3 + \frac{3}{2}R_2 \to \end{matrix} \begin{bmatrix} 1 & -\frac{1}{2} & \frac{3}{2} & 2 \\ 0 & 1 & -1 & 0 \\ 0 & 0 & 4 & 13 \end{bmatrix},$$

$$\begin{matrix} \\ \\ \frac{1}{4}R_3 \to \end{matrix} \begin{bmatrix} 1 & -\frac{1}{2} & \frac{3}{2} & 2 \\ 0 & 1 & -1 & 0 \\ 0 & 0 & 1 & \frac{13}{4} \end{bmatrix},$$

$$\begin{matrix} R_1 - \frac{3}{2}R_3 \to \\ R_2 + R_3 \to \\ \end{matrix} \begin{bmatrix} 1 & -\frac{1}{2} & 0 & -\frac{23}{8} \\ 0 & 1 & 0 & \frac{13}{4} \\ 0 & 0 & 1 & \frac{13}{4} \end{bmatrix},$$

$$\begin{matrix} R_1 + \frac{1}{2}R_2 \to \\ \\ \end{matrix} \begin{bmatrix} 1 & 0 & 0 & -\frac{5}{4} \\ 0 & 1 & 0 & \frac{13}{4} \\ 0 & 0 & 1 & \frac{13}{4} \end{bmatrix}.$$

From the final "reduced matrix" which is the augmented matrix of system (12), we read the solution $x = -\frac{5}{4}$, $y = \frac{13}{4}$, $z = \frac{13}{4}$ of the system (7).

We now do two additional examples to illustrate the two other cases that can occur, namely, systems with no solutions and systems with infinitely many solutions.

Example 2. Show that the system

$$x + 3y + 2z = 7,$$

(13)
$$2x + y - z = 5,$$

$$-x + 2y + 3z = 4$$

has no solution.

Solution. Reduce the augmented matrix as follows:

$$\begin{bmatrix} 1 & 3 & 2 & 7 \\ 2 & 1 & -1 & 5 \\ -1 & 2 & 3 & 4 \end{bmatrix},$$

$$\begin{matrix} \\ R_2 - 2R_1 \rightarrow \\ R_3 + R_1 \rightarrow \end{matrix} \begin{bmatrix} 1 & 3 & 2 & 7 \\ 0 & -5 & -5 & -9 \\ 0 & 5 & 5 & 11 \end{bmatrix},$$

$$\begin{matrix} \\ -R_2 \rightarrow \\ R_3 + R_2 \rightarrow \end{matrix} \begin{bmatrix} 1 & 3 & 2 & 7 \\ 0 & 5 & 5 & 9 \\ 0 & 0 & 0 & 2 \end{bmatrix}.$$

Without carrying the reduction any further, it is clear that this is the matrix of the system

$$x + 3y + 2z = 7,$$

(14)
$$5y + 5z = 9,$$

$$0z = 2,$$

which is equivalent to the system (13). Since the equation $0z = 2$ has no solution, the system (14) and therefore the original system (13) has no solution.

Example 3. Show that the following system has infinitely many solutions and find formulas that give all solutions of the system:

$$x_1 - 2x_2 - x_3 - x_4 = 3,$$

(15)
$$x_1 + 3x_2 + 2x_3 + x_4 = -2,$$

$$2x_1 + x_2 + x_3 = 1,$$

$$3x_1 + 4x_2 + 3x_3 + 3x_4 = -1.$$

Solution. We write the augmented matrix and reduce it as follows by elementary row transformations.

$$\begin{bmatrix} 1 & -2 & -1 & -1 & 3 \\ 1 & 3 & 2 & 1 & -2 \\ 2 & 1 & 1 & 0 & 1 \\ 3 & 4 & 3 & 3 & -1 \end{bmatrix},$$

$$\begin{matrix} \\ R_2 - R_1 \rightarrow \\ R_3 - 2R_1 \rightarrow \\ R_4 - 3R_1 \rightarrow \end{matrix} \begin{bmatrix} 1 & -2 & -1 & -1 & 3 \\ 0 & 5 & 3 & 2 & -5 \\ 0 & 5 & 3 & 2 & -5 \\ 0 & 10 & 6 & 6 & -10 \end{bmatrix},$$

$$\begin{matrix} \\ \frac{1}{5}R_2 \rightarrow \\ R_3 - R_2 \rightarrow \\ R_4 - 2R_2 \rightarrow \end{matrix} \begin{bmatrix} 1 & -2 & -1 & -1 & 3 \\ 0 & 1 & \frac{3}{5} & \frac{2}{5} & -1 \\ 0 & 0 & 0 & 0 & 0 \\ 0 & 0 & 0 & 0 & 0 \end{bmatrix},$$

$$\begin{matrix} R_1 + 2R_2 \rightarrow \\ \\ \\ \end{matrix} \begin{bmatrix} 1 & 0 & \frac{1}{5} & -\frac{1}{5} & 1 \\ 0 & 1 & \frac{3}{5} & \frac{2}{5} & -1 \\ 0 & 0 & 0 & 0 & 0 \\ 0 & 0 & 0 & 0 & 0 \end{bmatrix}.$$

The reduced matrix corresponds to the system of equations

$$x_1 + \tfrac{1}{5}x_3 - \tfrac{1}{5}x_4 = 1,$$
$$x_2 + \tfrac{3}{5}x_3 + \tfrac{2}{5}x_4 = -1$$

or

(16)

$$x_1 = 1 - \tfrac{1}{5}x_3 + \tfrac{1}{5}x_4,$$
$$x_2 = -1 - \tfrac{3}{5}x_3 - \tfrac{2}{5}x_4,$$

and it is clear that a solution may be found by assigning arbitrary values to x_3 and x_4 and computing x_1 and x_2 from (16). All solutions are found in this way—one for each pair of values assigned to x_3 and x_4. Thus, (16) is called a general solution of (15). Particular solutions are found by assigning numerical values to x_3 and x_4 in (16). Sample particular solutions are

$$x_1 = 1, \quad x_2 = -1, \quad x_3 = 0, \quad x_4 = 0$$

and

$$x_1 = 2, \quad x_2 = -8, \quad x_3 = 5, \quad x_4 = 10.$$

General solutions are not unique. For example, (16) could be solved for x_3 and x_4 in terms of x_1 and x_2. Any formula that gives all solutions in terms of parameters which may be assigned arbitrary values is called a general solution.

A general solution may also be given in vector form. For example, from (16) all solution vectors of (15) have the form

$$[1 - \tfrac{1}{5}x_3 + \tfrac{1}{5}x_4, -1 - \tfrac{3}{5}x_3 - \tfrac{2}{5}x_4, x_3, x_4],$$

where x_3, x_4 may be assigned arbitrary values. This general vector solution can also be written in the form

$$[1, -1, 0, 0] + x_3 [-\tfrac{1}{5}, -\tfrac{3}{5}, 1, 0] + x_4[\tfrac{1}{5}, -\tfrac{2}{5}, 0, 1].$$

Thus, all solutions of (15) are obtained by adding to the vector $[1, -1, 0, 0]$ any linear combination of the two vectors $[-\tfrac{1}{5}, -\tfrac{3}{5}, 1, 0]$ and $[\tfrac{1}{5}, -\tfrac{2}{5}, 0, 1]$.

Exercise 2.2

In Problems 1 through 20, find all solutions of the given systems or prove that solutions do not exist.

1. $x + y - z = 6,$
$2x + 5y - 2z = 10.$

2. $x + y = 0,$
$2x + 3y = 0.$

3. $x + y + z = 1,$
$2x - 3y + 7z = 0,$
$3x - 2y + 8z = 4.$

4. $x + y - 2z + t = 0,$
$2x + 2y - 5z + 3t = 0.$

5. $x - y + 2z = 1,$
$x + y + z = 2,$
$2x - y + z = 5.$

6. $x - y + 2z = 4,$
$3x + y + 4z = 6,$
$x + y + z = 1.$

7. $x + 3y + z = 2,$
$2x + 7y + 4z = 6,$
$x + y - 4z = 1.$

8. $x + 3y + z = 0,$
$2x + 7y + 4z = 0,$
$x + y - 4z = 0.$

9. $2x - y + 5z = 19$
$x + 5y - 3z = 4,$
$3x + 2y + 4z = 5.$

10. $2x - y + 5z = 19,$
$x + 5y - 3z = 4,$
$3x + 2y + 4z = 25.$

11. $x + 2y = 6,$
$3x + 7y = 14.$

12. $x - 3y = 2,$
$-2x + 6y = 12.$

13. $x - 3y = 2,$
$-2x + 6y = -4.$

14. $5x = 7.$

15. $0x = 7.$

16. $0x = 0.$

17. $x_1 + 2x_2 + \ x_3 = 0,$
       $2x_1 + 5x_2 + 4x_3 = 0,$
       $x_1 + 4x_2 + 6x_3 = 0.$

$\downarrow$ 18. $x_1 + 2x_2 - \ x_3 + 7x_4 = 4,$
         $2x_1 + 2x_2 + \ x_3 + 3x_4 = 4,$
         $x_1 - \ x_2 - \ x_3 + \ x_4 = 1,$
         $3x_1 + \ x_2 - 2x_3 + 4x_4 = 3.$

19. $x_1 + 2x_2 - 4x_3 + \ x_4 = 0,$
       $3x_1 + 2x_2 + \ x_3 + 2x_4 = 0,$
       $2x_1 \qquad + 5x_3 + \ x_4 = 0,$
       $4x_1 + 4x_2 - 3x_3 + 3x_4 = 0.$

20. $2x_1 + \ x_2 + 4x_3 - 2x_4 = 1,$
        $x_1 + 3x_2 - 2x_3 + 2x_4 = 4,$
       $3x_1 + 4x_2 + 2x_3 \qquad = 0,$
       $x_1 - 2x_2 + 6x_3 - 4x_4 = 7.$

21. Let A be any $m \times n$ matrix and consider the system of linear equations

$$AX = C.$$

Prove that a solution exists if and only if the vector C is in the subspace of $\mathscr{V}_m(\mathscr{F})$ spanned by the column vectors of A. *Hint.* Show that the vector AX is a linear combination of the column vectors of A.

22. Using the result of Problem 21, determine whether the vector $[9, -8, 3]$ is in the space spanned by the vectors $[2, 1, 4]$ and $[1, 3, 5]$.

23. Prove that a homogeneous system $AX = O$ has a nonzero solution vector if and only if the column vectors of A are linearly dependent.

24. If A is an $n \times n$ matrix with linearly independent column vectors, prove that the system

$$AX = C$$

always has a unique solution, whatever the vector C. *Hint.* Note that the column vectors of A constitute a basis for $\mathscr{V}_n(\mathscr{F})$.

2.4 Row-Equivalence and the Reduced Row-Echelon Form of a Matrix

In this section, we shall identify and exploit the basic principles involved in the method of solution of linear systems described in the last section.

Definition. *A matrix A is said to be* row-equivalent *to a matrix B if A can be transformed into B by a finite number of elementary row transformations. We write $A \sim B$ for A is row-equivalent to B.*

Theorem 2.6

(a) *If A is any matrix, $A \sim A$.*
(b) *If $A \sim B$, then $B \sim A$.*
(c) *If $A \sim B$ and $B \sim C$, then $A \sim C$.*

Proof. Parts (a) and (c) are both obvious from the definition of row-equivalence. To prove (b), it is only necessary to note that each elementary transformation ϵ can be "undone" by another elementary transformation ϵ^{-1}, which will be called its *inverse*. For example, the inverse of the interchange of two rows is the interchange of the same two rows again. If ϵ is the transformation that adds k times the jth row vector to the ith, then the inverse ϵ^{-1} of ϵ consists in adding $-k$ times the jth row vector to the ith. Finally, if ϵ is the transformation that multiplies a row by a nonzero scalar k, then ϵ^{-1} is the transformation that multiples the same row by $1/k$. Now, if A is transformed into B by a succession of elementary transformations $\epsilon_1, \epsilon_2, \ldots, \epsilon_m$, the elementary transformations $\epsilon_m^{-1}, \epsilon_{m-1}^{-1}, \ldots, \epsilon_1^{-1}$ will clearly transform B into A, and therefore (b) follows.

The three laws (a), (b), and (c) of Theorem 2.6 are called, respectively, the reflexive, symmetric, and transitive laws. Any relation, defined on the elements of a set, that is reflexive, symmetric, and transitive is called an *equivalence relation*. The simplest equivalence relation is, of course, *equality* but there are many others. Row-equivalence of matrices is, by Theorem 2.6, an equivalence relation, and we shall meet others later. Examples are given in Exercise 2.3, Problems 5 and 6.

Theorem 2.7. *Let A and B be the augmented matrices of two systems of linear equations in n unknowns. If $A \sim B$, then the two systems have the same solutions.*

Proof. If $A \sim B$ we can pass from the system with matrix A to the system with matrix B by a succession of operations of the type (a), (b), and (c) described in Section 2.3. It is obvious that operations of type (a) and (b) do not change the solutions of the system. The proof that operations of type (c) do not change the solutions is similar to that given in the proof of Theorem 1.1.

Definition. *A matrix R is called a* reduced row-echelon matrix *if it has the following properties.*

(1) *All zero row vectors (if any) occur below all nonzero row vectors.*

(2) *The first nonzero element in each row (reading from the left) is equal to 1.*

(3) *If the first nonzero element in the ith row is in the j_ith column, then every other element in the j_ith column is zero.*

(4) *If the first nonzero element in the ith row is in the j_ith column, then*

$$j_1 < j_2 < j_3 < \cdots.$$

The reader should verify that this definition provides a formal description of the "reduced" matrices obtained in the examples of Section 2.3 from which the general solutions of the corresponding system of equations were obtained.

Theorem 2.8. *Every matrix is row-equivalent to a reduced row-echelon matrix.*

Proof. The proof is a formalization of the reduction process used in Section 2.3. Let $A = [a_{ij}]$ be the given matrix. Suppose the first nonzero column of A is the j_1th column. By an interchange of rows of A (if necessary), we can find a matrix $B = [b_{ij}]$ row-equivalent to A with $b_{1j_1} \neq 0$ and all $b_{ij} = 0$ for $j < j_1$. Multiply the first row of B by $b_{1j_1}^{-1}$ and by adding suitable scalar multiples of the new first row to the other rows of B, we get $A \sim B \sim C$, where

$$C = \begin{bmatrix} 0 & \cdots & 0 & 1 & c_{1j_1+1} & \cdots & c_{1n} \\ 0 & \cdots & 0 & 0 & c_{2j_1+1} & \cdots & c_{2n} \\ \multicolumn{7}{c}{\dotfill} \\ 0 & \cdots & 0 & 0 & c_{mj_1+1} & \cdots & c_{mn} \end{bmatrix}.$$

Now let the j_2th column of C be the first column in which a nonzero element occurs in a row other than the first. Then, $j_1 < j_2$ and by reordering the rows, if necessary, we get a nonzero element c'_{2j_2} in the 2nd row, j_2th column. Divide the 2nd row by c'_{2j_2} and apply elementary row-transformations of type (γ) to obtain a matrix D, row equivalent to C, having the form

$$D = \begin{bmatrix} 0 & \cdots & 0 & 1 & c_{1j_1+1} & \cdots & 0 & d_{1j_2+1} & \cdots & d_{1n} \\ 0 & \cdots & 0 & 0 & 0 & \cdots & 1 & d_{2j_2+1} & \cdots & d_{2n} \\ 0 & \multicolumn{6}{c}{\dotfill} & 0 & d_{3j_2+1} & \cdots & d_{3n} \\ \multicolumn{10}{c}{\dotfill} \\ 0 & \multicolumn{6}{c}{\dotfill} & 0 & d_{mj_2+1} & \cdots & d_{mn} \end{bmatrix}.$$

Now let the j_3th be the first column of D in which a nonzero element occurs *below* the 2nd row, and repeat the process. It is clear that this process can be continued until we obtain a reduced row-echelon matrix R which is row-equivalent to A.

Exercise 2.3

1. Which of the following matrices are in reduced row-echelon form?

(a) $\begin{bmatrix} 1 & -3 \\ 0 & 1 \end{bmatrix}$.

(b) $\begin{bmatrix} 1 & 2 & 0 \\ 0 & 0 & 1 \\ 0 & 0 & 0 \end{bmatrix}$.

(c) $\begin{bmatrix} 0 & 0 & 0 \\ 0 & 1 & 0 \\ 0 & 0 & 1 \end{bmatrix}$.

(d) $\begin{bmatrix} 0 & 1 & 5 & 0 & 0 & 6 \\ 0 & 0 & 0 & 1 & 0 & 4 \\ 0 & 0 & 0 & 0 & 1 & 2 \\ 0 & 0 & 0 & 0 & 0 & 0 \end{bmatrix}$.

(e) $\begin{bmatrix} 2 & 0 & 0 & 0 \\ 0 & 5 & 0 & 0 \\ 0 & 0 & 9 & 0 \\ 0 & 0 & 0 & 1 \end{bmatrix}$.

2. To each matrix in Problem 1 which is not already in reduced row-echelon form, apply elementary row transformations to put it in this form.

3. Reduce each of the following to reduced row-echelon form:

(a) $\begin{bmatrix} 2 & 1 & 7 & 3 \\ 1 & 4 & 2 & 1 \\ 3 & 5 & 9 & 2 \end{bmatrix}$.

(b) $\begin{bmatrix} 1 & 4 & 5 \\ 2 & 1 & 7 \\ 1 & -10 & -1 \end{bmatrix}$.

(c) $\begin{bmatrix} 2 & -1 & 4 & 1 \\ 3 & 2 & 5 & -1 \\ 1 & 3 & 1 & -2 \\ 7 & 7 & 11 & -4 \end{bmatrix}$.

4. A reduced row-echelon matrix must satisfy four conditions which are stated in the definition. Write down four 4×5 matrices, each of which satisfies three of these conditions but violates the fourth, and such that each of the four matrices violates a different one of the four conditions.

5. If m is a fixed nonzero integer and a and b are two integers, we define the relation $a \equiv b \pmod{m}$, (read a is congruent to b modulo m) to mean that $a - b$ is divisible by m. Prove that congruence modulo m is an equivalence relation.

6. If two real numbers x and y are defined to be equivalent whenever $x - y$ is a rational number, prove that this equivalence is an equivalence relation.

7. Think of other equivalence relations defined among the rational, real, or complex numbers.

2.5 Rank of a Matrix

Let $A = [a_{ij}]$ be an $m \times n$ matrix with elements in the field $\mathscr{F}$. The row vectors $R_1, R_2, \ldots, R_m$ of A belong to $\mathscr{V}_n(\mathscr{F})$. The subspace $\mathscr{W}$ of $\mathscr{V}_n(\mathscr{F})$ spanned by the row vectors of A is called the *row space* of A.

Definition. *The dimension r of the row space of A is called the rank of A.*

Theorem 2.9. *Row-equivalent matrices have the same row space and, therefore, the same rank.*

Proof. Let A be any matrix and let $\mathscr{W}$ be its row space. If A' is obtained from A by applying elementary row transformations, then each row of A' is a linear combination of the rows of A. Hence, if $\mathscr{W}'$ is the row space of A', we have $\mathscr{W}' \subset \mathscr{W}$. However, A can also be obtained from A' by elementary row transformations and the same argument shows that $\mathscr{W} \subset \mathscr{W}'$ and, hence, $\mathscr{W} = \mathscr{W}'$. Now if $A \sim B$, B can be obtained from A by a succession of elementary transformations. Hence, A and B have the same row space and the same rank.

Theorem 2.10. *The nonzero row vectors of a reduced row-echelon matrix R are linearly independent and, hence, form a basis for the row space of R.*

Proof. Let R be of type $m \times n$. By the definition of a reduced row-echelon matrix, if R has r nonzero rows, there exist positive integers $j_1, j_2, \ldots, j_r$, all $\leq n$ such that $j_1 < j_2 < \cdots < j_r$ and R has a 1 in the ith row and j_ith column. All elements in the j_ith column except that in the ith row are 0. Hence, if $R_1, R_2, \ldots, R_r$ are the nonzero row vectors of R, the vector

$$X = c_1 R_1 + c_2 R_2 + \cdots + c_r R_r$$

has c_i as its j_ith coordinate $(i = 1, 2, \ldots, r)$. Thus, $X = O$ implies $c_1 = c_2 = \cdots = c_r = 0$ and $R_1, R_2, \ldots, R_r$ are linearly independent. Since by definition these vectors span the row space, they constitute a basis of the row space of R.

Corollary 1. *If A is any matrix and R is a reduced row-echelon matrix row-equivalent to A, then the nonzero row vectors of R constitute a basis for the row space of A.*

Corollary 2. *The rank of a matrix A is equal to the number of nonzero rows in reduced row-echelon matrix row-equivalent to A.*

Corollary 3. *If A is an $n \times n$ matrix of rank n, the reduced row-echelon matrix row-equivalent to A is the identity matrix I_n.*

Proof. This follows from Corollary 2 and the definition of a reduced row-echelon matrix.

In view of Corollary 1 the row reduction of a matrix gives a systematic procedure for finding a basis for the space spanned by a given set of vectors in $\mathscr{V}_n(\mathscr{F})$.

Example 1. Find a basis for the space spanned by the vectors $[1, 2, -3, 4]$, $[-2, 1, 7, -5]$, $[2, 5, -3, 4]$, and $[2, 10, -2, 7]$.

Solution. Reduce the matrix of which the rows are the given vectors as follows:

$$\begin{bmatrix} 1 & 2 & -3 & 4 \\ -2 & 1 & 7 & -5 \\ 2 & 5 & -3 & 4 \\ 2 & 10 & -2 & 7 \end{bmatrix},$$

$$\begin{matrix} R_1 \to \\ R_2 + 2R_1 \to \\ R_3 - 2R_1 \to \\ R_4 - 2R_1 \to \end{matrix} \begin{bmatrix} 1 & 2 & -3 & 4 \\ 0 & 5 & 1 & 3 \\ 0 & 1 & 3 & -4 \\ 0 & 6 & 4 & -1 \end{bmatrix},$$

$$\begin{matrix} R_1 \to \\ R_3 \to \\ R_2 - 5R_3 \to \\ R_4 - 6R_3 \to \end{matrix} \begin{bmatrix} 1 & 2 & -3 & 4 \\ 0 & 1 & 3 & -4 \\ 0 & 0 & -14 & 23 \\ 0 & 0 & -14 & 23 \end{bmatrix},$$

$$\begin{matrix} R_1 - 2R_2 \to \\ \\ \tfrac{1}{14}R_3 \to \\ R_4 - R_3 \to \end{matrix} \begin{bmatrix} 1 & 0 & -9 & 12 \\ 0 & 1 & 3 & -4 \\ 0 & 0 & 1 & -\frac{23}{14} \\ 0 & 0 & 0 & 0 \end{bmatrix},$$

$$\begin{matrix} R_1 + 9R_3 \to \\ R_2 - 3R_3 \to \\ \\ \end{matrix} \begin{bmatrix} 1 & 0 & 0 & -\frac{39}{14} \\ 0 & 1 & 0 & \frac{13}{14} \\ 0 & 0 & 1 & -\frac{23}{14} \\ 0 & 0 & 0 & 0 \end{bmatrix}.$$

Therefore a basis for the space is

$$[1, 0, 0, -\tfrac{39}{14}], \quad [0, 1, 0, \tfrac{13}{14}], \quad [0, 0, 1, -\tfrac{23}{14}]$$

and the dimension of the space is 3.

Example 2. Which, if any, of the vectors

$$X_1 = [2, 1, 4], \quad X_2 = [7, 3, -2], \quad X_3 = [7, 3, 5]$$

belong to the space spanned by $[4, 3, -1]$ and $[3, -2, 12]$?

Solution.

Method 1. By the standard reduction process (details left to the reader), we find

$$\begin{bmatrix} 4 & 3 & -1 \\ 3 & -2 & 12 \end{bmatrix} \sim \begin{bmatrix} 1 & 0 & 2 \\ 0 & 1 & -3 \end{bmatrix}.$$

Therefore a basis for the space is $[1, 0, 2]$, $[0, 1, -3]$ and every vector in the space can be written in the form

(17) $$a[1, 0, 2] + b[0, 1, -3] = [a, b, 2a - 3b].$$

Comparing X_1, X_2, X_3 with (17), we see that X_1 and X_2 do not belong to the space but X_3 does, because $5 = 2(7) - 3(3)$.

Method 2. The vector $[a, b, c]$ is in the space spanned by $[4, 3, -1]$ and $[3, -2, 12]$ if and only if there exist scalars x and y such that

$$[a, b, c] = x[4, 3, -1] + y[3, -2, 12]$$

or if and only if the system

$$4x + 3y = a,$$

$$3x - 2y = b,$$

$$-x + 12y = c$$

has a solution. The augmented matrix

$$\begin{bmatrix} 4 & 3 & a \\ -2 & 3 & b \\ -1 & 12 & c \end{bmatrix}$$

is row-equivalent to (details left to the student)

$$\begin{bmatrix} 1 & -12 & -c \\ 0 & 1 & (b + 3c)/34 \\ 0 & 0 & (2a - 3b - c) \end{bmatrix}.$$

Hence, solutions exist if and only if $2a - 3b - c = 0$ or $c = 2a - 3b$. Hence $[a, b, c]$ is in the space spanned by the given vectors if and only if $c = 2a - 3b$.

Exercise 2.4

1. Find a basis for the space spanned by each of the following sets of vectors:
   (a) $[2, 3, 5]$, $[-1, 5, 4]$, $[3, -2, 1]$.
   (b) $[1, 2, 3]$, $[4, 5, 6]$, $[7, 8, 9]$.
   (c) $[1, 4, 2, 5]$, $[3, 2, -4, 5]$, $[-2, 1, 5, -1]$, $[1, -3, -5, -2]$.
   (d) $[1, -1, 2, 2]$, $[3, 5, 2, -4]$, $[5, 3, 6, 1]$.

2. Find the rank of each of the following matrices:

   (a) $\begin{bmatrix} 2 & 1 & 7 & 3 \\ 1 & 4 & 2 & 1 \\ 3 & 5 & 9 & 2 \end{bmatrix}$.

   (b) $\begin{bmatrix} 1 & 4 & 5 \\ 2 & 1 & 7 \\ 1 & -10 & -1 \end{bmatrix}$.

   (c) $\begin{bmatrix} 2 & -1 & 4 & 1 \\ 3 & 2 & 5 & -1 \\ 1 & 3 & 1 & -2 \\ 7 & 7 & 11 & -4 \end{bmatrix}$.

   (d) $\begin{bmatrix} 1 & 1 & 0 & 0 \\ 0 & 1 & 1 & 0 \\ 0 & 0 & 1 & 1 \\ 1 & 2 & 2 & 1 \end{bmatrix}$.

   (e) $\begin{bmatrix} 1 & 2 & 3 \\ 4 & 5 & 6 \\ 7 & 8 & 9 \end{bmatrix}$.

4. Determine whether the vector $[5, -1, 6]$ is in the space spanned by the vectors $[1, 2, 4]$ and $[2, -1, 2]$.

5. Find necessary and sufficient conditions on a, b, c, d, in order that the vector $[a, b, c, d]$ be in the space spanned by $[1, 2, 3, -4]$ and $[2, 1, 5, -1]$.

6. Let $\mathscr{S}$ be the space spanned by the vectors $[1, 5, 2, -2]$, $[-3, 7, 4, -10]$, and $[3, 4, 1, 2]$ and let $\mathscr{T}$ be the space spanned by $[-2, 1, 1, -4]$ and $[2, 0, 5, -1]$. Find the dimensions of $\mathscr{S}$, $\mathscr{T}$ and $\mathscr{S} + \mathscr{T}$ and find bases for each of these spaces. Deduce (Theorem 1.10) the dimension of $\mathscr{S} \cap \mathscr{T}$ and find a basis for this space.

7. Let $\mathscr{S}$ be the space spanned by the vectors $[1, 2, 2, -3]$ and $[-1, 4, -2, 0]$ and let $\mathscr{T}$ be the space spanned by $[2, 0, -1, 4]$ and $[1, 8, 2, -6]$. Find the dimension of $\mathscr{S}$, $\mathscr{T}$, $\mathscr{S} + \mathscr{T}$, and $\mathscr{S} \cap \mathscr{T}$ and a basis for each of these spaces.

8. Find necessary and sufficient conditions that $[a, b, c, d, e]$ belong to the space spanned by:
   (a) $[1, 0, 0, 3, 2]$, $[0, 1, 0, 4, -3]$, and $[0, 0, 1, 2, 1]$.
   (b) $[1, 1, 0, 2, 5]$ and $[3, 4, -1, 7, 0]$.
   (c) $[1, 1, 2, 2, 3]$ and $[-1, 3, 1, 2, -1]$.
   (d) $[1, -1, 2, -2, 4]$, $[2, 0, 4, 1, 3]$, $[3, 7, 1, 2, 2]$.

2.6 The Solution Space of a Homogeneous System

For completeness and easy reference, we state formally the result established in Example 4, Section 1.5.

Theorem 2.11. *Let A be an $m \times n$ matrix with elements in $\mathcal{F}$. The set $\mathcal{S}$ of all solution vectors of the homogeneous system of linear equations*

(18)
$$AX = O,$$

with coefficient matrix A, is a subspace of $\mathcal{V}_n(\mathcal{F})$.

Proof. The proof given in Section 1.5 can be simplified by the use of matrix notation. If X_1 and X_2 are solutions of (18), then

$$A(X_1 + X_2) = AX_1 + AX_2 = O$$

so that $X_1 + X_2 \in \mathcal{S}$. Similarly, if $AX_1 = O$, then $A(kX_1) = k(AX_1) = O$ and $kX_1 \in \mathcal{S}$. Hence, $\mathcal{S}$ is a subspace of $\mathcal{V}_n(\mathcal{F})$.

As in Section 1.5, the space $\mathcal{S}$ is called the solution space of the system (18). It is also called the *null space* of the matrix A.

Theorem 2.12. *The solution space $\mathcal{S}$ of a homogeneous system $AX = O$ has dimension $n - r$ where n is the number of unknowns and r is the rank of the coefficient matrix A.*

Proof. Let R be a reduced row-echelon matrix which is row-equivalent to A. By Theorem 2.7, the system $RX = O$ has the same solution space as the system $AX = O$. But because R is a reduced row-echelon matrix, the equations $RX = O$ give the r unknowns $x_{j_1}, x_{j_2}, \ldots, x_{j_r}$ as linear functions of the remaining $n - r$ unknowns. To simplify the notation, assume that the unknowns are renumbered so that $j_1 = 1, j_2 = 2, \ldots, j_r = r$. We then have

(19)
$$x_1 = b_{11}x_{r+1} + \cdots + b_{n-r1}x_n,$$
$$x_2 = b_{12}x_{r+1} + \cdots + b_{n-r2}x_n,$$
$$\cdots\cdots\cdots\cdots\cdots\cdots\cdots\cdots$$
$$x_r = b_{1r}x_{r+1} + \cdots + b_{n-rr}x_n,$$

and arbitrary choice of values for $x_{r+1}, \ldots, x_n$ yields a solution for the system. Therefore we can find $n - r$ solution vectors of the form

$$S_1 = [b_{11}, \ldots, b_{1r}, 1, 0, 0, \ldots, 0],$$
$$S_2 = [b_{21}, \ldots, b_{2r}, 0, 1, 0, \ldots, 0],$$
$$\cdots\cdots\cdots\cdots\cdots\cdots\cdots\cdots$$
$$S_{n-r} = [b_{n-r1}, \ldots, b_{n-rr}, 0, 0, \ldots, 0, 1].$$

These are clearly linearly independent because the vector $S = c_1 S_1 + c_2 S_2 + \cdots +$ $c_{n-r} S_{n-r}$ has $c_1, c_2, \ldots, c_{n-r}$ for its last $n - r$ co-ordinates and, hence, $S = O$ implies $c_1 = c_2 = \cdots = c_{n-r} = 0$. Moreover, $S_1, \ldots, S_{n-r}$ span the solution space, for if

$$S = [s_1, s_2, \ldots, s_n]$$

is any solution vector of (18), then

$$T = S - s_{r+1} S_1 - s_{r+2} S_2 - \cdots - s_n S_{n-r}$$
$$= [t_1, t_2, \ldots, t_r, 0, 0, \ldots, 0]$$

is also a solution of (18) because it is a linear combination of solutions. Hence, the coordinates of T also satisfy (19) and, therefore, $t_1 = t_2 = \cdots = t_r = 0$. Thus, $T = O$ and $S = s_{r+1} S_1 + \cdots + s_n S_{n-r}$. Therefore, $S_1, \ldots, S_{n-r}$ span the solution space $\mathscr{S}$ and, being linearly independent, form a basis of $\mathscr{S}$. Hence, dim $\mathscr{S} = n - r$.

Corollary. *A homogeneous system*

$$AX = O$$

in n unknowns has only the zero solution $X = O$ if and only if rank $A = n$.

Proof. The system has only the zero solution if and only if the dimension of the solution space is zero, that is, if and only if $n - r = 0$ or $n = r$, the rank of A.

Example. Find a general solution and a basis for the solution space of the system

$$
\begin{aligned}
x_1 + 2x_2 \quad\quad + x_4 - x_5 &= 0, \\
2x_1 + 3x_2 + x_3 + 4x_4 - 3x_5 &= 0, \\
x_1 \quad\quad + 2x_3 + 5x_4 - 3x_5 &= 0.
\end{aligned}
$$

Solution. The reduced row-echelon matrix row-equivalent to the matrix of coefficients is (details left to the student)

$$
\begin{bmatrix}
1 & 0 & 2 & 5 & -3 \\
0 & 1 & -1 & -2 & 1 \\
0 & 0 & 0 & 0 & 0
\end{bmatrix}.
$$

The general solution is, therefore,

(20)
$$
\begin{aligned}
x_1 &= -2x_3 - 5x_4 + 3x_5, \\
x_2 &= \quad x_3 + 2x_4 - x_5,
\end{aligned}
$$

where x_3, x_4, x_5 are arbitrary. Using the method described in the proof of Theorem 2.12, a basis for the solution space is found to be

$$S_1 = [-2, 1, 1, 0, 0],$$
$$S_2 = [-5, 2, 0, 1, 0],$$
$$S_3 = [3, -1, 0, 0, 1].$$

Exercise 2.5

1. Find a basis for the solution space for each of the following systems:

(a) $3x_1 - x_2 + x_3 + 4x_4 = 0.$

(b) $x_1 + 2x_2 - x_3 + x_4 - 2x_5 = 0,$
$2x_1 + 5x_2 - 3x_3 - x_4 + x_5 = 0.$

(c) $x_1 - 2x_2 + x_3 = 0.$

(d) $x_1 + 5x_2 = 0,$
$3x_1 - 4x_2 = 0.$

(e) $3x_1 - x_2 + 2x_3 = 0,$
$5x_1 - x_2 + 3x_3 = 0.$

(f) $x_1 + x_2 - x_3 = 0,$
$3x_1 + 2x_2 + x_3 = 0,$
$2x_1 - x_2 + 3x_3 = 0.$

(g) $x_1 + x_2 - x_3 + 4x_4 - x_5 = 0,$
$2x_1 + x_2 + x_3 - x_4 + 2x_5 = 0,$
$x_1 - 2x_2 - 3x_3 + 2x_4 + x_5 = 0.$

2. Let $\mathscr{S}$ be any subspace of $\mathscr{V}_n(\mathscr{F})$ of dimension r. Prove that $\mathscr{S}$ is the solution space of a set of $n - r$ homogeneous linear equations in n unknowns. *Hint.* Choose a basis $B_1, B_2, \ldots, B_r$ of $\mathscr{S}$ and use either of the methods used in the solution of Example 2 Section 2.5 to find conditions that $[x_1, x_2, \ldots, x_n]$ is in the space spanned by $B_1, B_2, \ldots, B_r$.

2.7 The Column Space and Null Space of a Matrix

Let A be any $m \times n$ matrix and X any $n \times 1$ matrix, both with elements in $\mathscr{F}$. In what follows, we find it convenient to identify X with the vector of $\mathscr{V}_n(\mathscr{F})$ having the same coordinates relative to the standard basis. It clearly makes no difference to the algebra of vectors in $\mathscr{V}_n(\mathscr{F})$ whether we write these vectors as rows or columns, and since we want to consider vectors of the form AX, this necessitates writing X as a column vector. In spite of this, we still consider X as a vector in $\mathscr{V}_n(\mathscr{F})$.

With this understanding, consider the mapping

$$\tau : \mathscr{V}_n(\mathscr{F}) \to \mathscr{V}_m(\mathscr{F}),$$

defined by

(21) $$\tau(X) = AX.$$

Because A is $m \times n$ and X is $n \times 1$, AX is $m \times 1$ and, with our convention, belongs to $\mathscr{V}_m(\mathscr{F})$. If $X_1, X_2 \in \mathscr{V}_n(\mathscr{F})$ and $k \in \mathscr{F}$, we know that $A(X_1 + X_2) = AX_1 + AX_2$ and $A(kX_1) = kAX_1$. Hence we have

(22)
$$\tau(X_1 + X_2) = \tau(X_1) + \tau(X_2),$$
$$\tau(kX_1) = k\tau(X_1).$$

Because it satisfies (22), τ is an example of an important type of mapping called a *linear transformation*. Linear transformations will be studied in some detail in Chapter 5. For our present purposes, it is sufficient to deduce a few properties of the mapping τ defined by (21).

Lemma 2.13. *If A is an $m \times n$ matrix and X an $n \times 1$ matrix, then AX is a linear combination of the column vectors of A.*

Proof. By the definition of a matrix product,

$$AX = \begin{bmatrix} a_{11} & a_{12} & \cdots & a_{1n} \\ a_{21} & a_{22} & \cdots & a_{2n} \\ \hdotsfor{4} \\ a_{m1} & a_{m2} & \cdots & a_{mn} \end{bmatrix} \begin{bmatrix} x_1 \\ x_2 \\ \cdot \\ x_n \end{bmatrix} = \begin{bmatrix} a_{11}x_1 + a_{12}x_2 + \cdots + a_{1n}x_n \\ a_{21}x_1 + a_{22}x_2 + \cdots + a_{2n}x_n \\ \hdotsfor{1} \\ a_{m1}x_1 + a_{m2}x_2 + \cdots + a_{mn}x_n \end{bmatrix}$$

$$= x_1 \begin{bmatrix} a_{11} \\ a_{21} \\ \cdot \\ \cdot \\ \cdot \\ a_{m1} \end{bmatrix} + x_2 \begin{bmatrix} a_{12} \\ a_{22} \\ \cdot \\ \cdot \\ \cdot \\ a_{m2} \end{bmatrix} + \cdots + x_n \begin{bmatrix} a_{1n} \\ a_{2n} \\ \cdot \\ \cdot \\ \cdot \\ a_{mn} \end{bmatrix},$$

which proves the lemma.

Lemma 2.14. Im τ *is the subspace of $\mathscr{V}_m(\mathscr{F})$ spanned by the column vectors of A.*

Proof. Let $\mathscr{U}$ be the subspace of $\mathscr{V}_m(\mathscr{F})$ spanned by the column vectors of A. By Lemma 2.13, $AX \in \mathscr{U}$ for all X in $\mathscr{V}_n(\mathscr{F})$ and hence im $\tau \subset \mathscr{U}$. On the other hand, $A = AI_n$, where I_n is the $n \times n$ identity matrix, and hence by Theorem 2.5 the jth column of A is $AE_j = \tau(E_j)$ where E_j is the jth column of I_n. Thus the column vectors of A belong to im τ and $\mathscr{U} \subset$ im τ. Thus, $\mathscr{U} =$ im τ as required.

We have now associated with any $m \times n$ matrix A the following three vector spaces:

(1) The *row space of A*, the subspace $\mathscr{W}$ of $\mathscr{V}_n(\mathscr{F})$ spanned by the row vectors of A.

(2) The *column space of A*, the subspace $\mathscr{U}$ of $\mathscr{V}_m(\mathscr{F})$ spanned by the column vectors of A, which we have by Lemma 2.14 identified with the space

$$\operatorname{im} \tau = \{AX \mid X \in \mathscr{V}_n(\mathscr{F})\}.$$

(3) The solution space $\mathscr{N}$ of the homogeneous system $AX = O$ which, in this context, is renamed the *null space* of A and defined by

$$\mathscr{N} = \{X \in \mathscr{V}_n(\mathscr{F}) \mid AX = O\}.$$

Theorem 2.15. *For any $m \times n$ matrix A with column space $\mathscr{U}$ and null space $\mathscr{N}$,*

$$\dim \mathscr{N} + \dim \mathscr{U} = n.$$

Proof. Let $\dim \mathscr{N} = s$. By the corollary to Theorem 1.9, we can choose a basis

$$X_1, X_2, \ldots, X_s, \ldots, X_n$$

of $\mathscr{V}_n(\mathscr{F})$ of which the first s vectors $X_1 \cdots X_s$ form a basis of $\mathscr{N}$. Since every vector of $\mathscr{V}_n(\mathscr{F})$ has the form

$$X = \sum_{i=1}^{n} c_i X_i,$$

it follows that every vector of $\mathscr{U}$ (i.e., of $\operatorname{im} \tau$) has the form

$$AX = A\left(\sum_{i=1}^{n} c_i X_i \right) = \sum_{i=1}^{n} c_i (AX_i)$$

and hence $\mathscr{U}$ is spanned by the vectors AX_i $(i = 1, 2, \ldots, n)$. However, $AX_1 = AX_2 = \cdots = AX_s = O$ because $X_1, \ldots, X_s$ belong to $\mathscr{N}$. Therefore, $\mathscr{U}$ is spanned by the vectors

(23) $$AX_{s+1}, AX_{s+2}, \ldots, AX_n.$$

But these vectors can be shown to be linearly independent. For if

$$c_1(AX_{s+1}) + \cdots + c_{n-s}(AX_n) = O,$$

then

$$A(c_1 X_{s+1} + \cdots + c_{n-s} X_n) = O$$

and

$$c_1 X_{s+1} + \cdots + c_{n-s} X_n \in \mathscr{N}.$$

But a basis for $\mathcal{N}$ is $X_1, \ldots, X_s$, so that

$$c_1 X_{s+1} + \cdots + c_{n-s} X_n = b_1 X_1 + \cdots + b_s X_s$$

and the linear independence of $X_1, \ldots, X_n$ requires $c_1 = c_2 = \cdots = c_{n-s} = 0$
This proves that the vectors (23) are a basis for $\mathcal{U}$ and hence dim $\mathcal{U} = n - s$ and
dim $\mathcal{U}$ + dim $N = n$ as required.

Corollary 1. *For any matrix A the dimension of the column space is equal to the
dimension of the row space.*

Proof. By the theorem, dim $\mathcal{U} = n -$ dim $\mathcal{N}$. But by Theorem 2.12, dim $\mathcal{N} =$
$n - r$ where $r =$ dim $\mathcal{W}$. Hence,

$$\dim \mathcal{U} = n - (n - r) = r = \dim \mathcal{W}.$$

This important result is sometimes stated in terms of the *row rank* of A (the
dimension of the row space) and the *column rank* of A (the dimension of the column
space). The corollary states that the row rank and column rank of any matrix A are
equal. Their common value is the rank of A. The rank of a matrix A is, therefore
equal both to the maximum number of linearly independent rows of A and to the
maximum number of linearly independent columns of A.

Corollary 2. *If A is a square matrix, the column vectors of A are linearly independent if and only if the row vectors of A are linearly independent.*

Definition. *The dimension of the nullspace $\mathcal{N}$ of a matrix A is called the
nullity of A.*

Since by Corollary 1, dim $\mathcal{U} =$ rank A, Theorem 2.15 can be restated
For any $m \times n$ matrix A, rank A + nullity A = n.

Exercise 2.6

1. Find bases for the row space, the column space, and the null space of the following
matrices:

(a) $\begin{bmatrix} 2 & 1 & 5 \\ 3 & -2 & 4 \end{bmatrix}$.

(b) $\begin{bmatrix} 1 & -3 & -1 \\ 2 & 1 & 5 \\ 2 & -5 & -1 \end{bmatrix}$.

(c) $\begin{bmatrix} 1 & 2 & 4 & -3 \\ -1 & 3 & 2 & 5 \\ 5 & 0 & 8 & -19 \end{bmatrix}$.

(d) $\begin{bmatrix} 1 & 0 & 0 \\ 0 & 1 & 0 \\ 0 & 0 & 0 \end{bmatrix}$.

(e) $\begin{bmatrix} 1 & 0 & 0 \\ 0 & 0 & 0 \\ 0 & 0 & 0 \end{bmatrix}$.

(f) $\begin{bmatrix} 2 & 0 & 1 \\ 0 & 0 & 0 \\ -3 & 0 & 4 \end{bmatrix}$.

2. Does there exist a 3×3 matrix A of rank 2 such that $A^2 = O$? Give reasons for your answer.

3. If A is an $m \times n$ matrix of rank r and B an $n \times p$ matrix such that $AB = O$, prove that rank $B \leq n - r$. Does there always exist a matrix B of rank $n - r$ such that $AB = O$? *Hint.* Note that if $AB = O$, the columns of B belong to the null space of A.

4. Let A be any $m \times n$ matrix and let B be a matrix whose rows are a basis for $\mathcal{N}_A$, the null space of A. Prove that the null space of B is the row space $\mathcal{W}_A$ of A. *Hint.* Prove first that $\mathcal{W}_A \subset \mathcal{N}_B$ and then that dim $\mathcal{W}_A = $ dim $\mathcal{N}_B$. Then use Problem 4, Exercise 1.5.

2.8 Nonhomogeneous Systems

By Theorem 2.12, we know that a system

$$AX = O$$

of homogeneous linear equations in n unknowns has $n - r$ linearly independent solution vectors where $r = $ rank A. Moreover, we know how to find these solutions by reduction of A to row-echelon form as described in Section 2.4. We have also seen that the same reduction process applied to the augmented matrix can be used to find solutions of a nonhomogeneous system when these exist. We shall now prove the basic existence theorems for the nonhomogeneous case.

Let

(24) $$AX = C$$

by any system of linear equations in n unknowns, $x_1, x_2, \ldots, x_n$ with coefficient matrix A and augmented matrix B. Since (24) is equivalent to the vector equation

$$x_1 A_1 + x_2 A_2 + \cdots + x_n A_n = C,$$

where $A_1, A_2, \ldots, A_n$ are the column vectors of A, it is clear that a solution of (24) exists if and only if C is in the column space of A. Another way of stating this is that the columns of the augmented matrix B span the same space as the columns of A. In view of Corollary 1, Theorem 2.15, we have the following.

Theorem 2.16. *A system of linear equations*

$$AX = C$$

has solutions if and only if the rank of the augmented matrix $[A,C]$ is equal to the rank of the coefficient matrix A.

Next, we relate the solutions of the nonhomogeneous system $AX = C$ to the solutions of the homogeneous system $AX = O$ with the same coefficient matrix.

Theorem 2.17.

(a) *If S is any fixed solution vector of* (24) *and T is a solution vector of the homogeneous system $AX = O$, then $S + T$ is a solution vector of* (24).

(b) *Every solution vector of* (24) *has the form $S + T$ where S is a fixed solution vector of* (18) *and T a solution of $AX = O$.*

Proof.

(a) If $AS = C$ and $AT = O$, then

$$A(S + T) = AS + AT = C$$

and $S + T$ is a solution of (24).

(b) Now suppose $X = U$ is any solution, and S is a fixed solution, of (24). We have

$$A(U - S) = AU - AS = C - C = O.$$

Hence, $U - S$ is a solution T of $AX = O$ and $U = S + T$ as required.

Students who are familiar with the theory of linear differential equations should note the analogy between Theorem 2.17 and the theorem that states that the general solution of a nonhomogeneous linear differential equation is equal to the sum of a particular solution and the general solution of the corresponding homogeneous equation, usually called the complementary function.

The special case of systems of n equations in n unknowns is of particular interest and is dealt with in the next theorem.

Theorem 2.18. *If A is an $n \times n$ matrix, the following three statements are equivalent (that is, each implies the others):*

(a) *The rank of A is n.*

(b) *The system $AX = C$ has a unique solution for arbitrary vectors C.*

(c) *The system $AX = O$ has only the trivial solution $x_1 = x_2 = \cdots = x_n = 0$.*

Proof. It is sufficient to prove that

(a) $\Rightarrow$ (b) ["(a) implies (b)"], (b) $\Rightarrow$ (c) and (c) $\Rightarrow$ (a).

(1) (a) $\Rightarrow$ (b). If A has rank n the columns of A are linearly independent by Theorem 2.15, Corollary 2 and, hence, constitute a basis of $\mathscr{V}_n(\mathscr{F})$. It follows (Theorem 1.11) that every vector C can be written uniquely as a linear combination of the column vectors of A. Hence, $AX = C$ has a unique solution for X.

(2) (b) $\Rightarrow$ (c). If $AX = C$ has a unique solution S, then $AX = O$ has only the zero solution. For if $X = T \neq O$ were a solution of $AX = O$, then by Theorem 2.17, $X = S + T$ would be a second solution $\neq S$ of $AX = C$, contrary to the uniqueness of S.

(3) (c) $\Rightarrow$ (a). If $AX = O$ has only the solution $X = O$, then the column vectors of A are linearly independent and by Theorem 2.15, Corollary 2, the rank of A is n.

2.9 Nonsingular Matrices and Matrix Inversion

The following theorem provides useful information about the rank of the product of two matrices.

Theorem 2.19.   *Let A and B be any two matrices for which the product AB is defined.*

(a) *The row space of AB is a subspace of the row space of B.*

(b) *The column space of AB is a subspace of the column space of A.*

Proof.

(a) It follows from the definition of a matrix product that the row vectors of AB are linear combinations of the row vectors of B. For example, if $A = [a_{ij}]$ the ith row of AB is

$$a_{i1}B_1 + a_{i2}B_2 + \cdots + a_{in}B_n,$$

where $B_1, B_2, \ldots, B_n$ are the row vectors of B. Hence the row space of AB is a subspace of the row space of B.

(b) The column space of AB is the row space of

$$(AB)^T = B^T A^T,$$

and hence by (a) is a subspace of the row space of A^T, that is, of the column space of A.

Corollary.   *The rank of a product of two matrices is less than or equal to the rank of either factor.*

Proof.   If we write $\mathscr{W}_M$ and $\mathscr{U}_M$ for the row space and column space of a matrix M, then we have rank $AB = \dim \mathscr{W}_{AB} \leq \dim \mathscr{W}_B = $ rank B. By Theorem 2.15, Corollary 2, we also have rank $AB = \dim \mathscr{U}_{AB} \leq \dim \mathscr{U}_A = $ rank A.

Definition.   *An $n \times n$ matrix A is said to be* nonsingular *if rank $A = n$ and* singular *if rank $A < n$.*

Definition. *If A is an $n \times n$ matrix, an $n \times n$ matrix B such that $AB = BA = I$ is called an* inverse *of A.*

The next theorem tells us exactly which square matrices have inverses.

Theorem 2.20. *An $n \times n$ matrix A has an inverse if and only if it is non-singular.*

Proof. If A has an inverse B, then $AB = I$. The rank of the identity matrix I is clearly n so that rank $AB = n$. But the corollary to Theorem 2.19 then implies that rank $A = $ rank $B = n$ so that A is nonsingular.

Conversely, suppose A is nonsingular and hence of rank n. By Theorem 2.18, the system

$$AX = E_i,$$

where E_i is the ith column of the identity matrix, has a unique solution $X = B_i$. Let B be the $n \times n$ matrix with B_i for its ith column $(i = 1, 2, \ldots, n)$. By Theorem 2.5(b) it follows that $AB = I$.

By Theorem 2.15, Corollary 2, A^T is also nonsingular so by the same argument, there exists an $n \times n$ matrix C^T such that $A^T C^T = I$ and since $I^T = I$, this yields $CA = I$. Now,

$$C = CI = C(AB) = (CA)B = IB = B.$$

Hence, $C = B$, $BA = AB = I$, and B is an inverse of A.

Corollary 1. *The inverse of a nonsingular matrix is unique.*

Proof. If B and C are both inverses of A, then

$$B = BI = B(AC) = (BA)C = IC = C.$$

The inverse of a nonsingular matrix A will be denoted by A^{-1}.

Corollary 2. *If A is nonsingular, A^{-1} is nonsingular and $(A^{-1})^{-1} = A$.*

Proof. Since $AA^{-1} = I$ has rank n, A^{-1} has rank n by the corollary to Theorem 2.19. That $(A^{-1})^{-1} = A$ follows from $A^{-1}A = AA^{-1} = I$ and the uniqueness of the inverse.

Corollary 3. *If A and B are nonsingular $n \times n$ matrices, AB is nonsingular and $(AB)^{-1} = B^{-1}A^{-1}$.*

Proof. Clearly, $B^{-1}A^{-1}AB = ABB^{-1}A^{-1} = I$ so $B^{-1}A^{-1}$ is the inverse of AB which is nonsingular by Theorem 2.20.

Corollary 4. *If $A_1, A_2, \ldots, A_n$ are nonsingular $n \times n$ matrices, $(A_1 A_2 \cdots A_n)^{-1} = A_n^{-1} \cdots A_2^{-1} A_1^{-1}$. In particular,*

$$(A^m)^{-1} = (A^{-1})^m.$$

If A is nonsingular, $(A^{-1})^m$ will be written A^{-m}. If we define $A^0 = I$, it is clear that the index laws

$$A^m A^p = A^{m+p},$$

$$(A^m)^p = A^{mp},$$

hold for arbitrary integers m and p (positive, negative, or zero).

Corollary 5. *If A is nonsingular A^T is nonsingular and $(A^T)^{-1} = (A^{-1})^T$.*

The proof of Corollary 5 is left as an exercise.

We have shown that A^{-1} exists if A is nonsingular. We next describe a method of actually finding A^{-1}. If A is nonsingular, we have

$$AA^{-1} = I,$$

and it follows [Theorem 2.5(b)] that the ith column X_i of A^{-1} satisfies

(25) $$AX_i = E_i, \qquad (i = 1, 2, \ldots, n)$$

where E_i is the ith column of I. The solution of (25) is found by reducing the augmented matrix to reduced row-echelon form. Because A is nonsingular, it is row equivalent to I by Theorem 2.10, Corollary 3. For $i = 1$, the augmented matrix of (25) is

$$\begin{bmatrix} a_{11} & a_{12} & \cdots & a_{1n} & 1 \\ a_{21} & a_{22} & \cdots & a_{2n} & 0 \\ \cdots\cdots\cdots\cdots\cdots\cdots\cdots \\ a_{n1} & a_{n2} & \cdots & a_{nn} & 0 \end{bmatrix},$$

and its reduced form will be

$$\begin{bmatrix} 1 & 0 & \cdots & 0 & x_{11} \\ 0 & 1 & \cdots & 0 & x_{21} \\ \cdots\cdots\cdots\cdots\cdots \\ 0 & 0 & \cdots & 1 & x_{n1} \end{bmatrix},$$

of which the last column is X_1. Now for $i = 1, 2, \ldots, n$, (25) represents n systems of equations all with coefficient matrix A, the systems differing only in their constant terms. Hence, all n systems can be solved at once by reducing the matrix

$$
\begin{bmatrix}
a_{11} & a_{12} & \cdots & a_{1n} & 1 & 0 & \cdots & 0 \\
a_{21} & a_{22} & \cdots & a_{2n} & 0 & 1 & \cdots & 0 \\
\multicolumn{8}{c}{\cdots\cdots\cdots\cdots\cdots\cdots\cdots\cdots\cdots} \\
a_{n1} & a_{n2} & \cdots & a_{nn} & 0 & 0 & \cdots & 1
\end{bmatrix}.
$$

The reduced row-echelon form will be

$$
\begin{bmatrix}
1 & 0 & \cdots & 0 & x_{11} & x_{12} & \cdots & x_{1n} \\
0 & 1 & \cdots & 0 & x_{21} & x_{22} & \cdots & x_{2n} \\
\multicolumn{8}{c}{\cdots\cdots\cdots\cdots\cdots\cdots\cdots\cdots\cdots} \\
0 & 0 & \cdots & 1 & x_{n1} & x_{n2} & \cdots & x_{nn}
\end{bmatrix},
$$

of which the last n columns are the solutions of the n systems (25). It follows that the matrix $X = [x_{ij}]$ of these last n columns satisfies $AX = I$. Since the inverse A^{-1} is known to exist, we have $A^{-1}AX = A^{-1}I = A^{-1}$ or $X = A^{-1}$.

Let A and B be any two matrices having the same number of rows. If we use the notation $[A, B]$ for the matrix formed by writing the columns of A, in order, followed by the columns of B, in order, we can state our result as follows.

Theorem 2.21. *If A is a nonsingular $n \times n$ matrix and I the identity matrix of order n, the (unique) reduced row-echelon matrix row-equivalent to $[A, I]$ is $[I, A^{-1}]$.*

The only part of this theorem not already proved is the uniqueness of the reduced row-echelon matrix row equivalent to $[A, I]$. The definition of a reduced row-echelon matrix ensures that I is the only reduced row-echelon matrix row equivalent to a nonsingular matrix A. Since the last n columns in the reduced form of $[A, I]$ are the unique solutions of $AX = E_i$, it follows that the reduced form $[I, A^{-1}]$ is unique.

Example 1. Find A^{-1} if

$$
A = \begin{bmatrix}
1 & 2 & 1 \\
1 & 3 & 3 \\
1 & 3 & 4
\end{bmatrix}.
$$

Solution. Reduce the matrix $[A, I]$ as follows:

$$\begin{bmatrix} 1 & 2 & 1 & 1 & 0 & 0 \\ 1 & 3 & 3 & 0 & 1 & 0 \\ 1 & 3 & 4 & 0 & 0 & 1 \end{bmatrix},$$

$$\begin{matrix} \\ R_2 - R_1 \rightarrow \\ R_3 - R_1 \rightarrow \end{matrix} \begin{bmatrix} 1 & 2 & 1 & 1 & 0 & 0 \\ 0 & 1 & 2 & -1 & 1 & 0 \\ 0 & 1 & 3 & -1 & 0 & 1 \end{bmatrix},$$

$$\begin{matrix} \\ \\ R_3 - R_2 \rightarrow \end{matrix} \begin{bmatrix} 1 & 2 & 1 & 1 & 0 & 0 \\ 0 & 1 & 2 & -1 & 1 & 0 \\ 0 & 0 & 1 & 0 & -1 & 1 \end{bmatrix},$$

$$\begin{matrix} R_1 - R_3 \rightarrow \\ R_2 - 2R_3 \rightarrow \\ \end{matrix} \begin{bmatrix} 1 & 2 & 0 & 1 & 1 & -1 \\ 0 & 1 & 0 & -1 & 3 & -2 \\ 0 & 0 & 1 & 0 & -1 & 1 \end{bmatrix},$$

$$\begin{matrix} R_1 - 2R_2 \rightarrow \\ \\ \end{matrix} \begin{bmatrix} 1 & 0 & 0 & 3 & -5 & 3 \\ 0 & 1 & 0 & -1 & 3 & -2 \\ 0 & 0 & 1 & 0 & -1 & 1 \end{bmatrix}.$$

By Theorem 2.21

$$A^{-1} = \begin{bmatrix} 3 & -5 & 3 \\ -1 & 3 & -2 \\ 0 & -1 & 1 \end{bmatrix}.$$

Example 2. Use the inverse of the matrix A computed in Example 1 to find the solution of the system

$$x + 2y + z = 4,$$
$$x + 3y + 3z = -1,$$
$$x + 3y + 4z = 2.$$

Solution. The system can be written in the form

(26) $$AX = C,$$

where

$$X = \begin{bmatrix} x \\ y \\ z \end{bmatrix} \quad \text{and} \quad C = \begin{bmatrix} 4 \\ -1 \\ 2 \end{bmatrix}.$$

Since A is known to have an inverse from Example 1, we can multiply each side of (26) on the left by A^{-1} to get

$$X = A^{-1}C.$$

The solution of the system is, therefore,

$$X = A^{-1}C = \begin{bmatrix} 3 & -5 & 3 \\ -1 & 3 & -2 \\ 0 & -1 & 1 \end{bmatrix} \begin{bmatrix} 4 \\ -1 \\ 2 \end{bmatrix} = \begin{bmatrix} 23 \\ -11 \\ 3 \end{bmatrix},$$

or $x = 23$, $y = -11$, $z = 3$.

Exercise 2.7

1. Find the general solution of each of the following nonhomogeneous systems by reducing the augmented matrix to row-echelon form. In each case verify Theorem 2.17, namely, that the general solution is obtained by adding a particular solution to the general solution of the homogeneous system obtained by replacing the constant terms by zeros.

(a) $x + y - z = 6,$
$2x + 5y - 2z = 10.$

(b) $x + y = 2,$
$2x + 3y = 5.$

(c) $x + y + z = 1,$
$2x - 3y + 7z = 0,$
$3x - 2y + 8z = 4.$

(d) $x_1 + x_2 - 2x_3 + x_4 = 4,$
$2x_1 + 2x_2 - 5x_3 + 3x_4 = 2.$

(e) $x - y + 2z = 1,$
$x + y + z = 2,$
$2x - y + z = 5.$

(f) $x_1 - x_2 + 2x_3 = 4,$
$3x_1 + x_2 + 4x_3 = 6,$
$x_1 + x_2 + x_3 = 1.$

2. Find all solutions of the system

$$x + y - 2z + t = 4,$$
$$2x + 3y + z - t = 10$$

in terms of the appropriate number of arbitrary parameters. Check your result by substitution in the equations.

3. Interpret Theorem 2.17 geometrically in the case of each of the following systems.

(a) $ax + by = c$ (b) $ax + by + cz = d$

(c) $ax + by + cz = d$

     $ex + fy + gz = k$

4. If $X_1, X_2, \ldots, X_m$ are linearly independent vectors, prove that the vectors $Y_1, Y_2, \ldots, Y_m$, where

$$Y_i = \sum_{j=1}^{m} a_{ij}X_j$$

are linearly independent if and only if the matrix $A = [a_{ij}]$ is nonsingular.

5. Find the inverse of each of the following matrices:

(a) $\begin{bmatrix} 5 & 3 \\ 2 & 1 \end{bmatrix}$.   (b) $\begin{bmatrix} 4 & 1 \\ 5 & 2 \end{bmatrix}$.   (c) $\begin{bmatrix} 1 & 2 & 4 \\ 3 & 1 & 4 \\ 5 & 2 & 7 \end{bmatrix}$.

6. Find the inverse of the matrix

$$\begin{bmatrix} 1 & 2 & 1 \\ 2 & 5 & 2 \\ 1 & 3 & 3 \end{bmatrix}$$

and use it to find the solutions of each of the following systems:

(a) $x + 2y + z = 10$,   (b) $x + 2y + z = 2$,

    $2x + 5y + 2z = 14$, $2x + 5y + 2z = -1$,

    $x + 3y + 3z = 30$. $x + 3y + 3z = 6$.

7. Let $\mathscr{V}$ be a vector space over $\mathscr{F}$ with basis $X_1, X_2, \ldots, X_m$ and let

$$Y_i = \sum_{j=1}^{m} a_{ij}X_j.$$

What condition on the coefficients a_{ij} will ensure that $Y_1, Y_2, \ldots, Y_m$ is a basis of $\mathscr{V}$?

8. If A is an $n \times n$ matrix, prove that A is nonsingular if and only if the null space of A is the zero space.

9. If A and B are $n \times n$ matrices and A is nonsingular, prove that rank $AB =$ rank $B =$ rank BA.

10. A matrix A is called a *left zero divisor* if there exists a nonzero matrix B such that $AB = O$ and a *right zero divisor* if there exists a nonzero matrix C such that $CA = O$. Let A be an $m \times n$ matrix and prove that:

(a) If $m < n$, A is a left zero divisor.

(b) If $m > n$, A is a right zero divisor.

(c) If $m = n$, A is both a left and a right zero divisor if and only if A is singular.

11. If A and B are nonsingular $n \times n$ matrices, show by means of examples that we know nothing about the rank of $A + B$, that is, there exist nonsingular matrices A, B such that rank $(A + B) = r$ for any value of r such that $0 \le r \le n$.

12. Let A be the coefficient matrix and B the augmented matrix of a system of three equations in three unknowns. Discuss the intersections of the three planes represented by these three equations in each of the following cases.
    (a) rank $A = 3$, rank $B = 3$. (d) rank $A = 1$, rank $B = 2$.
    (b) rank $A = 2$, rank $B = 3$. (e) rank $A = 1$, rank $B = 1$.
    (c) rank $A = 2$, rank $B = 2$. (f) Is rank $A = 1$, rank $B = 3$ possible?

2.10 Elementary Matrices

In Section 2.3, we defined three types of elementary row transformations of a matrix which we designated as types (α), (β), and (γ). We now let ϵ represent any elementary row transformation and denote by $\epsilon(A)$ the matrix obtained from A by applying the elementary row transformation ϵ.

Definition. *An elementary matrix is any matrix of the form $\epsilon(I)$, in other words, any matrix obtained by applying an elementary row transformation to an identity matrix.*

The importance of elementary matrices results from the following theorem.

Theorem 2.22. *If A is any $m \times n$ matrix and ϵ is any elementary transformation, then*

$$(27) \qquad\qquad \epsilon(A) = \epsilon(I)A,$$

where I is the identity matrix of order m.

Proof. The proof consists in verifying (27) for each of the three types of elementary transformation. For types (α) and (β) this is immediate. For type (γ), if ϵ consists in adding k times the jth row to the ith row then $\epsilon(I)$ has 1's in the main diagonal, k in the ith row and jth column, and 0's elsewhere. Hence, $\epsilon(I)A$ differs from A only in its ith row which is equal to the ith row of A plus k times the jth row of A. This completes the proof.

Theorem 2.23. *Every elementary matrix is nonsingular and $[\epsilon(I)]^{-1} = \epsilon^{-1}(I)$. The transpose of an elementary matrix is an elementary matrix.*

Proof. By Theorem 2.9, $\epsilon(I)$ has the same rank as I and, therefore, is nonsingular. Since ϵ^{-1} is an elementary row transformation (Theorem 2.6), Theorem 2.22 gives

$$\epsilon^{-1}(I)\epsilon(I) = \epsilon^{-1}[\epsilon(I)] = I.$$

Similarly, $\epsilon(I)\epsilon^{-1}(I) = I$ and hence $\epsilon^{-1}(I) = [\epsilon(I)]^{-1}$. Finally, if ϵ is of type (α) or (β), $[\epsilon(I)]^T = \epsilon(I)$. If ϵ is of type (γ) and $\epsilon(I)$ is obtained by adding k times the jth row of I to the ith row, then $[\epsilon(I)]^T$ is obtained from I by adding k times the ith row to the jth row, and is therefore an elementary matrix.

Theorem 2.24. *If A is an $m \times n$ matrix which is row equivalent to B, there exists a nonsingular $m \times n$ matrix P such that $B = PA$.*

Proof. Since A can be transformed into B by a succession of elementary row transformations, we have by Theorem 2.22

$$B = E_1 E_2 \cdots E_s A,$$

where each E_i is an elementary $m \times m$ matrix. Hence, $B = PA$ where $P = E_1 E_2 \cdots E_s$ is nonsingular because each E_i is nonsingular.

Corollary. *If R is a reduced row-echelon matrix row-equivalent to A, then $R = PA$ where P is nonsingular.*

Theorem 2.25. *Every nonsingular $n \times n$ matrix is equal to a product of elementary $n \times n$ matrices.*

Proof. If A is a nonsingular $n \times n$ matrix it is row-equivalent to I by Theorem 2.10, Corollary 3. Hence, by Theorem 2.22, $E_1 E_2 \cdots E_s A = I$ and therefore $A = E_s^{-1} \cdots E_2^{-1} E_1^{-1}$, where each E_i and hence each E_i^{-1} is an elementary matrix.

Corollary. *A necessary and sufficient condition that two $m \times n$ matrices A and B are row-equivalent is that there exists a nonsingular $m \times m$ matrix P such that $B = PA$.*

Proof. The necessity is proved in Theorem 2.24 and the sufficiency follows from Theorem 2.25.

Exercise 2.8

1. Write down 4×4 elementary matrices that will induce the following elementary transformations in a 4×4 matrix A when used as left multipliers. Check your own answers.
   (a) Interchange of the 2nd and 4th rows of A.
   (b) Interchange of the 2nd and 3rd rows of A.
   (c) Multiplication of the 4th row of A by 5.
   (d) Addition of k times the 4th row of A to the 1st row of A.
   (e) Addition of k times the 1st row of A to the 4th row of A.

2. Reduce the matrix

$$A = \begin{bmatrix} 1 & 0 & 2 \\ 0 & 3 & -1 \\ 2 & 3 & 3 \end{bmatrix}$$

to a reduced row-echelon matrix R and write the elementary matrix corresponding to each elementary row transformation used. Hence, find a nonsingular matrix P such that $PA = R$.

3. Write the following matrices as products of elementary matrices

(a) $\begin{bmatrix} 1 & 3 \\ 2 & 8 \end{bmatrix}$.

(b) $\begin{bmatrix} 2 & -1 \\ 3 & 2 \end{bmatrix}$.

4. Choose any 4×4 matrix A and multiply it on the *right* by each of the five elementary matrices that you found in Problem 1. What is the effect of each of these multiplications on the matrix A?

5. Let E be the $n \times n$ matrix

$$\begin{bmatrix} 0 & 0 & \cdots & 0 & 1 \\ 0 & 0 & \cdots & 1 & 0 \\ \cdots\cdots\cdots\cdots\cdots \\ 1 & 0 & \cdots & 0 & 0 \end{bmatrix}.$$

What is the effect on an $n \times n$ matrix A of multiplying A by E (a) on the left, (b) on the right, (c) both on the left and the right.

Note. Problems 6 to 11 develop a connected theory. They should be worked in the order in which they are given.

6. Define elementary column transformations of a matrix A analogously to elementary row transformations. Prove that every elementary column transformation of A can be achieved by multiplying A on the *right* by an elementary matrix. *Hint.* This can be done either directly, as for rows, or by using transposes and Theorem 2.23.

7. Show that a reduced row-echelon $m \times n$ matrix R of rank r can be reduced by elementary column transformations to an $m \times n$ matrix of the form

$$C = \begin{bmatrix} 1 & 0 & & \cdots & & 0 \\ 0 & 1 & & \cdots & & 0 \\ \cdots\cdots\cdots\cdots\cdots\cdots\cdots \\ 0 & 0 & \cdots & 1 & \cdots & 0 \\ \cdots\cdots\cdots\cdots\cdots\cdots\cdots \\ 0 & 0 & & \cdots & & 0 \end{bmatrix},$$

in which the first r elements in the main diagonal are 1 and all other elements are 0.

8. Deduce from Problem 7 and Theorem 2.8 that every $m \times n$ matrix A of rank r can be reduced by elementary row and column transformations to the form C in Problem 7. The matrix C is then called the canonical form of A.

9. A matrix A is said to be row-column-equivalent (r-c-equivalent) to a matrix B if A can be transformed into B by a succession of elementary row and column transformations. Prove the following.

   (a) If A is a matrix, ϵ an elementary row transformation, and ϵ' an elementary column transformation, then

$$(\epsilon A)\epsilon' = \epsilon(A\epsilon').$$

   (b) r-c-equivalence is an equivalence relation.

   (c) Two $m \times n$ matrices A and B are r-c-equivalent if and only if they have the same canonical form and, hence, if and only if they have the same rank.

10. If A is any $m \times n$ matrix of rank r, prove that there exist a nonsingular $m \times m$ matrix P and a nonsingular $n \times n$ matrix Q such that $PAQ = C$, the canonical form of A.

11. Prove that two $m \times n$ matrices A and B are r-c-equivalent if and only if there exist a nonsingular $m \times m$ matrix P and a nonsingular $n \times n$ matrix Q such that $PAQ = B$.

2.11 Multiplication of Partitioned Matrices

If $A = [a_{ij}]$ is any matrix, the matrix that remains after some rows and/or columns of A have been deleted is called a submatrix of A. It is frequently convenient to partition a matrix into submatrices and to consider it as a matrix whose elements are themselves these submatrices. For example, the matrix

$$A = \left[\begin{array}{cc|cc|c} a_{11} & a_{12} & a_{13} & a_{14} & a_{15} \\ a_{21} & a_{22} & a_{23} & a_{24} & a_{25} \\ \hline a_{31} & a_{32} & a_{33} & a_{34} & a_{35} \\ a_{41} & a_{42} & a_{43} & a_{44} & a_{45} \end{array}\right]$$

may be partitioned as shown and written in the form

$$\begin{bmatrix} A_{11} & A_{12} & A_{13} \\ A_{21} & A_{22} & A_{23} \end{bmatrix},$$

where

$$A_{11} = \begin{bmatrix} a_{11} & a_{12} \\ a_{21} & a_{22} \end{bmatrix}, \quad A_{12} = \begin{bmatrix} a_{13} & a_{14} \\ a_{23} & a_{24} \end{bmatrix}, \quad A_{13} = \begin{bmatrix} a_{15} \\ a_{25} \end{bmatrix},$$

$$A_{21} = \begin{bmatrix} a_{31} & a_{32} \\ a_{41} & a_{42} \end{bmatrix}, \quad A_{22} = \begin{bmatrix} a_{33} & a_{34} \\ a_{43} & a_{44} \end{bmatrix}, \quad A_{23} = \begin{bmatrix} a_{35} \\ a_{45} \end{bmatrix}.$$

Now if B is any matrix with five rows, the product AB is defined. To be definite, suppose

$$
B = \begin{bmatrix} b_{11} & b_{12} & b_{13} \\ b_{21} & b_{22} & b_{23} \\ \hline b_{31} & b_{32} & b_{33} \\ b_{41} & b_{42} & b_{43} \\ \hline b_{51} & b_{52} & b_{53} \end{bmatrix} = \begin{bmatrix} B_{11} & B_{12} \\ B_{21} & B_{22} \\ B_{31} & B_{32} \end{bmatrix},
$$

and let B be partitioned in the manner shown so that its row partitions are spaced in the same way as the column partitions of A. Then we can multiply the two matrices in *partitioned form*. That is to say, we can write

$$
AB = \begin{bmatrix} A_{11}B_{11} + A_{12}B_{21} + A_{13}B_{31} & A_{11}B_{12} + A_{12}B_{22} + A_{13}B_{32} \\ A_{21}B_{11} + A_{22}B_{21} + A_{23}B_{31} & A_{21}B_{12} + A_{22}B_{22} + A_{23}B_{32} \end{bmatrix}
$$

as a partitioned form AB. By writing out the products in full, the student should verify, first, that all the matrix products and sums used in this multiplication are actually defined and, second, that the above product of the partitioned forms of A and B does actually give the product AB.

This rule for multiplying partitioned matrices is general and may be stated as follows. If

$$
A = \begin{bmatrix} A_{11} & A_{12} & \cdots & A_{1n} \\ \cdots\cdots\cdots\cdots\cdots\cdots \\ A_{m1} & A_{m2} & \cdots & A_{mn} \end{bmatrix} \quad \text{and} \quad B = \begin{bmatrix} B_{11} & B_{12} & \cdots & B_{1h} \\ \cdots\cdots\cdots\cdots\cdots\cdots \\ B_{n1} & B_{n2} & \cdots & B_{nh} \end{bmatrix},
$$

where A_{ij} is a matrix of type (r_i, s_j) and B_{jk} is a matrix of type (s_j, t_k), then

$$
AB = \left[\sum_{j=1}^{n} A_{ij}B_{jk} \right] (i = 1, \ldots, m; k = 1, \ldots, h).
$$

A formal proof of this rule will not be given but the student should test its truth by working several examples of the type indicated.

Example 1. By ordinary matrix multiplication

$$
\begin{bmatrix} 2 & -1 & 4 \\ \hline 5 & 1 & -2 \\ -1 & 4 & 1 \end{bmatrix} \begin{bmatrix} 1 & 3 & 1 \\ \hline 4 & 1 & 5 \\ 2 & 2 & 7 \end{bmatrix} = \begin{bmatrix} 6 & 13 & 25 \\ \hline 5 & 12 & -4 \\ 17 & 3 & 26 \end{bmatrix},
$$

whereas

$$[2][1] + [-1, 4]\begin{bmatrix} 4 \\ 2 \end{bmatrix} = [6],$$

$$[2][3, 1] + [-1, 4]\begin{bmatrix} 1 & 5 \\ 2 & 7 \end{bmatrix} = [13, 25],$$

$$\begin{bmatrix} 5 \\ -1 \end{bmatrix}[1] + \begin{bmatrix} 1 & -2 \\ 4 & 1 \end{bmatrix}\begin{bmatrix} 4 \\ 2 \end{bmatrix} = \begin{bmatrix} 5 \\ 17 \end{bmatrix},$$

and

$$\begin{bmatrix} 5 \\ -1 \end{bmatrix}[3 \quad 1] + \begin{bmatrix} 1 & -2 \\ 4 & 1 \end{bmatrix}\begin{bmatrix} 1 & 5 \\ 2 & 7 \end{bmatrix} = \begin{bmatrix} 12 & -4 \\ 3 & 26 \end{bmatrix}.$$

Example 2

$$\left[\begin{array}{cc|c} 3 & 4 & 0 \\ 2 & 1 & 0 \\ \hline -1 & -2 & 5 \end{array}\right]\left[\begin{array}{cc|c} 1 & 2 & 0 \\ 3 & 1 & 0 \\ \hline 5 & 3 & 3 \end{array}\right] = \left[\begin{array}{cc|c} 15 & 10 & 0 \\ 5 & 5 & 0 \\ \hline 18 & 11 & 15 \end{array}\right],$$

whereas

$$\begin{bmatrix} 3 & 4 \\ 2 & 1 \end{bmatrix}\begin{bmatrix} 1 & 2 \\ 3 & 1 \end{bmatrix} + \begin{bmatrix} 0 \\ 0 \end{bmatrix}[5 \quad 3] = \begin{bmatrix} 15 & 10 \\ 5 & 5 \end{bmatrix},$$

$$\begin{bmatrix} 3 & 4 \\ 2 & 1 \end{bmatrix}\begin{bmatrix} 0 \\ 0 \end{bmatrix} + \begin{bmatrix} 0 \\ 0 \end{bmatrix}[3] = \begin{bmatrix} 0 \\ 0 \end{bmatrix},$$

$$[-1, -2]\begin{bmatrix} 1 & 2 \\ 3 & 1 \end{bmatrix} + [5][5, 3] = [18 \quad 11],$$

and

$$[-1, -2]\begin{bmatrix} 0 \\ 0 \end{bmatrix} + [5][3] = [15].$$

Exercise 2.9

1. Compute the following matrix products as partitioned matrices (partitionings as indicated) and check by ordinary matrix multiplication.

(a) $\left[\begin{array}{cc|cc} 2 & 1 & 5 & 3 \\ 1 & 4 & 2 & -1 \\ \hline 3 & -1 & 2 & 2 \end{array}\right]\begin{bmatrix} 4 & 1 \\ 1 & 5 \\ -2 & 2 \\ 3 & 6 \end{bmatrix}.$

(b) $\begin{bmatrix} \begin{array}{cc|cc} 1 & 1 & 2 & 2 \\ 3 & 1 & 2 & 1 \\ \hline 1 & 0 & 5 & 2 \\ 1 & 4 & 3 & 1 \end{array} \end{bmatrix} \begin{bmatrix} \begin{array}{cc|cc} 1 & 2 & 1 & 2 \\ 1 & 1 & 3 & -5 \\ \hline 1 & 2 & 1 & 6 \\ 1 & 1 & 3 & 4 \end{array} \end{bmatrix}.$

(c) $\begin{bmatrix} \begin{array}{c|cc|cc} 2 & 0 & 0 & 0 & 0 \\ \hline 1 & 5 & 2 & 0 & 0 \\ 2 & -1 & 3 & 0 & 0 \\ \hline 1 & 4 & -1 & 3 & -1 \\ -1 & 5 & 2 & 1 & 6 \end{array} \end{bmatrix}^2.$

2. Show that the product of the $n \times n$ matrices

$$A = \begin{bmatrix} A_{11} & O & \cdots & O \\ A_{21} & A_{22} & \cdots & O \\ \hdotsfor{4} \\ A_{r1} & A_{r2} & \cdots & A_{rr} \end{bmatrix}, \qquad B = \begin{bmatrix} B_{11} & O & \cdots & O \\ B_{21} & B_{22} & \cdots & O \\ \hdotsfor{4} \\ B_{r1} & B_{r2} & \cdots & B_{rr} \end{bmatrix},$$

in which A_{ii}, B_{ii} are square matrices of order m_i and

$$\sum_{i=1}^{r} m_i = n,$$

has the form

$$AB = \begin{bmatrix} C_{11} & O & \cdots & O \\ C_{21} & C_{11} & \cdots & O \\ \hdotsfor{4} \\ C_{r1} & C_{r2} & \cdots & C_{rr} \end{bmatrix},$$

where $C_{ii} = A_{ii}B_{ii}$, $(i = 1, 2, \ldots, r)$.

CHAPTER THREE

DETERMINANTS

3.1 Permutations

The reader probably has some previous knowledge of determinants, at least those of order 2 and 3. In this chapter we shall define the determinant of an arbitrary square matrix of order n and derive its basic properties. It is first necessary to discuss permutations of finite sets.

Definition. *A permutation is a one-to-one mapping of a finite set onto itself.*

The elements of the finite set $\mathscr{S}$ will be designated by the first n integers so that

$$\mathscr{S} = \{1, 2, 3, \ldots, n\}.$$

If σ is any permutation, then $\sigma(1), \sigma(2), \ldots, \sigma(n)$ is simply a rearrangement of the integers $1, 2, \ldots, n$. Since the number of possible such rearrangements is $n!$, there are exactly $n!$ distinct permutations of $\mathscr{S}$. These include the identity mapping defined by $\sigma(i) = i$ $(i = 1, 2, \ldots, n)$, which will be denoted by I. Because $\sigma : \mathscr{S} \to \mathscr{S}$ is a one-to-one mapping of $\mathscr{S}$ onto $\mathscr{S}$, there exists an inverse mapping $\sigma^{-1} : \mathscr{S} \to \mathscr{S}$ which is also a permutation. The product of two permutations σ and τ is the composite mapping defined by $(\sigma\tau)(i) = \sigma[\tau(i)]$. If σ and τ are permutations of $\mathscr{S}$, so also is $\sigma\tau$. In particular, $\sigma\sigma^{-1} = \sigma^{-1}\sigma = I$.

A permutation σ is sometimes represented by the notation

$$\sigma = \begin{pmatrix} 1 & 2 & 3 & \cdots & n \\ \sigma(1) & \sigma(2) & \sigma(3) & \cdots & \sigma(n) \end{pmatrix}.$$

Example. If $n = 5$ the symbol

$$\sigma = \begin{pmatrix} 1 & 2 & 3 & 4 & 5 \\ 3 & 1 & 5 & 2 & 4 \end{pmatrix}$$

means that for each i, $\sigma(i)$ is the number immediately below i in the symbol for σ.

Now if

$$\tau = \begin{pmatrix} 1 & 2 & 3 & 4 & 5 \\ 5 & 4 & 1 & 2 & 3 \end{pmatrix},$$

we have

$$\sigma\tau = \begin{pmatrix} 1 & 2 & 3 & 4 & 5 \\ 4 & 2 & 3 & 1 & 5 \end{pmatrix}$$

because, for example, $\sigma\tau(1) = \sigma[\tau(1)] = \sigma(5) = 4$. Similarly,

$$\tau\sigma = \begin{pmatrix} 1 & 2 & 3 & 4 & 5 \\ 1 & 5 & 3 & 4 & 2 \end{pmatrix}.$$

A permutation σ of $\mathscr{S}$ is called a *transposition* if it interchanges two elements of $\mathscr{S}$ and leaves all others fixed. For example, if $\sigma(i) = j$, $\sigma(j) = i$, and $\sigma(k) = k$ for all k in $\mathscr{S}$ except i and j, then σ is a transposition and we write $\sigma = (ij)$. Thus, in the example given above, $\sigma\tau = (14)$ and $\tau\sigma = (25)$ are both transpositions.

We need two basic results which we state in the next two theorems.

Theorem 3.1. *Every permutation of a finite set $\mathscr{S}$ is the product of a finite number of transpositions.*

Theorem 3.2. *If a permutation σ can be written as a product of r transpositions and also as a product of s transpositions then r and s are either both even or both odd.*

We shall omit the proofs of these theorems. They may be found in [7], [9], or [6]. As a consequence of them, all permutations of $\mathscr{S}$ can be classified as *even* or *odd*. Thus, σ is said to be an *even permutation* if it is a product of an even number of transpositions or an *odd permutation* if it is a product of an odd number of transpositions.

Example 2. If σ and τ are the two permutations of Example 1, the student can verify that $\sigma = (24)(25)(23)(21)$ and is therefore even whereas $\tau = (53)(13)(24)$ and is therefore odd.

Theorem 3.3. *The product of two even permutations or of two odd permutations is even. The product of an even permutation and an odd permutation is odd. The identity permutation I is even. The inverse σ^{-1} is even if σ is even, odd if σ is odd.*

Proof. The first two statements are immediate consequences of the definition. That I is even follows from $I = (ij)^2$ where (ij) is any transposition. Thus, every transposition is its own inverse, and hence σ^{-1} is even if and only if σ is even.

Theorem 3.4. *If $\mathscr{S}$ is any finite set, exactly half the permutations of $\mathscr{S}$ are even.*

Proof. Let r be the number of even permutations and s the number of odd permutations. Let $\sigma_1, \sigma_2, \ldots, \sigma_r$ be the distinct even permutations. Let τ be any odd permutation. By Theorem 3.3, $\tau\sigma_1, \tau\sigma_2, \ldots, \tau\sigma_r$ are all odd. Moreover, they are all different for if $\tau\sigma_i = \tau\sigma_j$ we can multiply on the left by τ^{-1} and get $\sigma_i = \sigma_j$. Hence, $s \geq r$ since we have produced r different odd permutations. A similar argument shows that $r \geq s$ and hence $r = s$.

Exercise 3.1

1. Given

$$\sigma = \begin{pmatrix} 1 & 2 & 3 & 4 & 5 & 6 \\ 4 & 6 & 1 & 3 & 5 & 2 \end{pmatrix}, \qquad \tau = \begin{pmatrix} 1 & 2 & 3 & 4 & 5 & 6 \\ 3 & 1 & 2 & 5 & 6 & 4 \end{pmatrix},$$

compute $\sigma\tau$, $\tau\sigma$, σ^2, τ^2, τ^3, σ^{-1}, τ^{-1}, and $(\sigma\tau)^{-1}$.

2. If $S = \{s_1, s_2, \ldots, s_n\}$ and $r \leq n$, let σ be the permutation defined by $\sigma(s_i) = s_{i+1}$ for $i = 1, 2, \ldots, r - 1$, $\sigma(s_r) = s_1$, and $\sigma(s_j) = s_j$ for $j > r$. Then σ is called a *cycle* (or cyclic permutation) of order r. A cyclic permutation of $1, 2, 3, \ldots, r$ is usually written $\sigma = (123 \cdots r)$ rather than in the more cumbersome notation

$$\sigma = \begin{pmatrix} 1 & 2 & 3, \ldots, r-1 & r & r+1, \ldots, n \\ 2 & 3 & 4, \ldots, r & 1 & r+1, \ldots, n \end{pmatrix}$$

Prove that:

(a) If σ is cyclic of order r, then $\sigma^r = I$ but $\sigma^s \neq I$ if $s < r$.

(b) Every transposition is cyclic of order 2.

(c) Every cyclic permutation of order 3 is even.

(d) Every cyclic permutation of order 4 is odd.

(e) A cyclic permutation of order r is even if r is odd and odd if r is even.

3. If σ and τ are the permutations given in Problem 1, show that $\sigma = (143)(26)$ and $\tau = (132)(456)$.

4. If $\sigma = (12345)$, show that $\sigma^{-1} = (54321)$.

5. If $\sigma = \tau_1\tau_2 \cdots \tau_r$ where $\tau_1, \ldots, \tau_r$ are transpositions, show that

$$\sigma^{-1} = \tau_r\tau_{r-1} \cdots \tau_1.$$

3.2 Determinant of an $n \times n$ Matrix

Let A be any square matrix of order n with elements in the field $\mathscr{F}$ of scalars. It is possible to associate with A a number in $\mathscr{F}$ called the *determinant* of A (abbreviated det A) which has a number of interesting properties. We give the definition of det A first and will then derive its most important properties.

Let $A = [a_{ij}]$ and let σ be any permutation of the set $\{1, 2, \ldots, n\}$. Define the symbol $\epsilon(\sigma)$ by

$$\epsilon(\sigma) = 1 \qquad \text{if } \sigma \text{ is an even permutation,}$$
$$\epsilon(\sigma) = -1 \qquad \text{if } \sigma \text{ is an odd permutation.}$$

We now define the determinant of A by the equation

(1) $$\det A = \sum_{\sigma} \epsilon(\sigma)a_{1\sigma(1)}a_{2\sigma(2)} \cdots a_{n\sigma(n)},$$

where the summation extends over all $n!$ permutations σ of the set $\{1, 2, \ldots, n\}$.

The determinant of the matrix $[a_{ij}]$ is also written

$$\det A = \begin{vmatrix} a_{11} & a_{12} & \cdots & a_{1n} \\ a_{21} & a_{22} & \cdots & a_{2n} \\ \hdotsfor{4} \\ a_{n1} & a_{n2} & \cdots & a_{nn} \end{vmatrix},$$

with vertical lines replacing the square brackets used for the matrix A itself.

The student should check that for $n = 2$ and $n = 3$, our definition yields

$$\begin{vmatrix} a_{11} & a_{12} \\ a_{21} & a_{22} \end{vmatrix} = a_{11}a_{22} - a_{12}a_{21}$$

and

$$\begin{vmatrix} a_{11} & a_{12} & a_{13} \\ a_{21} & a_{22} & a_{23} \\ a_{31} & a_{32} & a_{33} \end{vmatrix} = a_{11}a_{22}a_{33} + a_{12}a_{23}a_{31} + a_{13}a_{21}a_{32} - a_{12}a_{21}a_{33} - a_{13}a_{22}a_{31} - a_{11}a_{23}a_{32}.$$

It follows from the definition (1) and from Theorem 3.4 that det A is the sum of $n!$ terms, half of which are preceded by a positive sign and half by a negative sign.

Moreover, each of these terms is a product of n factors a_{ij} of which exactly one comes from each row of A (because each initial subscript occurs exactly once) and exactly one comes from each column (because σ is a permutation and hence each second subscript occurs exactly once). The student should check this for the determinant of order 3.

3.3 Properties of Determinants

Because of the preceding remark, for any fixed j, $1 \leq j \leq n$, each term in det A contains exactly one factor which is an element of the jth column. Hence we have

$$(2) \qquad \det A = a_{1j}c_{1j} + a_{2j}c_{2j} + \cdots + a_{nj}c_{nj},$$

where for each i, j the coefficient c_{ij} depends only on the elements of A which do *not* occur either in the ith row or in the jth column. We state this fact formally, since several important properties of determinants follow from it.

Theorem 3.4. *If A is any $n \times n$ matrix and j is any one of the integers $1, 2, \ldots, n$, det A is a linear homogeneous function of the elements of the jth column of A.*

Corollary 1. *If $A_j = [a_{1j}, a_{2j}, \ldots, a_{nj}]^T$ is the jth column vector of A and $A_j = B_j + C_j$, then*

$$\det A = \det B + \det C,$$

where B and C are the matrices obtained from A by replacing the jth column by B_j and C_j, respectively.

Corollary 2. *If A' is the matrix obtained from A by replacing the jth column vector A_j by kA_j where k is any scalar, then det $A' = k$ det A.*

Corollary 3. *If one column vector of A is the zero vector, then det $A = 0$.*

These three corollaries all follow at once the theorem or from equation (2).

Theorem 3.5. *If A' is the matrix obtained from A by interchanging two columns, then det $A' = -$det A.*

Proof. If A' is obtained from A by interchanging the ith and the jth columns, we have by the definition (1) that

$$\det A' = \sum_{\sigma} \epsilon(\sigma)\, a_{1\tau(1)} a_{2\tau(2)} \cdots a_{n\tau(n)}$$

where $\tau = (ij)\sigma$. Hence $\epsilon(\tau) = -\epsilon(\sigma)$ and

$$\det A' = -\sum_\tau \epsilon(\tau)a_{1\tau(1)}a_{2\tau(2)} \cdots a_{n\tau(n)} = -\det A.$$

Corollary. *If two columns of A are identical, then* $\det A = 0$.

Proof. Interchanging the two identical columns, we get $-\det A = \det A$ and hence $\det A = 0$.

Theorem 3.6. *If k is any scalar and A' is the matrix obtained from A by adding k times the jth column vector to the ith column vector of A, then* $\det A' = \det A$.

Proof. By Theorem 3.4, Corollary 1 we have

$$\det A' = \det A + \det B,$$

where B is the matrix obtained from A by replacing the ith column by k times the jth column. By Theorem 3.4, Corollary 2, $\det B = k \det B'$ where B' has the ith and jth columns identical. Hence, by the corollary to Theorem 3.5, $\det B' = 0$ and $\det A' = \det A$.

Theorem 3.7. *If A is any square matrix,* $\det A^T = \det A$.

Proof. We rearrange the factors in each term of (1) so that the second subscripts are in natural order. Since, for example, the factor $a_{i\sigma(i)}$ in which $\sigma(i) = 1$ or $i = \sigma^{-1}(1)$ comes to the first place, we get

$$\det A = \sum_\sigma \epsilon(\sigma)a_{\tau(1)1}a_{\tau(2)2} \cdots a_{\tau(n)n},$$

where $\tau = \sigma^{-1}$. However $\epsilon(\sigma^{-1}) = \epsilon(\sigma)$ because if σ is written as a product of transpositions, then σ^{-1} is the product of the same transpositions in reverse order. Hence,

$$\det A = \sum_\tau \epsilon(\tau)a_{\tau(1)1}a_{\tau(2)2} \cdots a_{\tau(n)n} = \det A^T$$

because the second subscripts are the row indices of A^T.

Corollary. *Theorem 3.4 and its corollaries, Theorem 3.5 and its corollary, and Theorem 3.6 all remain valid when the word "column" is replaced by "row."*

Proof. These theorems, when applied to A^T, give the row form of the theorems for A.

Theorem 3.8. *If A is an n × n matrix and B is row equivalent to A, then* det $B = k$ det A where $k \neq 0$.

Proof. By Theorem 3.5 and the corollary to Theorem 3.7, an elementary row-transformation of type (α) merely changes the sign of det A. Similarly, by Theorem 3.4, Corollary 2, an elementary row-transformation of type (β) multiplies det A by a nonzero scalar. Finally, by Theorem 3.6 an elementary row-transformation of type (γ) leaves det A unchanged.

Corollary. *If R is the reduced row-echelon form of A, then* det A *is zero if and only if* det R *is zero.*

Definition. *An n × n matrix is said to be triangular if all elements below (or above) the main diagonal of A are zero.*

Theorem 3.9. *If A is a triangular matrix, then* det A *is equal to the product of the elements in the main diagonal of A.*

Proof. Let

$$A = \begin{bmatrix} a_{11} & a_{12} & a_{13} & \cdots & a_{1n} \\ 0 & a_{22} & a_{23} & \cdots & a_{2n} \\ 0 & 0 & a_{33} & \cdots & a_{3n} \\ \cdots\cdots\cdots\cdots\cdots\cdots\cdots \\ 0 & 0 & 0 & \cdots & a_{nn} \end{bmatrix}.$$

In writing out det A from the definition (1) the only nonzero term is $a_{11} a_{22} \cdots a_{nn}$. This is because to get a nonzero term, a_{11} must be chosen from the first column. Since a_{11} is in the first row, this leaves a_{22} as the only possible nonzero choice from the second column and so on. Thus, det $A = a_{11} a_{22} \cdots a_{nn}$ as required. If the zeros occur above the main diagonal instead of below, the same result follows from Theorem 3.7.

Corollary. *If I is the n × n identity matrix,* det $I = 1$.

The last two theorems lead to the following very important result.

Theorem 3.10. *If A is a square matrix of order n,* det $A = 0$ *if and only if* rank $A < n$, *that is, if and only if A is singular.*

Proof. Let R be the reduced-row echelon matrix row-equivalent to A. If rank $A < n$, then by Theorem 2.10, Corollary 2, R has at least one zero row vector and hence det $R = 0$. It follows from Theorem 3.8 that det $A = 0$. On the other hand, if rank $A = n$, then by Theorem 2.10, Corollary 3, $R = I$ and det $R = 1$. Hence, by Theorem 3.8, det $A \neq 0$. Thus, if det $A = 0$ we must have rank $A < n$.

We next investigate the determinant of a product of two matrices and show that it is equal to the product of the determinants of the two factors. We prove this first for the case in which one factor is an elementary matrix.

Lemma 3.11. *If E is an elementary matrix and B an arbitrary square matrix of the same order as E, then*

$$\det (EB) = \det E \det B.$$

Proof. We recall (Section 2.10) that $E = \epsilon(I)$ where ϵ is an elementary row transformation and that by Theorem 2.22

$$EB = \epsilon(I)B = \epsilon(B).$$

Now if E is of type (α) (see Section 2.3), we have $\det \epsilon(I) = -1$ and $\det \epsilon(B) = -\det B$ by Theorem 3.5. Thus, $\det EB = -\det B = (-1) \det B = \det E \det B$. If ϵ is of type β, we have $\det E = \det \epsilon(I) = k$ and $\det EB = \det \epsilon(B) = k \det B = \det E \det B$ by Theorem 3.4, Corollary 2. Finally if ϵ is of type (γ) we have by Theorem 3.9 that $\det E = \det \epsilon(I) = 1$ and hence $\det EB = \det \epsilon(B) = \det B = \det E \det B$ by Theorem 3.6.

Theorem 3.12. *If A and B are n by n matrices, then* $\det AB = \det A \det B$.

Proof. If either A or B is singular, then by the corollary to Theorem 2.19 the product AB is also singular. It follows from Theorem 3.10 that $\det AB = 0$ and either $\det A = 0$ or $\det B = 0$. Thus the theorem holds in this case.

Now assume that both A and B are nonsingular. By Theorem 2.25 we have $A = E_1 E_2 \cdots E_r$ where $E_1, E_2, \ldots, E_r$ are elementary matrices. Hence, by successive application of Lemma 3.11, we have

$$\begin{aligned}
\det AB &= \det (E_1 E_2 \cdots E_r B) \\
&= \det E_1 \det (E_2 \cdots E_r B) \\
&= \det E_1 \det E_2 \cdots \det E_r \det B \\
&= \det E_1 \det E_2 \cdots \det (E_{r-1} E_r) \det B \\
&= \det (E_1 E_2 \cdots E_r) \det B \\
&= \det A \det B.
\end{aligned}$$

3.4 Evaluation of Determinants

It is clear that if $n > 3$ the evaluation of a determinant of order n directly from the definition (1) is impractical because of the large number of multiplications necessary and the difficulty of writing down the $n!$ terms. However, if we reduce the matrix A to triangular form by means of elementary transformations, $\det A$ can be found by means of Theorem 3.9 provided we keep track of the changes in the determinant resulting from the elementary transformations. Moreover, a computer can easily be programmed to evaluate a determinant by this method. We illustrate with an example.

Example. Evaluate det A if

$$A = \begin{bmatrix} 2 & 1 & 9 & 1 \\ 4 & 3 & -1 & 2 \\ 1 & 4 & 3 & -2 \\ 3 & 2 & 1 & 4 \end{bmatrix}.$$

Solution

$$\det A = \begin{vmatrix} 2 & 1 & 9 & 1 \\ 4 & 3 & -1 & 2 \\ 1 & 4 & 3 & -2 \\ 3 & 2 & 1 & 4 \end{vmatrix} = - \begin{vmatrix} 1 & 4 & 3 & -2 \\ 4 & 3 & -1 & 2 \\ 2 & 1 & 9 & 1 \\ 3 & 2 & 1 & 4 \end{vmatrix} \begin{matrix} \leftarrow R_3 \\ \\ \leftarrow R_1{'} \\ \\ \end{matrix}$$

$$= - \begin{vmatrix} 1 & 4 & 3 & -2 \\ 0 & -13 & -13 & 10 \\ 0 & -7 & 3 & 5 \\ 0 & -10 & -8 & 10 \end{vmatrix} \begin{matrix} \\ \leftarrow R_2 - 4R_1 \\ \leftarrow R_3 - 2R_1{'} \\ \leftarrow R_4 - 3R_1 \end{matrix}$$

$$= 13 \begin{vmatrix} 1 & 4 & 3 & -2 \\ 0 & 1 & 1 & -\frac{10}{13} \\ 0 & -7 & 3 & 5 \\ 0 & -10 & -8 & 10 \end{vmatrix} \begin{matrix} \\ \leftarrow -\frac{1}{13}R_2 \\ {}' \\ \end{matrix}$$

$$= 13 \begin{vmatrix} 1 & 4 & 3 & -2 \\ 0 & 1 & 1 & \frac{10}{13} \\ 0 & 0 & 10 & -\frac{5}{13} \\ 0 & 0 & 2 & \frac{30}{13} \end{vmatrix} \begin{matrix} \\ \\ \leftarrow R_3 + 7R_2 {}' \\ \leftarrow R_4 + 10R_2 \end{matrix}$$

$$= 65 \begin{vmatrix} 1 & 4 & 3 & -2 \\ 0 & 1 & 1 & \frac{10}{13} \\ 0 & 0 & 2 & -\frac{1}{13} \\ 0 & 0 & 2 & \frac{30}{13} \end{vmatrix} \begin{matrix} \\ \\ \leftarrow \frac{1}{5}R_3 {}' \\ \end{matrix}$$

$$= 65 \begin{vmatrix} 1 & 4 & 3 & -2 \\ 0 & 1 & 1 & \frac{10}{13} \\ 0 & 0 & 2 & -\frac{1}{13} \\ 0 & 0 & 0 & \frac{31}{13} \end{vmatrix} \begin{matrix} \\ \\ {}' \\ \leftarrow R_4 - R_3 \end{matrix}$$

$$= (65)(2)(\tfrac{31}{13}) = 310.$$

Exercise 3.2

1. Evaluate the determinants of the following matrices:

(a) $\begin{bmatrix} 1 & 2 & 5 \\ 9 & 1 & 3 \\ 2 & 3 & 4 \end{bmatrix}$.

(b) $\begin{bmatrix} 2 & 1 & 3 \\ 3 & 2 & 1 \\ 2 & 2 & 5 \end{bmatrix}$.

(c) $\begin{bmatrix} 2 & 0 & 9 \\ 4 & 7 & 0 \\ 0 & 5 & 6 \end{bmatrix}$.

(d) $\begin{bmatrix} 1 & 5 & 2 & 4 \\ 3 & 2 & 1 & 9 \\ 1 & 7 & 6 & 8 \\ 2 & 2 & 3 & 2 \end{bmatrix}$.

(e) $\begin{bmatrix} 1 & 5 & 0 & 0 \\ 2 & 8 & 0 & 0 \\ 3 & 2 & 5 & 1 \\ 5 & 1 & 9 & 2 \end{bmatrix}$.

(f) $\begin{bmatrix} 2 & 0 & 0 & 0 \\ 5 & 9 & 1 & 6 \\ 5 & 1 & 3 & 5 \\ 2 & 8 & 2 & 1 \end{bmatrix}$.

2. Prove that

$$\begin{vmatrix} a_1 & 0 & 0 & 0 \\ b_1 & b_2 & b_3 & b_4 \\ c_1 & c_2 & c_3 & c_4 \\ d_1 & d_2 & d_3 & d_4 \end{vmatrix} = a_1 \begin{vmatrix} b_2 & b_3 & b_4 \\ c_2 & c_3 & c_4 \\ d_2 & d_3 & d_4 \end{vmatrix}.$$

Hint. Show that a_1 must occur as a factor of each term in the expansion and use the definition (1) to show that the total coefficient of a_1 is the third order determinant.

3. Let A be an $n \times n$ matrix partitioned as

$$A = \begin{bmatrix} B & O \\ C & D \end{bmatrix},$$

where B is an $r \times r$ *diagonal* matrix and D an arbitrary $n - r \times n - r$ matrix. Prove, by repeated application of the result of Problem 2, that

$$\det A = \det B \det D.$$

4. Let A be an $n \times n$ matrix partitioned as

$$A = \begin{bmatrix} B & O \\ O & D \end{bmatrix},$$

where B is an $r \times r$ and D an $n - r \times n - r$ matrix. Prove that $\det A = \det B \det D$. *Hint.* Verify that

$$A = \begin{bmatrix} I_r & O \\ O & D \end{bmatrix} \begin{bmatrix} B & O \\ O & I_{n-r} \end{bmatrix}$$

where I_r, I_{n-r} are identity matrices. Then use Theorem 3.12 and Problem 3.

5. Let A be an $n \times n$ matrix partitioned as

$$A = \begin{bmatrix} B & O \\ C & D \end{bmatrix},$$

where A is an $r \times r$ and D an $n - r \times n - r$ matrix. Prove that

$$\det A = \det B \det D.$$

Hint. If B is nonsingular, show that

$$A = \begin{bmatrix} I & O \\ CB^{-1} & D \end{bmatrix} \begin{bmatrix} B & O \\ O & I \end{bmatrix}$$

and use previous results. If B is singular, prove that A is singular too.

6. Evaluate $\det A$ if A is the $n \times n$ matrix

$$A = \begin{bmatrix} 0 & 0 & \cdots & 0 & 1 \\ 0 & 0 & \cdots & 1 & 0 \\ \multicolumn{5}{c}{\cdots\cdots\cdots\cdots} \\ 1 & 0 & \cdots & 0 & 0 \end{bmatrix}.$$

7. Let σ be any permutation of $1, 2, \ldots, n$. Let A be an $n \times n$ matrix and let $\sigma(A)$ be the matrix obtained from A by applying the permutation σ to the columns of A. Prove that

$$\det \sigma(A) = \epsilon(\sigma) \det A.$$

3.5 Expansion by Cofactors

Let $A = [a_{ij}]$ be a square matrix of order n, and let A_{ij} be the submatrix of A obtained by deleting the ith row and jth column.

Definition. *The* minor *of the element* a_{ij} *in A is* $\det A_{ij}$. *The* cofactor *of the element* a_{ij} *is* $(-1)^{i+j} \det A_{ij}$. *Thus the minor and the cofactor of* a_{ij} *are equal if* $i + j$ *is even but differ, in sign only, if* $i + j$ *is odd.*

Theorem 3.13. *For each j ($j = 1, 2, \ldots, n$),*

$$(3) \qquad \det A = a_{1j}c_{1j} + a_{2j}c_{2j} + \cdots + a_{nj}c_{nj},$$

where c_{ij} is the cofactor of a_{ij}, namely, $(-1)^{i+j} \det A_{ij}$.

Proof. We know from Theorem 3.4 that $\det A$ can be written in the form (3) where each c_{ij} depends only on the elements of A that occur neither in the jth column nor in the ith row. It remains only to show that the coefficient of a_{ij} in the expression (1) for $\det A$ is the cofactor c_{ij}.

We look first at the coefficient of a_{11} in (1). Since each term of $\det A$ contains one factor from each row and from each column of A and since a_{11} lies in the first row and first column, the remaining factors in any term containing a_{11} must be chosen in all possible ways (that is, one from each row and one from each column) from the matrix A_{11}. Thus the terms in det A which contain a_{11} as a factor are precisely the set of all terms of the form

$$\epsilon(\tau)a_{11}a_{2\sigma(2)} \cdots a_{n\sigma(n)},$$

where σ is a permutation of $2, 3, \ldots, n$, and τ is the permutation of $1, 2, \ldots, n$ defined by $\tau(1) = 1$ and $\tau(i) = \sigma(i)$ for $i = 2, 3, \ldots, n$. Since clearly $\epsilon(\tau) = \epsilon(\sigma)$, the total coefficient of a_{11} in det A is

$$\sum_{\sigma} \epsilon(\sigma)a_{2\sigma(2)} \cdots a_{n\sigma(n)},$$

which is exactly det A_{11}.

Now to find the coefficient of a_{ij} in the expression (1) for det A, we move the ith row of A into the first position by $i - 1$ successive interchanges. This does not change the relative order of the remaining $n - 1$ rows. We then bring the jth column to first position by $j - 1$ successive interchanges. The new matrix A' has a_{ij} in the top left-hand corner and its minor in A' is det A_{ij}, the same as its minor in A, because the relative order of the other rows and columns has not been changed. However, det $A' = (-1)^{i+j-2}$ det $A = (-1)^{i+j}$ det A because the number of row and column interchanges was $(i - 1) + (j - 1) = i + j - 2$. Thus the coefficient of a_{ij} in det A is $(-1)^{i+j}$ times its coefficient in det A' which, by Case 1, is det A_{ij}.

Corollary 1. *If $c_{ij} = (-1)^{i+j}$ det A_{ij} is the cofactor of a_{ij} in A, then for any fixed $i(1 \leq i \leq n)$,*

$$(4) \qquad \det A = a_{i1}c_{i1} + a_{i2}c_{i2} + \cdots + a_{in}c_{in}.$$

This follows at once by applying the theorem to det A^T and using Theorem 3.7. Equation (3) is called the expansion of det A on the jth column and equation (4) the expansion of det A on the ith row.

Corollary 2. *If $k \neq j$ and $i \neq h$, then*

$$(5) \qquad a_{1j}c_{1k} + a_{2j}c_{2k} + \cdots + a_{nj}c_{nk} = 0$$

and

$$(6) \qquad a_{i1}c_{h1} + a_{i2}c_{h2} + \cdots + a_{in}c_{hn} = 0.$$

Proof. The left-hand side of (5) is the expansion on the kth column of det A' where A' is the matrix obtained from A by replacing the kth column by the jth column. Since A' has two columns identical, det $A' = 0$ by the corollary to Theorem 3.5. A similar argument using rows yields equation (6).

Theorem 3.13 and its corollaries have two interesting applications. Let $A = [a_{ij}]$ be any $n \times n$ matrix, let $c_{ij} = (-1)^{i+j} \det A_{ij}$ be the cofactor in A of the element a_{ij}, and let $C = [c_{ij}]$. Equations (4) and (6) are equivalent to the matrix equation

$$(7) \qquad\qquad AC^T = (\det A)I$$

because the element in the ith row and jth column of AC^T is $\sum_{k=1}^{n} a_{ik} c_{jk}$ which by (6) is zero if $i \neq j$ and by (4) is $\det A$ if $i = j$. If A is nonsingular so that $\det A \neq 0$, equation (7) yields a formula for A^{-1}. We state the result formally.

Theorem 3.14. *If $A = [a_{ij}]$ is any nonsingular matrix, then*

$$A^{-1} = \frac{1}{\det A} C^T,$$

where $C = [c_{ij}]$ and c_{ij} is the cofactor of a_{ij} in A.

This formula can be used to compute A^{-1}, but the method described in Section 2.9 (Theorem 2.21) is preferable because it involves fewer calculations and is better suited to the use of computers. However, the formula of Theorem 3.14 is of interest for theoretical reasons.

As a second application of Theorem 3.13, we derive formulas for the solution of a system n linear equations in n unknowns with nonsingular coefficient matrix. Consider the system

$$
\begin{aligned}
a_{11}x_1 + a_{12}x_2 + \cdots + a_{1n}x_n &= b_1, \\
a_{21}x_1 + a_{22}x_2 + \cdots + a_{2n}x_n &= b_2, \\
&\cdots\cdots\cdots\cdots\cdots\cdots \\
a_{n1}x_1 + a_{n2}x_2 + \cdots + a_{nn}x_n &= b_n.
\end{aligned}
$$

(8)

Let $A = [a_{ij}]$ be nonsingular and let c_{ij} be the cofactor of a_{ij} in A. We multiply the first equation by c_{1j}, the second by $c_{2j}, \ldots$, and the nth by c_{nj}, and add the resulting equations. By (3) the coefficient of x_j in the resulting equation will be $\det A$ and by (5) the coefficient of x_k for $k \neq j$ will be zero. The right-hand side of this equation will be

$$b_1 c_{1j} + b_2 c_{2j} + \cdots + b_n c_{nj},$$

which is equal to $\det A_j$ where A_j is the matrix obtained from A by replacing the jth column by the column vector $[b_1, b_2, \ldots, b_n]^T$.

Thus, every solution of (8) is a solution of the system

$$dx_j = d_j \qquad (j = 1, 2, \ldots, n),$$

where $d = \det A$ and $d_j = \det A_j$. But since $d \neq 0$, rank $A = n$ and, by Theorem 2.18, the system (8) has a unique solution which is therefore given by

$$(9) \qquad\qquad x_j = \frac{d_j}{d}, \qquad (j = 1, 2, \ldots, n).$$

The solution of the system (8) by the formula (9) is known as Cramer's rule. It is mainly of theoretical interest, since in numerical cases the solution is more easily found by the row reduction process described in Section 2.3.

3.6 Determinants and Rank

We have seen that a square matrix is nonsingular if and only if its determinant is nonzero. The following theorem connects the rank of an arbitrary matrix with the order of its nonsingular submatrices.

Theorem 3.15. *The rank of any matrix A is equal to the order of the "largest" square submatrix of A with nonzero determinant. (Here, "largest" means "of greatest order.")*

Proof. Let r be the rank of A and let s be the order of the largest square submatrix with nonzero determinant. Let B be an $s \times s$ submatrix of A with $\det B \neq 0$. By Theorem 3.10, rank $B = s$. The s rows of A that contain the s rows of B are linearly independent, since otherwise the rows of B would be linearly dependent contrary to $\det B \neq 0$. Hence, $r \geq s$.

On the other hand, let C be a submatrix of A consisting of any $s + 1$ rows of A. If C contained $s + 1$ linearly independent columns, these columns would form an $(s + 1) \times (s + 1)$ submatrix D of rank $s + 1$ and, by Theorem 3.10, this would imply $\det D \neq 0$ contrary to the definition of s. Hence, rank $C < s + 1$ and *any* $s + 1$ rows of A are linearly dependent. Hence, $r \leq s$, and this, combined with the previous inequality, gives $s = r$.

We conclude this chapter with two remarks about the definition of $\det A$ and its extension to more general situations. Suppose that for each i and j chosen from $\{1, 2, \ldots, n\}$ we have a real-valued function $f_{ij} : \mathscr{R} \to \mathscr{R}$. Then, for each $x \in \mathscr{R}$, we have a matrix $[f_{ij}(x)]$ with elements in $\mathscr{R}$. If we write

$$d(x) = \det [f_{ij}(x)]$$

it is clear that $d : \mathscr{R} \to \mathscr{R}$ is also a real-valued function. Moreover, by equation (1), $d(x)$ is formed by adding certain products of the elements $f_{ij}(x)$. Hence, if each function f_{ij} is continuous in an interval (a, b), then d is continuous in (a, b) and if each f_{ij} is differentiable in (a, b), then d is differentiable in (a, b).

Finally, we note that the definition of determinant is applicable to a square matrix A even if its elements are not real or complex numbers, provided these elements can be multiplied and added and these operations obey the usual commutative, associative, and distributive laws. For example, if for $i, j = 1, 2, \ldots, n$, $p_{ij}(x)$ is a polynomial in the indeterminate x with coefficients in $\mathscr{F}$, then det $[p_{ij}(x)]$ is also a polynomial in x defined by (1) on putting $a_{ij} = p_{ij}(x)$.

Exercise 3.3

1. Find the value of the determinants of the following matrices.
   (a) By expanding on one row or column, using Theorem 3.13.
   (b) By using elementary row transformations to reduce the matrix to triangular form and applying Theorem 3.9.
   (c) By combining these two methods and using the result of Problem 2, Exercise 3.2.

   (i) $\begin{bmatrix} 2 & 5 & 5 & 3 \\ 7 & -8 & 2 & 3 \\ 1 & -1 & 4 & -2 \\ -3 & 9 & -1 & 3 \end{bmatrix}$.

   (ii) $\begin{bmatrix} 3 & 2 & 2 & 3 \\ 1 & -4 & 2 & 1 \\ 4 & 5 & -1 & 0 \\ -1 & -4 & 2 & 7 \end{bmatrix}$.

2. Let

$$d_n = \begin{vmatrix} 1 & 1 & \cdots & 1 \\ x_1 & x_2 & \cdots & x_n \\ x_1^2 & x_2^2 & \cdots & x_n^2 \\ \cdots\cdots\cdots\cdots\cdots\cdots \\ x_1^{n-1} & x_2^{n-1} & \cdots & x_n^{n-1} \end{vmatrix}$$

and prove that

$$d_n = (x_n - x_1)(x_n - x_2) \cdots (x_n - x_{n-1}) d_{n-1}.$$

Hence, deduce that

$$d_n = \prod_{\substack{i,j=1 \\ i>j}}^{n} (x_i - x_j).$$

The determinant d_n is called the Vandermonde determinant of order n.

3. If a, b, c, d are real numbers and

$$A = \begin{bmatrix} a & b & c & d \\ b & -a & d & -c \\ c & -d & -a & b \\ d & c & -b & -a \end{bmatrix},$$

prove that $\det A = 0$ implies $a = b = c = d = 0$. *Hint.* Compute AA^T and use Theorems 3.7 and 3.12.

4. Let $p_n(x) = \det A$ where

$$A = \begin{bmatrix} x & 0 & 0 & \cdots & & 0 & c_n \\ 1 & x & 0 & \cdots & & 0 & c_{n-1} \\ 0 & 1 & x & \cdots & & 0 & c_{n-2} \\ \multicolumn{7}{c}{\cdots\cdots\cdots\cdots\cdots\cdots\cdots\cdots} \\ 0 & 0 & 0 & \cdots & 1 & x & c_2 \\ 0 & 0 & 0 & \cdots & 0 & 1 & x + c_1 \end{bmatrix}.$$

Prove that $p_n(x) = xp_{n-1}(x) + (-1)^{n-1}c_n$ and hence

$$\det A = x^n + c_1 x^{n-1} - c_2 x^{n-2} + \cdots + (-1)^{n-1}c_n.$$

5. Let $\varphi_{ij}(x)$ $(i, j = 1, 2, \ldots, n)$ be n^2 functions of x which are differentiable in some interval (a, b). Let $A = [\varphi_{ij}(x)]$ and let $\varphi(x) = \det A$. Prove that the derivative $\varphi'(x)$ is equal to the sum of the n determinants $\det A_j$ $(j = 1, 2, \ldots, n)$ where A_j is the matrix obtained from A by replacing the elements of the jth row (or column) by their derivatives. *Hint.* Apply the rule for differentiating a product of n functions to each term in the expansion [Equation (1)] of $\det A$. Then group the resulting terms.

6. Let $\varphi_1(x), \ldots, \varphi_n(x)$ be n solutions of the linear differential equation, with constant coefficients

$$\frac{d^n y}{dx^n} + a_1 \frac{d^{n-1}y}{dx^{n-1}} + \cdots + a_n y = 0,$$

and denote by W the determinant

$$\begin{vmatrix} \varphi_1(x) & \varphi_2(x) & \cdots & \varphi_n(x) \\ \varphi_1'(x) & \varphi_2'(x) & \cdots & \varphi_n'(x) \\ \multicolumn{4}{c}{\cdots\cdots\cdots\cdots\cdots\cdots\cdots} \\ \varphi_1^{(n-1)}(x) & \varphi_2^{(n-1)}(x) & \cdots & \varphi_n^{(n-1)}(x) \end{vmatrix}$$

whose ith row consists of the $(i-1)$th derivatives of $\varphi_1(x), \ldots, \varphi_n(x)$. Prove that

$$\frac{dW}{dx} = -a_1 W,$$

and hence that $W = W_0 e^{-a_1 x}$, where W_0 is the value of W when $x = 0$. [The determinant W is called the *Wronskian* of the solutions $\varphi_1(x), \ldots, \varphi_n(x)$ of the differential equation. Our formula for W shows that if $W = 0$ for any value of x, then $W_0 = 0$ and $W = 0$ for all x. The condition $W = 0$ is necessary and sufficient for the linear dependence of the solutions $\varphi_1(x), \ldots, \varphi_n(x)$.]

7. Let

$$A = \begin{bmatrix} x_1 & x_2 & \cdots & x_n \\ a_{11} & a_{12} & \cdots & a_{1n} \\ a_{21} & a_{22} & \cdots & a_{2n} \\ \cdots\cdots\cdots\cdots\cdots\cdots\cdots \\ a_{n-11} & a_{n-12} & \cdots & a_{n-1n} \end{bmatrix}$$

and let c_i be the cofactor of x_i in A $(i = 1, 2, \ldots, n)$. Prove that $[c_1, c_2, \ldots, c_n]$ is a solution of the system

$$\sum_{j=1}^{n} a_{ij}x_j = 0 \qquad (i = 1, 2, \ldots, n - 1).$$

8. Use the result of Problem 7 to write solutions of the systems:

(a) $2x - 5y + 3z = 0,$
   $3x + y - 4z = 0.$

(b) $x \qquad + 2z + w = 0,$
   $2x - y \qquad + 3w = 0,$
   $3x + 4y + 2z \qquad = 0.$

| INNER PRODUCTS

4.1 The Dot Product in $\mathscr{V}_3(\mathscr{R})$

Let

$$X = [x_1, x_2, x_3],$$

$$Y = [y_1, y_2, y_3]$$

be any two vectors in $\mathscr{V}_3(\mathscr{R})$. In Section 0.7 we defined the inner product $X \cdot Y$ and proved (Theorem 0.7) that it is given by the equation

(1) $$X \cdot Y = x_1 y_1 + x_2 y_2 + x_3 y_3.$$

The length $\|X\|$ of the vector X and the angle θ between the vectors X and Y are given in terms of inner products by the formulas

$$\|X\| = \sqrt{X \cdot X},$$

(2) $$\cos \theta = \frac{X \cdot Y}{\|X\| \, \|Y\|}.$$

The second of these formulas yields the condition $X \cdot Y = 0$ for the orthogonality of X and Y.

Because it is desirable at this stage to give a more general definition of an inner product, we shall in future refer to the inner product defined by (1) as either the

144

dot product or the *standard inner product* in $\mathscr{V}_3(\mathscr{R})$. Our program is first to general-ize the dot product (and the concepts depending on it) in an obvious way to the *n*-dimensional space $\mathscr{V}_n(\mathscr{R})$. Later, we shall define an abstract inner product on an arbitrary vector space $\mathscr{V}$ over $\mathscr{R}$ by assuming the basic properties of $X \cdot Y$ as postulates. If $\mathscr{V}$ is finite dimensional, we can then determine all possible inner products and these turn out to be related algebraically to the dot product in a rather simple way.

We shall continue to use geometric language such as "length of a vector," "orthogonal vectors," or "mutually orthogonal subspaces," in spite of the fact that when $n > 3$ there is no obvious physical representation of our vectors in geo-metric terms. When $n > 3$ our concepts, definitions, and theorems are essentially algebraic. Nevertheless, the geometric language is retained, not only because the theorems have geometric as well as algebraic content when $n = 2$ or 3, but because it actually adds to our understanding and appreciation of these theorems if we think of them in geometric terms as well as algebraic.

4.2 The Standard Inner Product in $\mathscr{V}_n(\mathscr{R})$

The obvious analogy with inner products in $\mathscr{V}_3(\mathscr{R})$, discussed in Section 0.7, suggests the following definition.

Definition. *If* $X = [x_1, x_2, \ldots, x_n]$, $Y = [y_1, y_2, \ldots, y_n]$ *are any two vectors of* $\mathscr{V}_n(\mathscr{R})$, *the standard inner product (or dot product) of* X *and* Y *is defined by*

$$X \cdot Y = x_1 y_1 + x_2 y_2 + \cdots + x_n y_n.$$

Theorem 4.1. *The standard inner product* $X \cdot Y$ *in* $\mathscr{V}_n(\mathscr{R})$ *has the following basic properties. For all vectors* X, Y *in* $\mathscr{V}_n(\mathscr{R})$ *and all scalars* k:

(a) $X \cdot Y$ *is a scalar,* $X \cdot X \geq 0$, *and* $X \cdot X = 0$ *implies* $X = O$.
(b) $X \cdot Y = Y \cdot X$.
(c) $X \cdot (Y_1 + Y_2) = X \cdot Y_1 + X \cdot Y_2$.
(d) $X \cdot (kY) = k(X \cdot Y)$.

Since each item in this theorem is an obvious consequence of the definition of $X \cdot Y$, we omit any formal proof. We point out, however, that the properties $X \cdot X \geq 0$ and $X \cdot X = 0$ only if $X = O$ are true only because the field of scalars is $\mathscr{R}$ so that the coordinates of X are real numbers. Inner products in $\mathscr{V}_n(\mathscr{C})$ will be considered in Chapter 8, but throughout this chapter, the field of scalars will be the field $\mathscr{R}$ of all real numbers.

We illustrate the computational implications of Theorem 4.1 by a few examples.

Example 1.

$$(X + Y) \cdot (X + Y) = (X + Y) \cdot X + (X + Y) \cdot Y$$
$$= X \cdot X + 2(X \cdot Y) + Y \cdot Y$$

by successive application of (b) and (c).

Example 2.

$$X \cdot (2Y + 3Z) = X \cdot (2Y) + X \cdot (3Z) \qquad \text{by (c)}$$
$$= 2(X \cdot Y) + 3(X \cdot Z) \qquad \text{by (d)}.$$

Example 3.

$$(X + 2Y) \cdot (3X - Y) = X \cdot [3X + (-1)Y] + 2Y[3X + (-1)Y]$$
$$= 3X \cdot X - X \cdot Y + 6Y \cdot X - 2Y \cdot Y$$
$$= 3X \cdot X + 5X \cdot Y - 2Y \cdot Y.$$

Again, by analogy with the corresponding concepts in $\mathscr{V}_3(\mathscr{R})$ (see Section 0.7), we define length and orthogonality of vectors.

Definition. *The length $\|X\|$ of a vector X in $\mathscr{V}_n(\mathscr{R})$ is defined by*

(3) $$\|X\| = (X \cdot X)^{1/2} = (x_1^2 + x_2^2 + \cdots + x_n^2)^{1/2}.$$

A vector X is orthogonal to Y if $X \cdot Y = 0$.

Because $Y \cdot X = X \cdot Y$, if X is orthogonal to Y then also Y is orthogonal to X and we say simply that X and Y are orthogonal. It should be noted that according to this definition, the zero vector is orthogonal to every vector of $\mathscr{V}_n(\mathscr{R})$.

In $\mathscr{V}_2(\mathscr{R})$ or $\mathscr{V}_3(\mathscr{R})$ the angle θ between two vectors X and Y is given by (2). Since $|\cos \theta| \leq 1$ for all θ, it follows from (2) that the inequality

(4) $$|X \cdot Y| \leq \|X\| \, \|Y\|$$

holds for any two vectors in $\mathscr{V}_2(\mathscr{R})$ or in $\mathscr{V}_3(\mathscr{R})$. We now show that (4) holds for any two vectors in $\mathscr{V}_n(\mathscr{R})$.

Theorem 4.2. *If X and Y are any two vectors in $\mathscr{V}_n(\mathscr{R})$, then $|X \cdot Y| \leq \|X\| \, \|Y\|$. This is known as the Cauchy-Schwarz inequality.*

Proof. If either $X = O$ or $Y = O$, then $X \cdot Y = \|X\| \, \|Y\| = 0$ and (4) holds with the equality sign. Hence, we assume $X \neq O$ and $Y \neq O$. Let $Z = hX - kY$ where h and k are real numbers. Then, by Theorem 4.1,

(5) $$Z \cdot Z = h^2 \|X\|^2 - 2hk(X \cdot Y) + k^2 \|Y\|^2 \geq 0$$

for all real values of h and k. If we put $h = \|Y\|$ and $k = \|X\|$, equation (5) becomes

(6) $$2\|X\|^2 \|Y\|^2 \geq 2\|X\| \|Y\| (X \cdot Y).$$

Since $X \neq O$ and $Y \neq O$, $\|X\| \|Y\| > 0$ by Theorem 4.1(a). Hence, dividing each side of (6) by $2\|X\| \|Y\|$, we get

$$\|X\| \|Y\| \geq X \cdot Y.$$

If we now repeat this argument using $Z = hX + kY$, we get $\|X\| \|Y\| \geq -X \cdot Y$ and hence $\|X\| \|Y\| \geq |X \cdot Y|$ as required.

The next theorem shows that the length of a vector of $\mathscr{V}_n(\mathscr{R})$, as defined in (3), satisfies the basic algebraic properties that we associate with length in $\mathscr{V}_2(\mathscr{R})$ or $\mathscr{V}_3(\mathscr{R})$.

Theorem 4.3. *If X and Y are vectors in $\mathscr{V}_n(\mathscr{R})$ and k is any scalar:*
(a) $\|X\| \geq 0$ *and* $\|X\| = 0$ *if and only if* $X = O$.
(b) $\|kX\| = |k| \|X\|$, *for any scalar k.*
(c) $\|X + Y\| \leq \|X\| + \|Y\|$.

Proof. (a) and (b) follow directly from (3) and the fact that the field of scalars is the real numbers. To prove (c) we have

$$\|X + Y\|^2 = (X + Y) \cdot (X + Y)$$
$$= X \cdot X + 2X \cdot Y + Y \cdot Y$$
$$= \|X\|^2 + 2X \cdot Y + \|Y\|^2$$
$$\leq \|X\|^2 + 2|X \cdot Y| + \|Y\|^2$$
$$\leq \|X\|^2 + 2\|X\| \|Y\| + \|Y\|^2 \qquad \text{by Theorem 4.2}$$
$$= (\|X\| + \|Y\|)^2$$

and (c) follows on taking positive square roots.

The inequality in Theorem 4.3(c) is called the *triangle inequality* because in two or three dimensions, it is equivalent to the fact that the sum of two sides of a triangle is greater than the third side.

Theorem 4.4. *If a vector X in $\mathscr{V}_n(\mathscr{R})$ is orthogonal to each of the vectors $Y_1, Y_2, \ldots, Y_r$, then X is orthogonal to every vector in the subspace $\mathscr{S}$ of $\mathscr{V}_n(\mathscr{R})$ spanned by $Y_1, \ldots, Y_r$.*

Proof. If $Y = c_1 Y_1 + c_2 Y_2 + \cdots + c_r Y_r$ is any vector of $\mathcal{S}$, we have

$$X \cdot Y = X \cdot (c_1 Y_1 + \cdots + c_r Y_r)$$
$$= c_1 X \cdot Y_1 + \cdots + c_r X \cdot Y_r = 0$$

because $X \cdot Y_i = 0$ for each i.

Definition. *A vector X is said to be orthogonal to a subspace $\mathcal{S}$ of $\mathcal{V}_n(\mathcal{R})$ if X is orthogonal to every vector in $\mathcal{S}$. Two subspaces $\mathcal{S}$ and $\mathcal{T}$ are said to be orthogonal if every vector of $\mathcal{S}$ is orthogonal to every vector of $\mathcal{T}$.*

Corollary. *If a subspace $\mathcal{S}$ of $\mathcal{V}_n(\mathcal{R})$ has a basis $B_1, B_2, \ldots, B_r$, a vector X is orthogonal to $\mathcal{S}$ if and only if*

$$X \cdot B_i = 0, \qquad (i = 1, 2, \ldots, r).$$

Example 4. Find a vector of unit length which is orthogonal to the vector $A = [2, -1, 6]$.

Solution. Let $X = [x, y, z]$ be the required vector so that

$$A \cdot X = 2x - y + 6z = 0.$$

Any solution of this equation, for example,

$$X = [2, -2, -1],$$

gives a vector orthogonal to A. However,

$$\|X\| = (2^2 + (-2)^2 + 1^2)^{1/2} = 3.$$

Hence the vector $\frac{1}{3}X = [\frac{2}{3}, -\frac{2}{3}, -\frac{1}{3}]$ has length 1 and is orthogonal to A.

Example 5. Find two mutually orthogonal vectors each of which is orthogonal to the vector $A = [4, 2, 3]$.

Solution. Find any nonzero solution vector of the equation

$$4x + 2y + 3z = 0,$$

for example, $B = [3, -3, -2]$. We now require a third vector $C = [x, y, z]$ orthogonal to both A and B. This means C must be a solution vector of the system

$$4x + 2y + 3z = 0,$$
$$3x - 3y - 2z = 0,$$

which is found to be $C = [5, 17, -18]$. Thus, B and C are orthogonal to each other and to A. The solution is, of course, by no means unique.

Exercise 4.1

1. Find two linearly independent vectors, each of which is orthogonal to the vector $[1, 1, 2]$.

2. Find two mutually orthogonal vectors, each of which is orthogonal to $[5, 2, -1]$.

3 Find two mutually orthogonal vectors of unit length, each of which is orthogonal to $[2, -1, 3]$.

4. Find three linearly independent vectors in $\mathscr{V}_4(\mathscr{R})$, each of which is orthogonal to $[1, 2, 1, 5]$.

5. Find three mutually orthogonal vectors of $\mathscr{V}_4(\mathscr{R})$, each of which is orthogonal to $[3, 2, 1, 1]$.

6. Find a nonzero vector in $\mathscr{V}_4(\mathscr{R})$ which is orthogonal to each of the three vectors $[1, 1, 1, -1]$, $[2, 1, 1, 1]$, $[1, 2, 0, 1]$.

7. Prove that given any r vectors in $\mathscr{V}_n(\mathscr{R})$, if $r < n$, it is always possible to find a nonzero vector which is orthogonal to each of the r vectors and, therefore, orthogonal to the space spanned by them.

8. Let A by any matrix with real elements. Prove that the null space of A is orthogonal to the row space of A.

9. Prove that for any two vectors X, Y of $\mathscr{V}_n(\mathscr{R})$, $|X \cdot Y| = \|X\| \, \|Y\|$ if and only if X and Y are linearly dependent.

10. Prove that if $\|X + Y\| = \|X\| + \|Y\|$, then X and Y are linearly dependent. Give an example to show that the converse of this statement is false.

11. Let X and Y be vectors in $\mathscr{V}_3(\mathscr{R})$. Prove algebraically that if $X + Y$ is orthogonal to $X - Y$, then $\|X\| = \|Y\|$. Interpret this result geometrically.

4.3 Orthogonal Complements and Projections in $\mathscr{V}_n(\mathscr{R})$

Definition. *If $\mathscr{S}$ is any subspace of $\mathscr{V}_n(\mathscr{R})$ the set $\mathscr{S}^\perp$ of all vectors of $\mathscr{V}_n(\mathscr{R})$ orthogonal to $\mathscr{S}$ is called the orthogonal complement of $\mathscr{S}$.*

This definition can also be stated in the form

$$(7) \qquad \mathscr{S}^\perp = \{X \in \mathscr{V}_n(\mathscr{R}) \,|\, X \cdot S = 0 \text{ for all } S \in \mathscr{S}\}.$$

Theorem 4.5. *If $\mathscr{S}$ is a subspace of $\mathscr{V}_n(\mathscr{R})$ of dimension r, then $\mathscr{S}^\perp$ is a subspace of $\mathscr{V}_n(\mathscr{R})$ of dimension $n - r$. Moreover, $\mathscr{S} \cap \mathscr{S}^\perp = O$, $\mathscr{S} + \mathscr{S}^\perp = \mathscr{V}_n(\mathscr{R})$ and $(\mathscr{S}^\perp)^\perp = \mathscr{S}$.*

Proof.

1. Since $O \in S^\perp$, we know that $S^\perp$ is nonempty. If $X, Y \in \mathscr{S}^\perp$ and $k \in \mathscr{R}$ we have, for all $S \in \mathscr{S}$, $(X + Y) \cdot S = X \cdot S + Y \cdot S = 0$ and $(kX) \cdot S = k(X \cdot S) = 0$. Hence, $\mathscr{S}^\perp$ is a subspace of $\mathscr{V}_n(\mathscr{R})$.

2. Since dim $\mathscr{S} = r$ let $B_1, B_2, \ldots, B_r$ be a basis of $\mathscr{S}$, let $B_i = [b_{i1}, b_{i2}, \ldots, b_{in}]$, and let $B = [b_{ij}]$. By the corollary to Theorem 4.4, $X \in \mathscr{S}^\perp$ if and only if $B_i \cdot X = 0$ for $i = 1, 2, \ldots, r$. But this is equivalent to the condition that X is a solution vector of the system of equations

$$BX = O$$

with coefficient matrix B. In other words $\mathscr{S}^\perp$ is simply the null space of the matrix B and, therefore (Theorem 2.12), has dimension $n - r$.

3. If $X \in \mathscr{S} \cap \mathscr{S}^\perp$, then $X \cdot X = 0$ and $X = O$ by Theorem 4.1(a). Hence, $\mathscr{S} \cap \mathscr{S}^\perp = O$.

4. Since $\mathscr{S} \cap \mathscr{S}^\perp = O$ we have, by Theorem 1.10,

$$\dim (\mathscr{S} + \mathscr{S}^\perp) = \dim \mathscr{S} + \dim \mathscr{S}^\perp$$
$$= r + (n - r) = n.$$

Hence, $\mathscr{S} + \mathscr{S}^\perp = \mathscr{V}_n(\mathscr{R})$ by Problem 4, Exercise 1.5.

5. Finally, it is clear that by definition $\mathscr{S} \subset (\mathscr{S}^\perp)^\perp$. But since dim $\mathscr{S}^\perp = n - r$, dim $(\mathscr{S}^\perp)^\perp = n - (n - r) = r = \dim \mathscr{S}$. Hence, $\mathscr{S} = (\mathscr{S}^\perp)^\perp$ by Problem 4, Exercise 1.5.

Corollary. *Every vector X of $\mathscr{V}_n(\mathscr{R})$ can be written in a unique way in the form $X = S + T$ where $S \in \mathscr{S}$ and $T \in \mathscr{S}^\perp$.*

Proof. Because $\mathscr{V}_n(\mathscr{R}) = \mathscr{S} + \mathscr{S}^\perp$ it follows that X can be written in the desired form. If

$$X = S + T = S_1 + T_1,$$

where $S, S_1 \in \mathscr{S}$ and $T, T_1 \in \mathscr{S}^\perp$, then $S - S_1 = T_1 - T$. But $S - S_1 \in \mathscr{S}$ and $T_1 - T \in \mathscr{S}^\perp$ and hence both these vectors belong to $\mathscr{S} \cap \mathscr{S}^\perp$. Since $\mathscr{S} \cap \mathscr{S}^\perp = O$, we must have $S_1 = S$ and $T_1 = T$.

Definition. *If $\mathscr{V}$ is a vector space with subspaces $\mathscr{S}$ and $\mathscr{T}$ such that $\mathscr{S} \cap \mathscr{T} = O$ and $\mathscr{V} = \mathscr{S} + \mathscr{T}$, then $\mathscr{V}$ is called the* direct sum *of $\mathscr{S}$ and $\mathscr{T}$ and we write $\mathscr{V} = \mathscr{S} \oplus \mathscr{T}$.*

The same proof used in the Corollary to Theorem 4.5 shows that if $\mathscr{V}$ is the direct sum of subspaces $\mathscr{S}$ and $\mathscr{T}$, then every vector X of $\mathscr{V}$ has a unique representation in the form $X = S + T$ where $S \in \mathscr{S}$ and $T \in \mathscr{T}$.

Definition. *If $\mathscr{V} = \mathscr{S} \oplus \mathscr{T}$ and if a vector X of $\mathscr{V}$ is written $X = S + T$, $S \in \mathscr{S}$, $T \in \mathscr{T}$, then S is called the projection of X on the subspace $\mathscr{S}$ in the direction of the subspace $\mathscr{T}$. If $\mathscr{V} = \mathscr{V}_n(\mathscr{R})$ and $\mathscr{T} = \mathscr{S}^\perp$, then S and T are called the orthogonal projections of X on the subspaces $\mathscr{S}$ and $\mathscr{T}$.*

Example 1. Let $\mathscr{V} = \mathscr{V}_3(\mathscr{R})$ and let $\mathscr{S}$ be the subspace spanned by a single nonzero vector so that $\mathscr{S}$ is the set of all vectors lying in a fixed line l through the origin O (Figure 4.1). The orthogonal complement $\mathscr{S}^\perp$ of $\mathscr{S}$ is therefore the space of all vectors orthogonal to $\mathscr{S}$ or all vectors lying in the plane p which is perpendicular to l and passes through O. Now let X be an arbitrary vector in $\mathscr{V}_3(\mathscr{R})$ with geometric representation $\overrightarrow{OP}$. To write X in the form $S + T$, where $S \in \mathscr{S}$ and $T \in \mathscr{S}^\perp$, we must construct a parallelogram of which $\overrightarrow{OP}$ is a diagonal and whose sides $\overrightarrow{OQ}$ and $\overrightarrow{OR}$ lie in the line l and in the plane p, respectively. Since S and T are orthogonal, it is clear that the parallelogram is a rectangle. Moreover, its sides are the orthogonal projections, in the ordinary geometric sense, of the vector $\overrightarrow{OP}$ onto

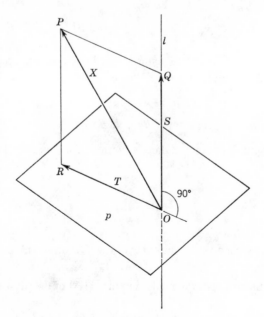

Figure 4.1

the line l and the plane p. These two projections are the geometric representations of the vectors S and T for which $X = S + T$.

Example 2. Let $\mathscr{S}$ be the subspace of $\mathscr{V}_3(\mathscr{R})$ spanned by the vector $S = [2 \ -1, 6]$. Find the projections of the vector $X = [4, 1, 2]$ on $\mathscr{S}$ and $\mathscr{S}^\perp$.

Solution. Since every vector of $\mathscr{S}$ is a scalar multiple of S, we require

(8) $$X = kS + T,$$

where $T \in S^\perp$ and hence $S \cdot T = 0$. From (8)

$$X \cdot S = k(S \cdot S) + S \cdot T = k(S \cdot S)$$

or

$$8 - 1 + 12 = k(4 + 1 + 36)$$

whence $k = \frac{19}{41}$, and $kS = [\frac{38}{41}, -\frac{19}{41}, \frac{114}{41}]$. Again, from (8),

$$T = X - (\tfrac{19}{41})S = [\tfrac{44}{41}, \tfrac{60}{41}, -\tfrac{32}{41}].$$

Example 3. Find the orthogonal projection of $X = [1, 2, 0, -2]$ onto the subspace $\mathscr{S}$ of $\mathscr{V}_4(\mathscr{R})$ spanned by $A = [2, 2, 0, -1]$ and $B = [1, 2, -1, 3]$, and also onto $\mathscr{S}^\perp$.

Outline of Solution. Let

(9) $$X = hA + kB + T,$$

where h, k are scalars and $T \in \mathscr{S}^\perp$. Then $hA + kB$ and T are the required projections. To find h and k we have, since $A \cdot T = B \cdot T = 0$,

$$A \cdot X = h(A \cdot A) + k(A \cdot B),$$
$$B \cdot X = h(A \cdot B) + k(B \cdot B).$$

Substitute for A, B, and X and solve these equations for h and k. Then substitute in (9) to find T.

Exercise 4.2

1. Find a basis for the orthogonal complement in $\mathscr{V}_3(\mathscr{R})$ of the subspace spanned by:
   (a) $[4, 7, -2]$.   (b) $[1, 8, 3]$.

2. Find a basis for the orthogonal complement in $\mathscr{V}_3(\mathscr{R})$ of the subspace spanned by the two vectors:
   (a) $[1, 2, -1], [3, 1, 4]$.   (b) $[2, 2, 5], [1, -7, 4]$.

3. Find the projections of the vector $[3, 4, 1]$ onto the space spanned by $[1, 1, 1]$ and on its orthogonal complement.

4. Find the projections of $[5, 1, 2]$ onto the plane $x - 2y + 3z = 0$ and onto the line through the origin perpendicular to this plane.

5. Complete the solution of Example 3.

6. Find the orthogonal projection of the vector $[2, 1, -3]$ onto the plane $2x - 3y + 5z = 0$.

7. Find the orthogonal projection of the vector $[2, -1, 2]$ onto the space spanned by the vector $[1, -1, 3]$.

8. Find a basis for the solution space of the equations

$$x + 2y - z - w = 0,$$
$$3x - y + z + 4w = 0$$

and also for the orthogonal complement of this space.

9. Find the projections of the vector $[1, 4, 2, 2]$ onto the two orthogonally complementary spaces of Problem 9.

4.4 Volume in $\mathscr{V}_n(\mathscr{R})$

The concept of length of a vector has already been generalized to $\mathscr{V}_n(\mathscr{R})$. We now seek similar generalizations of the concepts of area and volume. We start by deriving a simple formula for the area of a parallelogram.

Theorem 4.6. *If* $X = [x_1, x_2, x_3]$, $Y = [y_1, y_2, y_3]$ *are any two vectors of* $\mathscr{V}_3(\mathscr{R})$, *the area of the parallelogram of which X and Y are adjacent sides is equal to the positive square root of the determinant*

(10)
$$\begin{vmatrix} X \cdot X & X \cdot Y \\ Y \cdot X & Y \cdot Y \end{vmatrix}.$$

Proof. Let θ be the angle between X and Y and let a be the area of the parallelogram. If we consider X as base, the altitude of the parallelogram is (Figure 4.2) $\|Y\| \sin \theta$ and $a = \|X\| \|Y\| \sin \theta$. Hence,

$$\begin{aligned} a^2 &= \|X\|^2 \|Y\|^2 \sin^2 \theta = \|X\|^2 \|Y\|^2 (1 - \cos^2 \theta) \\ &= \|X\|^2 \|Y\|^2 - \|X\|^2 \|Y\|^2 \cos^2 \theta \\ &= \|X\|^2 \|Y\|^2 - (X \cdot Y)^2 \\ &= \begin{vmatrix} X \cdot X & X \cdot Y \\ Y \cdot X & Y \cdot Y \end{vmatrix}, \end{aligned}$$

and the theorem follows.

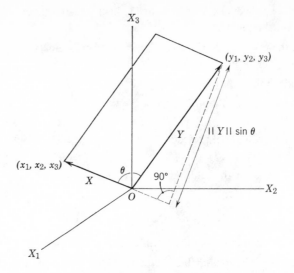

Figure 4.2

Corollary. *If $X = [x_1, x_2]$, $Y = [y_1, y_2]$ are any two vectors in $\mathscr{V}_2(\mathscr{R})$, the area of the parallelogram of which X and Y are adjacent edges is equal to the positive square root of the determinant* (10) *and also to the absolute value of the determinant*

$$\begin{vmatrix} x_1 & x_2 \\ y_1 & y_2 \end{vmatrix}.$$

Proof. The first statement follows from the theorem by identifying the vectors X and Y with the vectors $[x_1, x_2, 0]$, $[y_1, y_2, 0]$ in $\mathscr{V}_3(\mathscr{R})$.

The second statement follows because

$$\begin{bmatrix} X \cdot X & X \cdot Y \\ Y \cdot X & Y \cdot Y \end{bmatrix} = \begin{bmatrix} x_1 & x_2 \\ y_1 & y_2 \end{bmatrix} \begin{bmatrix} x_1 & y_1 \\ x_2 & y_2 \end{bmatrix}$$

and, hence, by Theorems 3.12 and 3.7,

$$\begin{vmatrix} X \cdot X & X \cdot Y \\ Y \cdot X & Y \cdot Y \end{vmatrix} = \begin{vmatrix} x_1 & x_2 \\ y_1 & y_2 \end{vmatrix}^2.$$

Next, we find similar formulas for the volume of a parallelepiped in $\mathscr{V}_3(\mathscr{R})$.

Theorem 4.7. *Let $X = [x_1, x_2, x_3]$, $Y = [y_1, y_2, y_3]$, $Z = [z_1, z_2, z_3]$ be any three linearly independent vectors in $\mathscr{V}_3(\mathscr{R})$. The volume of the parallelepiped of which X, Y, Z are adjacent edges is equal to the positive square root of the* *determinant*

$$\begin{vmatrix} X \cdot X & X \cdot Y & X \cdot Z \\ Y \cdot X & Y \cdot Y & Y \cdot Z \\ Z \cdot X & Z \cdot Y & Z \cdot Z \end{vmatrix},$$

which is, in turn, equal to the absolute value of the determinant

$$\begin{vmatrix} x_1 & x_2 & x_3 \\ y_1 & y_2 & y_3 \\ z_1 & z_2 & z_3 \end{vmatrix}.$$

Proof. The volume of this parallelepiped is the product of the area of a base, say, the parallelogram defined by X and Y, and the altitude, which is the length of the projection of Z onto a line through the origin perpendicular to the plane of X and Y. Thus, the altitude is the length of the projection of Z onto the orthogonal complement in $\mathscr{V}_3(\mathscr{R})$ of the space spanned by X and Y.

To find this projection W of Z, we write

$$Z = W + bX + cY,$$

where, by the corollary to Theorem 4.5, W, b, c are uniquely determined. Forming inner products of Z with X, Y, and W, respectively, we get, since $X \cdot W = Y \cdot W = 0$.

(11)
$$X \cdot Z = b \, \|X\|^2 + c(X \cdot Y),$$

$$Y \cdot Z = b(Y \cdot X) + c \, \|Y\|^2,$$

$$W \cdot Z = \|W\|^2.$$

By Theorem 4.6, the area a of the base of the parallelepiped is given by (10) so that, on solving the first two equations of (11) for b and c by Cramer's rule, we get

$$a^2 b = \begin{vmatrix} X \cdot Z & X \cdot Y \\ Y \cdot Z & Y \cdot Y \end{vmatrix} = - \begin{vmatrix} X \cdot Y & X \cdot Z \\ Y \cdot Y & Y \cdot Z \end{vmatrix},$$

$$a^2 c = \begin{vmatrix} X \cdot X & X \cdot Z \\ Y \cdot X & Y \cdot Z \end{vmatrix}.$$

If v is the required volume, we have, using the third equation of (11),

$$v^2 = a^2 \, \|W\|^2 = a^2(W \cdot Z) = a^2 Z \cdot (Z - bX - cY) = a^2(Z \cdot Z - bX \cdot Z - cY \cdot Z)$$

Now, substituting for a^2, b, and c, we find

$$v^2 = X \cdot Z \begin{vmatrix} X \cdot Y & X \cdot Z \\ Y \cdot Y & Y \cdot Z \end{vmatrix} - Y \cdot Z \begin{vmatrix} X \cdot X & X \cdot Z \\ Y \cdot X & Y \cdot Z \end{vmatrix} + Z \cdot Z \begin{vmatrix} X \cdot X & X \cdot Y \\ Y \cdot X & Y \cdot Y \end{vmatrix}$$

$$= \begin{vmatrix} X \cdot X & Y \cdot X & Z \cdot X \\ X \cdot Y & Y \cdot Y & Z \cdot Y \\ X \cdot Z & Y \cdot Z & Z \cdot Z \end{vmatrix} = \begin{vmatrix} x_1 & x_2 & x_3 \\ y_1 & y_2 & y_3 \\ z_1 & z_2 & z_3 \end{vmatrix} \begin{vmatrix} x_1 & y_1 & z_1 \\ x_2 & y_2 & z_2 \\ x_3 & y_3 & z_3 \end{vmatrix} = \begin{vmatrix} x_1 & x_2 & x_3 \\ y_1 & y_2 & y_3 \\ z_1 & z_2 & z_3 \end{vmatrix}^2 ,$$

and the proof is complete.

The results of Theorems 4.6 and 4.7 can be generalized to n-dimensional space. First, we define what is meant by an r-dimensional parallelepiped in $\mathscr{V}_n(\mathscr{R})$. To this end we note that the parallelogram whose sides are the two vectors X and Y in $\mathscr{V}_3(\mathscr{R})$ "contains" all vectors of the form $aX + bY$, where $0 \leq a \leq 1$ and $0 \leq b \leq 1$. By this we mean that all these vectors have geometric representations whose terminal points lie inside or on the boundary of the parallelogram. In fact, a necessary and sufficient condition that a vector have its terminal point in or on the parallelogram is that it be of the above form. Similarly, if X, Y, Z are three linearly independent vectors in 3-space, the parallelepiped whose edges are the vectors X, Y, and Z can be characterized as that portion of space that contains all vectors of the form $aX + bY + cZ$, where $0 \leq a \leq 1$, $0 \leq b \leq 1$, and $0 \leq c \leq 1$. The vertices of this parallelepiped are the terminal points of the eight vectors $e_1 X + e_2 Y + e_3 Z$, where each e_i is either zero or one. It is now clear how to formulate the definition of an r-dimensional parallelepiped.

Definition. *Any r linearly independent vectors X_1, $X_2, \ldots, X_r$ of $\mathscr{V}_n(\mathscr{R})$ are said to define an r-dimensional parallelepiped which consists of all vectors of the form*

$$a_1 X_1 + a_2 X_2 + \cdots + a_r X_r,$$

where $0 \leq a_i \leq 1$ $(i = 1, 2, \ldots, r)$.

Our next task is to define the r-volume of an r-dimensional parallelepiped in such a way that in $\mathscr{V}_3(\mathscr{R})$ 1-volume will mean length, 2-volume, area, and 3-volume, volume in the ordinary sense. We give the definition by induction on r. If X is any nonzero vector in $\mathscr{V}_n(\mathscr{R})$, its 1-volume is defined to be its length, namely, $\sqrt{X \cdot X}$. Now suppose that the $(r - 1)$-volume of an $(r - 1)$-dimensional parallelepiped has been defined. Let Π be the r-dimensional parallelepiped defined by the r linearly independent vectors X_1, $X_2, \ldots, X_r$. By the "base" of Π we shall understand the $(r - 1)$-dimensional parallelepiped defined by $X_1, X_2, \ldots, X_{r-1}$, and by the "altitude" of Π we shall mean the length of the projection of X_r onto the orthogonal complement in $\mathscr{V}_n(\mathscr{R})$ of the space spanned by the vectors $X_1, X_2, \ldots, X_{r-1}$. By our induction assumption the $(r - 1)$-volume of the base is already defined. Therefore we define the r-volume of Π to be the product of the $(r - 1)$-volume of its base and its altitude. We shall now use this definition to derive a formula for the r-volume of an r-dimensional parallelepiped which will include Theorems 4.6 and 4.7 as special cases.

Theorem 4.8. *The r-volume of the r-dimensional parallelepiped II defined by the r linearly independent vectors $X_1, X_2, \ldots, X_r$ of $\mathscr{V}_n(\mathscr{R})$ is the positive square root of the determinant*

(12)
$$\begin{vmatrix} X_1 \cdot X_1 & X_1 \cdot X_2 & \cdots & X_1 \cdot X_2 \\ X_2 \cdot X_1 & X_2 \cdot X_2 & \cdots & X_2 \cdot X_r \\ \cdots\cdots\cdots\cdots\cdots\cdots\cdots\cdots\cdots \\ X_r \cdot X_1 & X_r \cdot X_2 & \cdots & X_r \cdot X_r \end{vmatrix}.$$

Proof. The theorem is true for $r = 1$ by the definition of length (or 1-volume) of a vector. We assume the truth of the theorem for an $(r - 1)$-dimensional parallelepiped so that, if the $(r - 1)$-volume of the base of Π is b, we have

$$b^2 = \begin{vmatrix} X_1 \cdot X_1 & X_1 \cdot X_2 & \cdots & X_1 \cdot X_{r-1} \\ X_2 \cdot X_1 & X_2 \cdot X_2 & \cdots & X_2 \cdot X_{r-1} \\ \cdots\cdots\cdots\cdots\cdots\cdots\cdots\cdots\cdots \\ X_{r-1} \cdot X_1 & X_{r-1} \cdot X_2 & \cdots & X_{r-1} \cdot X_{r-1} \end{vmatrix}.$$

To find the length of the projection of X_r onto the orthogonal complement of the space generated by $X_1, X_2, \ldots, X_{r-1}$, we write

(13)
$$X_r = Y + a_1 X_1 + a_2 X_2 + \cdots + a_{r-1} X_{r-1},$$

where Y is the required projection and, therefore, $Y \cdot X_i = 0$ for $i = 1, 2, \ldots, r - 1$. It follows that

(14)
$$Y \cdot Y = Y \cdot X_r,$$

and

(15)
$$\begin{aligned} a_1 X_1 \cdot X_1 + a_2 X_1 \cdot X_2 + \cdots + a_{r-1} X_1 \cdot X_{r-1} &= X_1 \cdot X_r, \\ a_1 X_2 \cdot X_1 + a_2 X_2 \cdot X_2 + \cdots + a_{r-1} X_2 \cdot X_{r-1} &= X_2 \cdot X_r, \\ \cdots\cdots\cdots\cdots\cdots\cdots\cdots\cdots\cdots\cdots\cdots\cdots\cdots\cdots \\ a_1 X_{r-1} \cdot X_1 + a_2 X_{r-1} \cdot X_2 + \cdots + a_{r-1} X_{r-1} \cdot X_{r-1} &= X_{r-1} \cdot X_r. \end{aligned}$$

The determinant of the coefficients of equations (15) is b^2. Solving for $a_1, a_2, \ldots, a_{r-1}$, we get by Cramer's rule,

$$b^2 a_1 = (-1)^{r-2} b_1, \ b^2 a_2 = (-1)^{r-3} b_2, \ \ldots, \ b^2 a_{r-1} = b_{r-1},$$

where $b_1, b_2, \ldots, b_{r-1}$ are the minors of the first $r-1$ elements in the last row of (12), and b^2 is the minor of the last element of this row. Substituting in (13), we get

$$b^2 Y = b^2(-a_1 X_1 - a_2 X_2 - \cdots - a_{r-1} X_{r-1} + X_r)$$
$$= (-1)^{r-1} b_1 X_1 + (-1)^{r-2} b_2 X_2 + \cdots + (-1) b_{r-1} X_{r-1} + b^2 X_r,$$

whence from (14)

$$b^2 \| Y \|^2 = b^2 Y \cdot X_r$$
$$= (-1)^{r-1}[b_1 X_r \cdot X_1 - b_2 X_r \cdot X_2 + \cdots + (-1)^{r-1} b^2 X_r \cdot X_r].$$

The right-hand side of this equation is exactly the expansion of (12) by the elements of the last row, and the left-hand side is $b^2 \| Y \|^2$, the square of the r-volume of our r-dimensional parallelepiped. The theorem is therefore proved. We can also conclude that the determinant (12) is nonnegative.

If a different set of $r-1$ of the vectors $X_1, X_2, \ldots, X_r$ had been chosen to define the base of the r-dimensional parallelepiped, the resulting expression for v^2 would have been the determinant (12) with a certain permutation applied to the subscripts on the X's. Because this determinant is symmetric about the main diagonal, the effect of such a permutation of subscripts is to permute the rows and columns of (7) is precisely the same manner. It follows that the determinant itself would be unchanged, and the volume is therefore independent of the base chosen.

Corollary. *The n-volume of the n-dimensional parallelepiped in $\mathcal{V}_n(\mathcal{R})$ defined by the vectors*

$$X_i = [x_{i1}, x_{i2}, \ldots, x_{in}] \qquad (i = 1, 2, \ldots, n)$$

is the absolute value of det X *where*

$$X = \begin{bmatrix} x_{11} & x_{12} & \cdots & x_{1n} \\ x_{21} & x_{22} & \cdots & x_{2n} \\ \cdots\cdots\cdots\cdots\cdots \\ x_{n1} & x_{n2} & \cdots & x_{nn} \end{bmatrix}.$$

Proof. The determinant of XX^T is exactly the determinant (12) and is equal to the square of the determinant of X.

If det X is positive, the vectors $X_1, X_2, \ldots, X_n$ are said to be *positively oriented;* if the determinant is negative, the vectors are said to be *negatively oriented*. Note that the orientation of a set of vectors depends on the order in which they are written.

Exercise 4.3

1. Find the area of the parallelogram whose vertices are:
   (a) $(0, 0)$, $(1, 3)$, $(-2, 1)$, and $(-1, 4)$.
   (b) $(2, 4)$, $(4, 5)$, $(5, 2)$, and $(7, 3)$.
   (c) $(-1, 3)$, $(1, 5)$, $(3, 2)$, and $(5, 4)$.
   (d) $(0, 0, 0)$, $(1, -2, 2)$, $(3, 4, 2)$, and $(4, 2, 4)$.
   (e) $(2, 2, 1)$, $(3, 0, 6)$, $(4, 1, 5)$, and $(1, 1, 2)$.

2. Find the volume of the parallelepipeds whose adjacent edges are the vectors:
   (a) $[1, 1, 2]$, $[3, -1, 0]$, $[5, 2, -1]$.
   (b) $[1, 1, 0]$, $[1, 0, 1]$, $[0, 1, 1]$.

3. Prove both algebraically and geometrically that the parallelogram with edges X and Y has the same area as the parallelogram with edges X and $Y + kX$ for any scalar k.

4. Prove both algebraically and geometrically that the volume of the parallelepiped in $\mathscr{V}_3(\mathscr{R})$ with edges X, Y, and Z is equal to the volume of the parallelepiped with edges X, Y, $Z + hX + kY$ for any scalars h and k.

5. Find the 3-volume of the three-dimensional parallelepipeds in $\mathscr{V}_4(\mathscr{R})$ defined by the vectors:
   (a) $[2, 1, 0, -1]$, $[3, -1, 5, 2]$, $[0, 4, -1, 2]$.
   (b) $[1, 1, 0, 0]$, $[0, 2, 2, 0]$, $[0, 0, 3, 3]$.

6. Find the 2-volume of the parallelogram in $\mathscr{V}_4(\mathscr{R})$ two of whose edges are the vectors $[1, 3, -1, 6]$ and $[-1, 2, 4, 3]$.

7. Show that the parallelepiped in $\mathscr{V}_3(\mathscr{R})$ defined by the vectors $[2, 2, 1]$, $[1, -2, 2]$, and $[-2, 1, 2]$ is a cube. Find the volume of this cube.

8. Prove that if the vectors $X_1, X_2, \ldots, X_r$ are mutually orthogonal, the r-volume of the parallelepiped defined by them is equal to the product of their lengths.

9. Prove that r vectors $X_1, X_2, \ldots, X_r$ of $\mathscr{V}_n(\mathscr{R})$ are linearly dependent if and only if the determinant (12) is equal to zero.

10. Prove that two vectors X_1 and X_2 in the plane are positively oriented if and only if the sense of rotation from X_1 to X_2, through an angle less than $180°$, is counterclockwise.

4.5 Abstract Inner Product Spaces over the Real Field

Let $\mathscr{V}$ be any vector space over the real field $\mathscr{R}$, and let X, Y be any two vectors in $\mathscr{V}$.

Definition. *An inner product on a real vector space $\mathscr{V}$ is a function that associates with any pair of vectors X, Y in $\mathscr{V}$ a real number denoted by (X, Y) such that the following postulates are satisfied.*

IP1. $(X, X) \geq 0$ for all $X \in \mathscr{V}$ and $(X, X) = 0$ if and only if $X = O$.
IP2. $(X, Y) = (Y, X)$ for all $X, Y \in \mathscr{V}$.
IP3. $(X, kY) = k(X, Y)$ for all $X, Y \in \mathscr{V}$ and $k \in \mathscr{R}$.
IP4. $(X, Y_1 + Y_2) = (X, Y_1) + (X, Y_2)$ for all X, Y_1, Y_2 in $\mathscr{V}$.

We give several examples.

Example 1. In $\mathscr{V}_n(\mathscr{R})$ the standard inner product $X \cdot Y$ is an inner product in the sense of our definition. In fact, the postulates IP1 through IP4 are exactly the properties proved for $X \cdot Y$ in Theorem 4.1.

Example 2. In $\mathscr{V}_2(\mathscr{R})$ let $X = [x_1, x_2]$, $Y = [y_1, y_2]$ and define

$$(X, Y) = x_1 y_1 + 2x_1 y_2 + 2x_2 y_1 + 5x_2 y_2.$$

If we put $Y = X$,

$$(X, X) = x_1^2 + 4x_1 x_2 + 5x_2^2 = (x_1 + 2x_2)^2 + x_2^2.$$

Since x_1, x_2 are real numbers, it follows that $(X, X) \geq 0$ for all X and $(X, X) = 0$ if and only if $x_1 = x_2 = 0$, and hence $X = O$. Thus, IP1 holds and the other three postulates are easily verified.

Example 3. Let $\mathscr{V}$ be the space of all real-valued continuous functions defined on the interval $0 \leq x \leq 1$. If $f, g \in \mathscr{V}$, define

$$(f, g) = \int_0^1 f(x)g(x)\, dx.$$

Basic properties of the definite integral ensure that IP2, 3, 4 hold and that $(f, f) \geq 0$ for all $f \in \mathscr{V}$. The continuity of f ensures that

$$\int_0^1 [f(x)]^2\, dx > 0$$

if $f \neq 0$ and hence $(f, f) = 0$ implies $f = 0$.

Theorem 4.9. *(The Cauchy-Schwarz inequality) If (X, Y) is any inner product defined on a vector space $\mathscr{V}$ over $\mathscr{R}$, then for any two vectors X, Y in $\mathscr{V}$,*

$$|(X, Y)| \leq (X, Y)^{1/2}(Y, Y)^{1/2}.$$

Proof. The proof given for Theorem 4.2 applies with (X, Y) replacing $X \cdot Y$ because it makes use only of Theorem 4.1 which is now replaced by IP1 through IP4.

Definition. *A vector space $\mathscr{V}$ over the real field $\mathscr{R}$ on which is defined an inner product (X, Y) is called a* real inner product space. *A finite dimensional real inner product space is called a* Euclidean space.

The reason for the term Euclidean space is that the inner product enables us to define the length of a vector in such a way that Theorem 4.3 continues to hold.

Definition. *If $\mathscr{V}$ is a real inner product space with inner product (X, Y), the length of a vector X in $\mathscr{V}$, denoted by $\|X\|$, is defined by*

$$\|X\| = (X, X)^{1/2}.$$

Theorem 4.10. *In any real inner product space $\mathscr{V}$ the following hold for all vectors X, Y and all real numbers k.*

(a) $\|X\| \geq 0$ *and* $\|X\| = 0$ *if and only if* $X = O$.
(b) $\|kX\| = |k| \, \|X\|$.
(c) $\|X + Y\| \leq \|X\| + \|Y\|$.

Proof. The proof of Theorem 4.3 applies with $X \cdot Y$ replaced with (X, Y) and Theorem 4.2 by Theorem 4.9.

Theorems 4.9 and 4.10(c), when applied to the inner product space of continuous functions on a closed interval $[a, b]$ (see Example 3), give the following inequalities for any such continuous functions f and g:

(16) $$\left| \int_a^b f(x)g(x) \, dx \right| \leq \left\{ \int_a^b [f(x)]^2 \, dx \right\}^{1/2} \left\{ \int_a^b [g(x)]^2 \, dx \right\}^{1/2},$$

(17) $$\left\{ \int_a^b [f(x) + g(x)]^2 \, dx \right\}^{1/2} \leq \left\{ \int_a^b [f(x)]^2 \, dx \right\}^{1/2} + \left\{ \int_a^b [g(x)]^2 \, dx \right\}^{1/2}.$$

If the student has doubts about the power of the abstract axiomatic approach to inner products, let him try to prove (16) and (17) directly from the definition of an integral.

4.6 Orthonormal Bases

We assume throughout this section that $\mathscr{V}$ is a Euclidean vector space with inner product (X, Y).

Definition. *A vector X is said to be* orthogonal *to a vector Y if and only if* $(X, Y) = 0$.

It follows from IP2 that if X is orthogonal to Y, then Y is orthogonal to X and we say simply that X and Y are orthogonal. By IP4, $(X, Y) + (X, O) = (X, Y)$ and hence for all X, $(X, O) = (O, X) = 0$. Thus the zero vector is orthogonal to every vector of $\mathscr{V}$.

Definition. *A vector X is* orthogonal to a subspace $\mathscr{S}$ of $\mathscr{V}$ if X is orthogonal *to every vector in* $\mathscr{S}$. *A subspace* $\mathscr{T}$ of $\mathscr{V}$ is orthogonal to a subspace $\mathscr{S}$ if *every vector of* $\mathscr{T}$ *is orthogonal to every vector of* $\mathscr{S}$.

As for vectors, so for subspaces, $\mathscr{S}$ orthogonal to $\mathscr{T}$ implies $\mathscr{T}$ orthogonal to $\mathscr{S}$. Theorem 4.4 clearly holds in the following abstract form.

Theorem 4.11. *If a vector X in* $\mathscr{V}$ *is orthogonal to each of the vectors* $Y_1, \ldots, Y_r$, *then X is orthogonal to the subspace spanned by* $Y_1, \ldots, Y_r$. *If a subspace* $\mathscr{S}$ *has a basis* $B_1, \ldots, B_r$, *then X is orthogonal to* $\mathscr{S}$ *if and only if* $(X, B_i) = 0$, $(i = 1, 2, \ldots, r)$.

The vectors $X_1, X_2, \ldots, X_r$ are said to be *mutually orthogonal* if for $i \neq j$ $(X_i, X_j) = 0$, $(i, j = 1, 2, \ldots, r)$.

Theorem 4.12. *Any set of mutually orthogonal nonzero vectors in* $\mathscr{V}$ *is linearly independent.*

Proof. Suppose $X_1, X_2, \ldots, X_r$ are nonzero mutually orthogonal vectors so that $(X_i, X_j) = 0$ if $i \neq j$. If

$$c_1X_1 + c_2X_2 + \cdots + c_2X_2 = O,$$

then

$$0 = (X_i, O) = (X_i, c_1X_1 + c_2X_2 + \cdots, c_rX_r)$$
$$= c_1(X_i, X_1) + c_2(X_i, X_2) + \cdots + c_r(X_i, X_r) \qquad \text{by IP2 and IP3}$$
$$= c_i(X_i, X_i).$$

But since $X_i \neq O$, $(X_i, X_i) \neq 0$ by IP1. Hence, $c_i = 0$ $(i = 1, 2, \ldots, r)$ and $X_1, \ldots, X_r$ are linearly independent.

Lemma 4.13. *If* $\mathscr{S}$ *is a proper subspace of a Euclidean vector space* $\mathscr{V}$, *then* $\mathscr{V}$ *contains a nonzero vector orthogonal to* $\mathscr{S}$.

Proof. Let $S_1, S_2, \ldots, S_r$ be a basis of $\mathscr{S}$. Since $\mathscr{S}$ is a proper subspace of $\mathscr{V}$, we can choose $S_{r+1} \in \mathscr{V}$ such that $S_{r+1} \notin \mathscr{S}$. We let

$$(18) \qquad\qquad T = \sum_{j=1}^{r+1} a_j S_j$$

and seek values of the scalars $a_1, a_2, \ldots, a_{r+1}$ such that T is orthogonal to $\mathscr{S}$. By Theorem 4.11 this requires only that $(S_i, T) = 0$, $(i = 1, 2, \ldots, r)$ or

$$\sum_{j=1}^{r+1} a_j (S_i, S_j) = 0 \qquad (i = 1, 2, \ldots, r).$$

Since this is a system of r equations in $r + 1$ unknowns, it has a nonzero solution for $a_1, a_2, \ldots, a_{r+1}$ and substitution in (18) gives a vector T orthogonal to $\mathscr{S}$. Moreover, $T \neq 0$ because $a_1, a_2, \ldots, a_{r+1}$ are not all zero and $S_1, S_2, \ldots, S_{r+1}$ are linearly independent by Lemma 1.6.

Definition. *A vector X in $\mathscr{V}$ such that $\|X\| = 1$ is called a unit vector.*

If X is any nonzero vector of $\mathscr{V}$ and $k = \dfrac{1}{\|X\|}$, then kX is a unit vector. For example, if $X = [2, 1, 3, 1, 1]$ in $\mathscr{V}_5(\mathscr{R})$, $\|X\| = (16)^{1/2} = 4$ and hence $[\frac{1}{2}, \frac{1}{4}, \frac{3}{4}, \frac{1}{4}, \frac{1}{4}]$ is a unit vector.

Definition. *A basis of a Euclidean vector space that consists of mutually orthogonal unit vectors is called an* orthonormal basis.

It is clear from the previous remark that any basis consisting of mutually orthogonal vectors yields an orthonormal basis when each basis vector is multiplied by the reciprocal of its length.

Theorem 4.14. *Every nonzero Euclidean space has an orthonormal basis.*

Proof. Let $\mathscr{V}$ be a nonzero Euclidean vector space of dimension n. Choose any nonzero vector X_1 in $\mathscr{V}$, and let $\mathscr{S}_1$ be the subspace spanned by X_1. If $1 < n$, $\mathscr{S}_1$ is a proper subspace and, by Lemma 4.13, $\mathscr{V}$ contains a nonzero vector X_2 orthogonal to $\mathscr{S}_1$. Thus $\{X_1, X_2\}$ is a mutually orthogonal set of nonzero vectors. Now suppose a mutually orthogonal set $\{X_1, X_2, \ldots, X_r\}$ of nonzero vectors has been found. Then the subspace $\mathscr{S}_r$ spanned by these r vectors has dimension r (Theorem 4.12). If $r < n$ there exists (by Lemma 4.13) a nonzero vector X_{r+1} orthogonal to $\mathscr{S}_r$ and, therefore, to each of $X_1, X_2, \ldots, X_r$. Obviously this process continues until we obtain n nonzero mutually orthogonal vectors $X_1, X_2, \ldots, X_n$ which are linearly independent by Theorem 4.12 and hence a basis of $\mathscr{V}$. Now if $k_i = \|X_i\|^{-1}$, then $k_1 X_1, k_2 X_2, \ldots, k_n X_n$ is an orthonormal basis for $\mathscr{V}$.

Theorem 4.14 ensures that every Euclidean vector space has an orthonormal basis. Such a basis can actually be constructed from an arbitrary basis by an orthogonalization process which will now be described. Suppose that $X_1, X_2, \ldots, X_n$ is any basis of $\mathscr{V}$, and consider the vectors $Y_1, Y_2, \ldots, Y_n$ defined by

$$Y_1 = X_1,$$

$$Y_2 = X_2 - \frac{(X_2, X_1)}{(X_1, X_1)} Y_1,$$

$$Y_3 = X_3 - \frac{(X_3, Y_2)}{(Y_2, Y_2)} Y_2 - \frac{(X_3, Y_1)}{(Y_1, Y_1)} Y_1,$$

$$\cdots \cdots \cdots \cdots \cdots \cdots \cdots \cdots$$

$$Y_n = X_n - \frac{(X_n, Y_{n-1})}{(Y_{n-1}, Y_{n-1})} Y_{n-1} - \cdots - \frac{(X_n, Y_1)}{(Y_1, Y_1)} Y_1.$$

It is easy to verify successively that $(Y_2, Y_1) = 0$, $(Y_3, Y_1) = (Y_3, Y_2) = 0$, and in general, $(Y_i, Y_{i-1}) = (Y_i, Y_{i-2}) = \cdots = (Y_i, Y_1) = 0$. Hence the vectors Y_i are mutually orthogonal. Moreover, successive substitution shows that each Y_i has the form

$$X_i - (\text{a linear combination of } X_{i-1}, \ldots, X_1),$$

which cannot be O since the X_i are linearly independent. Hence, by Theorem 4.12, $Y_1, Y_2, \ldots, Y_n$ are linearly independent and therefore constitute a basis of $\mathscr{V}$. Now, if we replace each Y_i by the corresponding unit vector

$$E_i = \frac{Y_i}{\|Y_i\|},$$

we have an orthonormal basis $E_1, E_2, \ldots, E_r$ for $\mathscr{V}$. The construction just described is known as the *Gram-Schmidt orthogonalization process*.

Theorem 4.15. *If $E_1, E_2, \ldots, E_s$ ($1 \leq s < n$) are mutually orthogonal unit vectors in a space $\mathscr{V}$ of dimension n, there exist unit vectors $E_{s+1}, \ldots, E_n$ in $\mathscr{V}$ such that $E_1, \ldots, E_n$ is an orthonormal basis of $\mathscr{V}$.*

Proof. By Theorem 1.8, $\mathscr{V}$ has a basis of the form

$$E_1, \ldots, E_s, X_{s+1}, \ldots, X_n.$$

Application of the Gram-Schmidt orthogonalization process will leave $E_1, \ldots, E_s$ unchanged. Hence the final orthonormal basis will include $E_1, \ldots, E_s$. This result could also be proved by successive application of Lemma 4.13.

4.7 Inner Product Space Isomorphisms

In Section 1.8 we defined an isomorphism of a vector space $\mathscr{V}$ onto a vector space $\mathscr{W}$ (over the same field of scalars) as a one-to-one mapping $\sigma : \mathscr{V} \to \mathscr{W}$, of $\mathscr{V}$ onto $\mathscr{W}$, having the properties that $\sigma(V_1 + V_2) = \sigma V_1 + \sigma V_2$ and $\sigma(kV_1) = k(\sigma V_1)$ for all V_1, V_2 in $\mathscr{V}$ and all scalars k. Now suppose $\mathscr{V}$ and $\mathscr{W}$ are both vector spaces over the real field $\mathscr{R}$ and are both inner product spaces. We denote the inner product in $\mathscr{V}$ by $(\ ,\ )_1$ and in $\mathscr{W}$ by $(\ ,\ )_2$.

Definition. *An isomorphism* $\sigma : \mathscr{V} \to \mathscr{W}$ *between inner product spaces is called an inner product space isomorphism* (IPS-*isomorphism*) *if it satisfies, for all* V_1, V_2 *in* $\mathscr{V}$,

$$(\sigma V_1, \sigma V_2)_2 = (V_1, V_2)_1.$$

Theorem 4.16. *Let* $\mathscr{V}$ *be any Euclidean vector space of dimension n with inner product* $(\ ,\ )$. *Then* $\mathscr{V}$ *is* IPS-*isomorphic to* $\mathscr{V}_n(\mathscr{R})$ *with standard inner product.*

Proof. By Theorem 4.14 $\mathscr{V}$ has an orthonormal basis $F_1, F_2, \ldots, F_n$ characterized by

$$(19) \qquad \begin{aligned} (F_i, F_j) &= 0 \quad \text{if } i \neq j, \\ (F_i, F_i) &= 1 \quad (i = 1, 2, \ldots, n). \end{aligned}$$

Let $V_1 = x_1 F_1 + x_2 F_2 + \cdots + x_n F_n$ and let $\alpha : \mathscr{V} \to \mathscr{V}_n(\mathscr{R})$ be the vector space isomorphism discussed in Section 1.8, namely,

$$\alpha(V_1) = [x_1, x_2, \ldots, x_n].$$

Now, if $V_2 = y_1 F_1 + y_2 F_2 + \cdots + y_n F_n$ we have by IP2, 3, and 4,

$$(V_1, V_2) = (\sum x_i F_i, \sum y_i F_i) = \sum_{i=1}^{n} \sum_{j=1}^{n} x_i y_j (F_i, F_j)$$

$$= \sum_{i=1}^{n} x_i y_i \qquad\qquad \text{by (19)}$$

$$= \alpha(V_1) \cdot \alpha(V_2),$$

the standard inner product of $\alpha(V_1)$ and $\alpha(V_2)$ in $\mathscr{V}_n(\mathscr{R})$.

4.8 Inner Products as Bilinear Forms

Let $\mathscr{V}$ be a Euclidean vector space of dimension n with inner product $(\ , \ )$, and let $B_1, B_2, \ldots, B_n$ be any basis of $\mathscr{V}$. By IP2, 3, and 4, we have for any two vectors

$$(20) \qquad V_1 = \sum_{i=1}^{n} x_i B_i, \qquad V_2 = \sum_{j=1}^{n} y_j B_j,$$

$$(21) \qquad (V_1, V_2) = \sum_{i=1}^{n} \sum_{j=1}^{n} x_i y_j (B_i, B_j).$$

If the n^2 inner products (B_i, B_j) are known, equation (21) gives the inner product of any two vectors in terms of their coordinates relative to the basis $B_1, \ldots, B_n$ and the n^2 inner products (B_i, B_j).

We now ask to what extent, given a basis $B_1, \ldots, B_n$, we can define (B_i, B_j) arbitrarily and expect (21) to define an inner product in $\mathscr{V}$. To seek an answer, we arbitrarily choose n^2 real numbers b_{ij} and let $(B_i, B_j) = b_{ij}$. Then if V_1, V_2 are the vectors (20) we have

$$(22) \qquad (V_1, V_2) = \sum_{i=1}^{n} \sum_{j=1}^{n} b_{ij} x_i y_j$$

and we ask what restrictions must be imposed on the coefficients b_{ij} in order that (22) define an inner product in $\mathscr{V}$.

Because the right-hand side of (22) is a linear homogeneous function of $y_1, y_2, \ldots, y_n$, it follows at once that both IP3 and IP4 hold whatever the choice of b_{ij}. On the other hand, IP2 will hold if and only if $(B_i, B_j) = (B_j, B_i)$, that is, $b_{ij} = b_{ji}$. Finally, to satisfy IP1 it is necessary that

$$(V_1, V_1) = \sum_{i=1}^{n} \sum_{j=1}^{n} b_{ij} x_i x_j \geq 0$$

for every choice of $x_1, x_2, \ldots, x_n$.

We now introduce some terminology that will enable us to state our results succinctly.

Definition. *An expression of the form*

$$(23) \qquad \sum_{i=1}^{n} \sum_{j=1}^{n} b_{ij} x_i y_j \qquad\qquad b_{ij} \in \mathscr{R}$$

is called a real bilinear form *in the 2n variables* $x_1, x_2, \ldots, x_n, y_1, y_2, \ldots, y_n$. *The n × n matrix* $B = [b_{ij}]$ *is called the* matrix of the bilinear form.

Definition. *A square matrix* $B = [b_{ij}]$ *is said to be* symmetric *if for all* $i, j, b_{ij} = b_{ji}$, *that is, if* $B^T = B$. *A bilinear form is said to be* symmetric *if its matrix is symmetric.*

Definition. *An expression of the form*

(24)
$$\sum_{i=1}^{n} \sum_{j=1}^{n} a_{ij} x_i x_j, \qquad\qquad a_{ij} \in \mathcal{R}$$

is called a real quadratic form *in the n variables* $x_1, x_2, \ldots, x_n$.

Since it is assumed in (24) that $x_i x_j = x_j x_i$, we can assign half the total coefficient of the term in $x_i x_j$ to $x_i x_j$ and half to $x_j x_i$. In other words, every quadratic form can be written in the form (24) where $[a_{ij}]$ is a uniquely determined symmetric matrix.

Example 1. The quadratic form $x_1^2 - 18x_1 x_2 + 5x_2^2$ can be written

$$x_1^2 - 9x_1 x_2 - 9x_2 x_1 + 5x_2^2$$

and, therefore, its matrix is

$$A = \begin{bmatrix} 1 & -9 \\ -9 & 5 \end{bmatrix}.$$

Definition. *A real quadratic form* (24) *is said to be* nonnegative definite *if it takes only nonnegative values when arbitrary real values are substituted for* $x_1, x_2, \ldots, x_n$. *A nonnegative definite quadratic form which takes the value zero only when* $x_1 = x_2 = \cdots = x_n = 0$ *is said to be* positive definite.

Example 2. The quadratic form of Example 1 is not nonnegative definite because it takes the value -12 when $x_1 = x_2 = 1$.

Example 3. The quadratic form $x_1^2 - 4x_1 x_2 + 5x_2^2$ is positive definite because it can be written in the form

$$(x_1 - 2x_2)^2 + x_2^2,$$

which is nonnegative for all real values of x_1, x_2, and is zero only if $x_1 = x_2 = 0$.

Example 4. The quadratic form

$$x_1^2 + x_2^2 + 2x_3^2 - 2x_1 x_3 - 2x_2 x_3$$

is nonnegative definite but not positive definite because it can be written in the form

$$(x_1 - x_3)^2 + (x_2 - x_3)^2,$$

which is nonnegative for all real values of x_1, x_2, x_3 but is zero for nonzero values, for example, $x_1 = x_2 = x_3 = 2$.

Definition. *A real symmetric matrix $A = [a_{ij}]$ is said to be positive (nonnegative) definite if the associated quadratic form*

$$\sum_{i=1}^{n} \sum_{j=1}^{n} a_{ij} x_i x_j$$

is positive (nonnegative) definite. Similarly, a symmetric bilinear form is said to be positive (nonnegative) definite if its matrix is positive (nonnegative) definite.

Bilinear and quadratic forms can be written conveniently and compactly in matrix notation. For example, if $B = [b_{ij}]$ is an $n \times n$ matrix and if

$$X = [x_1, x_2, \ldots, x_n], \qquad Y = \begin{bmatrix} y_1 \\ y_2 \\ \cdot \\ \cdot \\ \cdot \\ y_n \end{bmatrix},$$

then XBY is a 1×1 matrix whose single element is the bilinear form (23). Since a 1×1 matrix can for most purposes be identified with its single element, we shall write

$$XBY = \sum_{i=1}^{n} \sum_{j=1}^{n} b_{ij} x_i y_j$$

and shall regularly use the notation XBY for this bilinear form. Similarly, if $A = [a_{ij}]$ is an $n \times n$ symmetric matrix, XAX^T is a 1×1 matrix whose single element is the quadratic form (24). We therefore write

$$XAX^T = \sum_{i=1}^{n} \sum_{j=1}^{n} a_{ij} x_i x_j,$$

and we shall normally use the notation XAX^T (or $X^T A X$ if X is a column vector) for the quadratic form with (symmetric) matrix A. We now summarize the results of this section in a formal statement.

Theorem 4.17. *Let $\mathscr{V}$ be a real vector space of finite dimension n, and let $B_1, B_2, \ldots, B_n$ be any basis of $\mathscr{V}$. Then every inner product that can be defined in $\mathscr{V}$ has the form $(V_1, V_2) = XAY$ where X, Y are the coordinate vectors of V_1, V_2 relative to the given basis and $A = [a_{ij}] = [(B_i, B_j)]$ is a positive definite symmetric matrix. Conversely, every positive definite symmetric bilinear form in the B-coordinates of the vectors V_1, V_2 defines an inner product in $\mathscr{V}$.*

Exercise 4.4

1. Verify that the vectors $[\frac{1}{3}, -\frac{2}{3}, -\frac{2}{3}]$, $[\frac{2}{3}, -\frac{1}{3}, \frac{2}{3}]$, and $[\frac{2}{3}, \frac{2}{3}, -\frac{1}{3}]$ form an orthonormal basis for $\mathcal{V}_3(\mathscr{R})$ relative to the standard inner product.

2. In $\mathcal{V}_2(\mathscr{R})$ let (X, Y) be the inner product of Example 2, Section 4.5. Show that the vectors $[1, 0]$ and $[2, -1]$ form an orthonormal basis for $\mathcal{V}_2(\mathscr{R})$ relative to this inner product.

3. In $\mathcal{V}_2(\mathscr{R})$, using the same inner product as in Problem 2, prove that $X_1 = [-\frac{2}{5}, \frac{3}{5}]$ is a unit vector and find a vector X_2 such that X_1, X_2 is an orthonormal basis for $\mathcal{V}_2(\mathscr{R})$.

4. Which of the following define inner products in $\mathcal{V}_2(\mathscr{R})$? Give reasons. (Assume $X = [x_1, x_2]$, $Y = [y_1, y_2]$.)
   (a) $(X, Y) = 2x_1y_1 + 5x_2y_2$.
   (b) $(X, Y) = x_1^2 - 2x_1y_2 - 2x_2y_1 + y_1^2$.
   (c) $(X, Y) = x_1y_1 - 2x_1y_2 - 2x_2y_1 + 4x_2y_2$.
   (d) $(X, Y) = x_1y_1 + 5x_1y_2 + 5x_2y_1 - x_2y_2$.
   (e) $(X, Y) = x_1y_1 + 3x_1y_2 + 3x_2y_1 + 10x_2y_2$.

5. Given the basis $[2, 0, 1]$, $[3, -1, 5]$, and $[0, 4, 2]$ for $\mathcal{V}_3(\mathscr{R})$, construct from it by the Gram-Schmidt process an orthonormal basis relative to the standard inner product.

6. Let $\mathscr{P}$ be the vector space over $\mathscr{R}$ consisting of all polynomials in x of degree ≤ 2 with real coefficients. Define an inner product in $\mathscr{P}$ by

$$(p, q) = \int_0^1 p(x)q(x)\, dx.$$

   (a) Verify that this does define an inner product.
   (b) Apply the Gram-Schmidt process to the basis 1, x, x^2 of $\mathscr{P}$ to obtain an orthonormal basis, relative to this inner product.
   (c) Find the coordinates relative to the orthonormal basis found in (b) of an arbitrary vector (that is, polynomial) $ax^2 + bx + c$.
   (d) Use (c) and the method of Theorem 4.16 to define an IPS-isomorphism σ between $\mathscr{P}$ and $\mathcal{V}_3(\mathscr{R})$ with standard inner product, and verify that Theorem 4.16 holds in this case.

7. Let $\mathcal{M}$ be the vector space over $\mathscr{R}$ consisting of all $m \times n$ matrices with elements in $\mathscr{R}$, and let $A = [a_{ij}]$, $B = [b_{ij}]$ be any two elements of $\mathcal{M}$. Prove that

$$(A, B) = \sum_{j=1}^{n} \sum_{i=1}^{m} a_{ij}b_{ij}$$

   defines an inner product in the space $\mathcal{M}$.

8. If $\mathscr{M}$, A, B, and (A, B) have the same meaning as in Problem 7, show that AB^T is an $m \times m$ matrix and that (A, B) is equal to the sum of the elements in the main diagonal of AB^T.

Note. If $C = [c_{ij}]$ is an $n \times n$ matrix the sum of its diagonal elements $\sum\limits_{i=1}^{n} c_{ii}$ is called the *trace* of C and written tr C. Thus, Problem 7 states that

$$(A, B) = \text{tr } (AB^T)$$

defines an inner product in the space $\mathscr{M}$ of all real $m \times n$ matrices.

9. Write down the matrix of each of the following bilinear forms and verify that each can be written as a matrix product XAY.
   (a) $x_1 y_1 + x_2 y_2 + 2x_3 y_2 - 5x_2 y_3$.
   (b) $2x_1 y_1 + 3x_1 y_2 + 5x_1 y_3 - 2x_2 y_2 + 7x_3 y_1 + 4x_3 y_3$.
   (c) $x_1 y_1 + x_2 y_2 + x_3 y_3 + x_4 y_4$.
   (d) $x_1 y_1 - 5x_2 y_3 + 2x_3 y_2$.

10. Write down the (symmetric!) matrix of each of the following quadratic forms and verify that they can be written as matrix products XAX^T
    (a) $x_1^2 + 5x_2^2 - 7x_3^2$. (b) $2x_1 x_2 + 6x_1 x_3 - 4x_2 x_3$.
    (c) $x_1^2 + 2x_2^2 - 5x_3^2 - x_1 x_2 + 4x_2 x_3 - 3x_3 x_1$.

11. If

$$A = \begin{bmatrix} 2 & 1 & 5 \\ 1 & 3 & -2 \\ 5 & -2 & 4 \end{bmatrix},$$

write out in full the quadratic form XAX^T and the bilinear form XAY.

12. Let $\mathscr{V}$ be an n-dimensional vector space over $\mathscr{R}$ with an inner product (,). Let $\mathscr{S}$ be a subspace of $\mathscr{V}$ of dimension r. Denote by $\mathscr{S}^{\perp}$ the set

$$\{X \in \mathscr{V} \mid (X, S) = 0 \text{ for all } S \in \mathscr{S}\}.$$

   (a) Prove that $\mathscr{S}^{\perp}$ is a subspace of $\mathscr{V}$.
   (b) Prove that dim $\mathscr{S} = n - r$. *Hint.* Use Lemma 4.13.
   (c) Prove that $\mathscr{S} \cap \mathscr{S}^{\perp} = O$ and $\mathscr{S} + \mathscr{S}^{\perp} = \mathscr{V}$.
   (d) Prove that every vector X in $\mathscr{V}$ can be written in one and only one way in the form $X = S + T$ where $S \in \mathscr{S}$ and $T \in \mathscr{S}^{\perp}$. The vectors S and T are called the *orthogonal projections* of X on the subspaces $\mathscr{S}$ and $\mathscr{S}^{\perp}$ (relative to the given inner product).

Note. In view of Problem 12, the discussion of volume in $\mathscr{V}_n(\mathscr{R})$ given in Section 4.4 could be extended almost word for word to an arbitrary n-dimensional Euclidean vector space $\mathscr{V}$ with arbitrary inner product (,).

13. Let $\mathscr{P}$ be the inner product space of Problem 6. Let $\mathscr{S}_1$ be the set of all polynomials of degree zero together with the zero polynomial, $\mathscr{S}_2$ the set of all scalar

multiples of x, and $\mathscr{S}_3$ the set of all polynomials $a + bx$, a, $b \in \mathscr{R}$. Prove that $\mathscr{S}_1$, $\mathscr{S}_2$, $\mathscr{S}_3$ are subspaces and find the subspaces $\mathscr{S}_1^\perp$, $\mathscr{S}_2^\perp$, and $\mathscr{S}_3^\perp$ relative to the inner product of Problem 6.

14. Find the orthogonal projections of $1 - 2x + 3x^2$:
    (a) onto $\mathscr{S}_1$ and $\mathscr{S}_1^\perp$.
    (b) onto $\mathscr{S}_2$ and $\mathscr{S}_2^\perp$ (notation as in Problem 13).

4.9 Change of Basis

Let $\mathscr{V}$ be an n-dimensional vector space over the field $\mathscr{F}$ of scalars and let E_1, $E_2, \ldots, E_n$ and $F_1, F_2, \ldots, F_n$ be two bases of $\mathscr{V}$. We wish to investigate the relationship between the coordinate vectors in $\mathscr{V}_n(\mathscr{F})$, relative to the two bases, of a fixed vector V in $\mathscr{V}$. This problem was discussed in Section 1.8 for the spaces $\mathscr{V}_2(\mathscr{R})$ and $\mathscr{V}_3(\mathscr{R})$ where it is essentially the problem of transformation of co-ordinates. The student may find it advantageous to reread Section 1.8 before continuing.

Since $E_1, E_2, \ldots, E_n$ is a basis for $\mathscr{V}$, we can write

$$F_1 = p_{11}E_1 + p_{21}E_2 + \cdots + p_{n1}E_n,$$
(25)
$$\ldots\ldots\ldots\ldots\ldots\ldots\ldots\ldots\ldots\ldots\ldots\ldots$$
$$F_n = p_{1n}E_1 + p_{2n}E_2 + \cdots + p_{nn}E_n.$$

Definition. *The matrix* $P = [p_{ij}]$ *(i the row index as usual) defined by* (25) *is called the* transition matrix *from the E-basis to the F-basis.*

Note that the transition matrix P is the transpose of the matrix of coefficients in (25). That is, the elements of ith row of P are the coefficients of E_i in (25).

Since $F_1, F_2, \ldots, F_n$ is a basis, the matrix $P = [p_{ij}]$ is nonsingular because a linear relation among the columns of P would imply the same linear relation among $F_1, F_2, \ldots, F_n$. (See Exercise 2.6, Problem 7.)

Theorem 4.18. *If a vector V in $\mathscr{V}$ has coordinate (column) vector X relative to the E-basis and coordinate vector X' relative to the F-basis, then $X = PX'$ where P is the transition matrix from the E-basis to the F-basis.*

Proof. We have by (25)

$$V = x_1E_1 + x_2E_2 + \cdots + x_nE_n$$
$$= x_1'F_1 + x_2'F_2 + \cdots + x_n'F_n$$
$$= \left(\sum_{j=1}^{n} p_{1j}x_j'\right)E_1 + \left(\sum_{j=1}^{n} p_{2j}x_j'\right)E_2 + \cdots + \left(\sum_{j=1}^{n} p_{nj}x_j'\right)E_n.$$

Since the coordinates of V relative to the E-basis are uniquely determined, we must have

$$x_i = \sum_{j=1}^{n} p_{ij} x_j' \qquad\qquad (i = 1, 2, \ldots, n)$$

or $X = PX'$ as required.

Corollary 1. *If P is the transition matrix from the E-basis to the F-basis and Q is the transition matrix from the F-basis to a third basis $G_1, G_2, \ldots, G_n$, the transition matrix from the E-basis is to the G-basis is PQ.*

Proof. If X'' is the coordinate vector of V relative to the G-basis, we have $X = PX'$, $X' = QX''$ and, therefore, $X = PQX''$.

Corollary 2. *If P is the transition matrix from the E-basis to the F-basis, then P^{-1} is the transition matrix from the F-basis to the E-basis.*

Proof. Put $G_i = E_i$ and by Corollary 1, we have $PQ = I$.

Next, we investigate the change from one orthonormal basis to another. Let $\mathscr{V}$ be any Euclidean space with inner product $(\ ,\ )$ and two orthonormal bases $E_1, E_2, \ldots, E_n$ and $F_1, F_2, \ldots, F_n$. We introduce a widely used symbol δ_{ij}, called a *Kronecker delta*, and defined for $i, j = 1, 2, \ldots, n$ by

$$\delta_{ij} = 1 \qquad \text{if } i = j,$$

$$\delta_{ij} = 0 \qquad \text{if } i \neq j,$$

or, in matrix language, by $[\delta_{ij}] = I$. Because the two bases are orthonormal, we have

(26) $$(E_i, E_j) = (F_i, F_j) = \delta_{ij}.$$

Now, if $P = [p_{ij}]$ is the transition matrix from the E-basis to the F-basis, we have by (25)

$$(F_i, F_j) = (p_{1i} E_1 + p_{2i} E_2 + \cdots + p_{ni} E_n, p_{1j} E_1 + p_{2j} E_2 + \cdots + p_{nj} E_n)$$

$$= p_{1i} p_{1j} + p_{2i} p_{2j} + \cdots + p_{ni} p_{nj}, \qquad \text{since } (E_i, E_j) = \delta_{ij}$$

$$= \delta_{ij}, \qquad\qquad\qquad\qquad\qquad \text{by (26)}.$$

Thus the transition matrix P has the property that its column vectors

$$[p_{1i}, p_{2i}, \ldots, p_{ni}]^T$$

are mutually orthogonal unit vectors of $\mathscr{V}_n(\mathscr{R})$ relative to the standard inner product. This fact can also be expressed by the matrix equation

(27) $$P^T P = I,$$

which is equivalent to

$$\sum_{k=1}^{n} p_{ki} p_{kj} = \delta_{ij}.$$

Since (27) implies $P^T = P^{-1}$ and hence

$$PP^T = I,$$

it follows that the rows of P are also mutually orthogonal unit vectors of $\mathscr{V}_n(\mathscr{R})$ relative to the standard inner product.

Definition. *A matrix P, with real elements, such that $P^T P = I$ is called an orthogonal matrix.*

Theorem 4.19. *If $E_1, \ldots, E_n$ is an orthonormal basis of a Euclidean space $\mathscr{V}$ and if $F_1, \ldots, F_n$ where*

$$F_i = \sum_{i=1}^{n} p_{ij} E_i$$

is a second basis of $\mathscr{V}$, the F-basis is orthonormal if and only if the transition matrix P from the E- to the F-basis is orthogonal.

Proof. We have already shown that if the F-basis is orthonormal, P is an orthogonal matrix. But if P is orthogonal we again have, since $(E_i, E_j) = \delta_{ij}$,

$$(F_i, F_j) = p_{1i} p_{1j} + p_{2i} p_{2j} + \cdots + p_{ni} p_{nj}$$

$$= \delta_{ij}$$

because P is orthogonal. Hence the F-basis is orthonormal.

Now let $E_1, \ldots, E_n$ and $F_1, \ldots, F_n$ be orthonormal bases of $\mathscr{V}$ and let P be the transition matrix from the E- to the F-basis. Let V, W be vectors in $\mathscr{V}$ with coordinate vectors X, Y relative to the E-basis and X', Y' relative to the F-basis. These coordinate vectors are considered as column vectors and satisfy $X = PX'$, $Y = PY'$ by Theorem 4.18. Since the standard inner product $X \cdot Y$ can be written as a matrix product

$$X^T Y = x_1 y_1 + x_2 y_2 + \cdots + x_n y_n,$$

we have

$$X^T Y = (PX')^T PY' = X'^T P^T PY' = X'^T Y'$$

because P is orthogonal and $P^T P = I$. Hence we have $X \cdot Y = X' \cdot Y'$ for the coordinate vectors relative to any two orthonormal bases. Thus, we have proved the following.

Theorem 4.20. *In an n-dimensional Euclidean vector space the dot product of the coordinate vectors of any two vectors of $\mathscr{V}$ is invariant under transformation from one orthonormal basis to another.*

This theorem, which we have proved via the orthogonal transition matrix, can actually be regarded as a corollary of Theorem 4.16 in which it was shown that

$$\alpha(V_1) \cdot \alpha(V_2) = (V_1, V_2)$$

where $\alpha(V)$ is the coordinate vector of V relative to *any* orthonormal basis of $\mathscr{V}$. Consequently, the dot product is the same for all such bases.

Theorem 4.21. *If P and Q are orthogonal matrices, then PQ, P^T, and P^{-1} are orthogonal and $\det P = \pm 1$.*

Proof. The first three statements are obvious consequences of the definition. Since $P^T P = I$, $\det P^T P = (\det P^T)(\det P) = (\det P)^2 = 1$ and hence $\det P = \pm 1$.

Definition. *An orthogonal matrix P is said to be* proper *if* $\det P = 1$, improper *if* $\det P = -1$.

Clearly, P^{-1} is proper or improper according as P is proper or improper. Moreover, if P and Q are both proper or both improper, then PQ is proper but if one of P, Q is proper and one improper, then PQ is improper.

4.10 Rotation and Reflection of Coordinate Axes in $\mathscr{V}_n(\mathscr{R})$

It was pointed out in Section 1.8 that each basis of a vector space $\mathscr{V}$ provides a way of assigning coordinates to the vectors in $\mathscr{V}$. A change of basis in $\mathscr{V}_n(\mathscr{R})$ is essentially a change of coordinate axes leaving the origin fixed. An orthonormal basis relative to the standard inner product in $\mathscr{V}_n(\mathscr{R})$, of course, corresponds to a rectangular coordinate system in which the same unit is used on each axis. We pass from one such coordinate system to another via an orthogonal transition matrix.

We recall (Section 4.4) that an ordered set of linearly independent vectors

(28) $$E_i = [e_{i1}, e_{i2}, \ldots, e_{in}] \qquad (i = 1, 2, \ldots, n)$$

in $\mathscr{V}_n(\mathscr{R})$ are said to be positively or negatively oriented according as det $[e_{ij}]$ is positive or negative.

Theorem 4.22. *Two bases $E_1, \ldots, E_n$ and $F_1, \ldots, F_n$ of $\mathscr{V}_n(\mathscr{R})$ have the same orientation if and only if the transition matrix P from the E-basis to the F-basis has positive determinant.*

Proof. Suppose the E-basis is given by (28) and the F-basis by

$$F_i = [f_{i1}, f_{i2}, \ldots, f_{in}].$$

If the transition matrix is $P = [p_{ij}]$,

(29) $$F_i = p_{1i}E_1 + p_{2i}E_2 + \cdots + p_{ni}E_n$$

and, equating the jth coordinates in (29), we get

$$f_{ij} = p_{1i}e_{1j} + p_{2i}e_{2j} + \cdots + p_{ni}e_{nj}$$

whence, if $F = [f_{ij}]$, $E = [e_{ij}]$,

$$F = P^T E,$$

and

$$\det F = (\det P)(\det E).$$

Hence, det F and det E have the same sign if and only if det $P > 0$. In particular, if the transition matrix P is proper orthogonal, not only are the length and orthogonality of the basis vectors preserved but so, too, is the orientation of the basis.

Definition. *A transformation of coordinates $X = PX'$ in $\mathscr{V}_n(\mathscr{R})$ with proper orthogonal transition matrix P is called a* rotation of coordinate axes *or a* rotation of basis vectors.

In order to justify this definition of a rotation of axes in n-space, we must show that it coincides in the case $n = 3$ with our usual concept of a rotation of axes in space. A rotation of axes in three-dimensional space is usually defined as a continuous rigid motion of the coordinate axes which leaves the origin fixed. It is easy to see that such a transformation cannot change the orientation of the basis vectors. For let

$$E_i = [e_{i1}, e_{i2}, e_{i3}] \qquad (i = 1, 2, 3)$$

be an orthonormal basis for $\mathscr{V}_3(\mathscr{R})$ which can be transformed by a rotation into a new normal orthogonal basis $F_i = [f_{i1}, f_{i2}, f_{i3}]$. Let $X_i = (x_{i1}, x_{i2}, x_{i3})$ be the coordinates of the basis vectors at any intermediate position so that each x_{ij} varies continuously from e_{ij} to f_{ij}. Since det $[x_{ij}]$ is a continuous function of the co-ordinates x_{ij}, it assumes every value between det $[e_{ij}]$ and det $[f_{ij}]$. Hence, if det $[e_{ij}]$ and det $[f_{ij}]$ differ in sign, det $[x_{ij}]$ must assume the value zero at some intermediate position in the course of the rotation from the E-basis to the F-basis. This is impossible, since in a rigid motion the basis vectors remain orthogonal and therefore linearly independent. Thus the orientation of the basis vectors cannot change in a continuous displacement of the basis vectors unless at some stage one of the basis vectors passes through the plane of the other two. A rotation of axes is therefore a transformation from one orthonormal basis to a similarly oriented one. Hence, its transition matrix is proper orthogonal by Theorems 4.19 and 4.22.

Conversely, given an orthonormal basis E_1, E_2, E_3 in $\mathcal{V}_3(\mathcal{R})$, every change of basis with proper orthogonal transition matrix leads by Theorems 4.19 and 4.22 to a similarly oriented orthonormal basis F_1, F_2, F_3. We shall show that any such transformation is a rotation of axes. It is clear that E_1 and E_2 can be brought into coincidence with F_1 and F_2 by a continuous rigid motion of E_1, E_2, and E_3, and that E_3 must then coincide either with F_3 or with $-F_3$. But, since F_1, F_2, F_3 have the same orientation as E_1, E_2, E_3, it follows that F_1, F_2, $-F_3$ have the opposite orientation. Hence, since a continuous rigid motion cannot change the orientation, the rigid motion that brings E_1 and E_2 into coincidence with F_1 and F_2 must also bring E_3 into coincidence with F_3.

Hence, every transformation from an orthonormal basis via a proper orthogonal transition matrix is a rotation of axes. Thus, in $\mathcal{V}_3(\mathcal{R})$ the transformations (from an orthonormal basis) with proper orthogonal transition matrices are exactly the rotations.

Every improper orthogonal matrix P can clearly be converted into a proper orthogonal matrix R by changing the sign of all elements in the first column, which is equivalent to multiplying it on the right by the improper orthogonal matrix

$$Q = \begin{bmatrix} -1 & 0 & \cdots & 0 \\ 0 & 1 & \cdots & 0 \\ \cdots\cdots\cdots\cdots \\ 0 & 0 & \cdots & 1 \end{bmatrix}.$$

This is the matrix of the transformation from the basis $F_1, F_2, \ldots, F_n$ to the basis $-F_1, F_2, \ldots, F_n$, and may be thought of as a reflection of the F_1 basis vector (or coordinate axis) in the subspace spanned by $F_2, F_3, \ldots, F_n$ (for example, reflection of the x-axis in the yz-plane). We have, then, $PQ = R$ or, since $Q^2 = I$, $P = RQ$. Hence the transformation with transition matrix P is equivalent to a rotation with transition matrix R followed by a reflection with transition matrix Q, or a change in sign of one of the basis vectors.

In many applications, we are concerned with change from the standard basis

$$E_1 = [1, 0, 0, \ldots, 0],$$
$$E_2 = [0, 1, 0, \ldots, 0],$$
$$\ldots \ldots \ldots \ldots \ldots$$
$$E_n = [0, 0, 0, \ldots, 1]$$

of $\mathscr{V}_n(\mathscr{R})$ to a new basis (usually orthonormal) $F_1, F_2, \ldots, F_n$. Now if

$$F_i = [p_{1i}, p_{2i}, \ldots, p_{ni}] = p_{1i}E_1 + p_{2i}E_2 + \cdots + p_{ni}E_n,$$

it follows from (25) that the transition matrix from the E-basis to the F-basis is the matrix P of which $F_1, F_2, \ldots, F_n$ are the columns.

Example 1. Find the coordinates of the point $(2, 1, 5)$ relative to a coordinate system based on the basis vectors

$$F_1 = [1, 0, 1], \qquad F_2 = [2, 1, 3], \qquad \text{and} \qquad F_3 = [2, -1, 2].$$

Solution. The transition matrix from the standard basis to the F-basis is

$$P = \begin{bmatrix} 1 & 2 & 2 \\ 0 & 1 & -1 \\ 1 & 3 & 2 \end{bmatrix}.$$

The new coordinates of the vector X are given by $X = PX'$ or $X' = P^{-1}X$. Computing P^{-1}, we find

$$(30) \qquad P^{-1} = \begin{bmatrix} 5 & 2 & -4 \\ -1 & 0 & 1 \\ -1 & -1 & 1 \end{bmatrix}$$

and hence the new coordinates of (x_1, x_2, x_3) are

$$x_1' = 5x_1 + 2x_2 - 4x_3,$$
$$x_2' = -x_1 \qquad\qquad x_3,$$
$$x_3' = -x_1 - x_2 + x_3.$$

In particular, the F-coordinates of $(2, 1, 5)$ are $(-8, 3, 2)$.

Example 2. Find an equation of the surface

$$(31) \qquad 6x_1^2 + x_2^2 + 5x_3^2 + 4x_1x_2 - 11x_1x_3 - 4x_2x_3 = 0$$

relative to the F-coordinate system of Example 1.

Solution 1. Let (x_1', x_2', x_3') be the F-coordinates of the point (x_1, x_2, x_3). By Theorem 4.18, $X = PX'$ or

$$x_1 = x_1' + 2x_2' + 2x_3',$$
$$x_2 = \quad\ x_2' - \ x_3',$$
$$x_3 = x_1' + 3x_2' + 2x_3'.$$

By substituting these in equation (31) and reducing, we find that (details left to the student),

(32)
$$-x_1'x_2' + x_3'^2 = 0.$$

Thus, x_1, x_2, x_3 satisfy (31) if and only if (x_1', x_2', x_3') satisfy (32) and hence (32) is an equation of the surface in the F-coordinate system.

Solution 2. The left-hand side of (31) is the quadratic form $X^T A X$ where

$$A = \begin{bmatrix} 6 & 2 & -\frac{11}{2} \\ 2 & 1 & -2 \\ -\frac{11}{2} & -2 & 5 \end{bmatrix}.$$

Since $X = PX'$, $X^T = X'^T P^T$ and $X^T A X = X'^T (P^T A P) X'$ where

$$P^T A P = \begin{bmatrix} 1 & 0 & 1 \\ 2 & 1 & 3 \\ 2 & -1 & 2 \end{bmatrix} \begin{bmatrix} 6 & 2 & -\frac{11}{2} \\ 2 & 1 & -2 \\ -\frac{11}{2} & -2 & 5 \end{bmatrix} \begin{bmatrix} 1 & 2 & 2 \\ 0 & 1 & -1 \\ 1 & 3 & 2 \end{bmatrix} = \begin{bmatrix} 0 & -\frac{1}{2} & 0 \\ -\frac{1}{2} & 0 & 0 \\ 0 & 0 & 1 \end{bmatrix}$$

and hence $X^T A X = X'^T P^T A P X' = x_3'^2 - x_1'x_2'$ and equation (31) transforms into (32).

Example 3. Find the transition matrix from the F-basis of Example 1 to the basis

$$G_1 = [1, 1, 0], \qquad G_2 = [1, 0, 2], \qquad G_3 = [2, 1, 1].$$

Solution. Let E_1, E_2, E_3 be the standard basis for $\mathscr{V}_3(\mathscr{R})$. The transition matrix from the E-basis to the F-basis is P (see Example 1) and, therefore, the transition matrix from the F-basis to the E-basis is P^{-1}. The transition matrix from the E-basis to the G-basis is

$$Q = \begin{bmatrix} 1 & 1 & 2 \\ 1 & 0 & 1 \\ 0 & 2 & 1 \end{bmatrix}$$

and hence the transition matrix from the F-basis to the G-basis is

$$P^{-1}Q = \begin{bmatrix} 5 & 2 & -4 \\ -1 & 0 & 1 \\ -1 & -1 & 1 \end{bmatrix} \begin{bmatrix} 1 & 1 & 2 \\ 1 & 0 & 1 \\ 0 & 2 & 1 \end{bmatrix} = \begin{bmatrix} 7 & -3 & 8 \\ -1 & 1 & -1 \\ -2 & 1 & -2 \end{bmatrix}.$$

Exercise 4.5

1. Write the transition matrices for the following changes in basis vectors:
   (a) From the basis $[1, 0]$, $[0, 1]$ to the basis $[2, 1]$, $[3, 2]$ in $\mathscr{V}_2(\mathscr{R})$.
   (b) From the basis $[2, -1]$, $[1, 3]$ to the basis $[3, 5]$, $[-2, 3]$ in $\mathscr{V}_2(\mathscr{R})$.
   (c) From the standard basis E_1, E_2, E_3 in $\mathscr{V}_3(\mathscr{R})$ to the basis E_2, E_3, E_1.
   (d) From the standard basis in $\mathscr{V}_3(\mathscr{R})$ to the basis

   $$F_1 = [2, -1, 0], \qquad F_2 = [1, 1, 1], \qquad F_3 = [-3, 0, 4].$$

   (e) From the F-basis in (d) to the standard basis.
   (f) From the F-basis in (d) to the basis

   $$G_1 = [1, 1, 0], \qquad G_2 = [1, 0, 1], \qquad G_3 = [0, 1, 1].$$

2. Choose unit vectors F_1, F_2 along the lines $x + 2y = 0$ and $x - 2y = 0$ as basis of $\mathscr{V}_2(\mathscr{R})$. Find:
   (a) The transition matrix from the standard basis to the F-basis.
   (b) The coordinates of the points $(1, 0)$, $(0, 1)$, and $(2, 5)$ relative to the coordinate system based on the F-basis.
   (c) An equation of the circle $x^2 + y^2 = r^2$ in the coordinate system based on the F-basis.

3. Find an equation of the hyperbola

   $$\frac{x^2}{a^2} - \frac{y^2}{b^2} = 1$$

   relative to a coordinate system using its asymptotes as coordinate axes and the same unit of length as in the original coordinate system.

4. Find the coordinates of the vector $[2, 1, 3, 4]$ of $\mathscr{V}_4(\mathscr{R})$ relative to the basis vectors $F_1 = [1, 1, 0, 0]$, $F_2 = [1, 0, 1, 1]$, $F_3 = [2, 0, 0, 2]$, $F_3 = [0, 0, 2, 2]$.

5. Find an equation of the curve

   $$8x^2 - 10xy - 3y^2 = 12$$

   relative to the coordinate system based on the vectors

   $$F_1 = [\tfrac{1}{14}, -\tfrac{2}{7}], \qquad F_2 = [\tfrac{3}{14}, \tfrac{1}{7}].$$

6. Find an orthonormal basis of $\mathcal{V}_3(\mathcal{R})$ of which $[\frac{1}{3}, \frac{2}{3}, \frac{2}{3}]$ is the first basis vector, and write the transition matrix from the standard basis to this new basis.

7. Prove that every proper orthogonal second-order matrix has the form

$$\begin{bmatrix} \cos\theta & -\sin\theta \\ \sin\theta & \cos\theta \end{bmatrix}$$

and every improper orthogonal second order matrix has the form

$$\begin{bmatrix} \cos\theta & \sin\theta \\ \sin\theta & -\cos\theta \end{bmatrix},$$

where θ is a suitably chosen angle.

8. Verify that the transformations of coordinates from the standard basis of $\mathcal{V}_3(\mathcal{R})$ to each of the following bases is orthogonal, and write the transition matrix in each case.
   (a) $F_1 = [\frac{2}{3}, \frac{2}{3}, \frac{1}{3}]$, $F_2 = [-\frac{2}{3}, \frac{1}{3}, \frac{2}{3}]$, $F_3 = [\frac{1}{3}, -\frac{2}{3}, \frac{2}{3}]$.
   (b) $G_1 = [\frac{2}{7}, \frac{3}{7}, \frac{6}{7}]$, $G_2 = [\frac{6}{7}, \frac{2}{7}, -\frac{3}{7}]$, $G_3 = [\frac{3}{7}, -\frac{6}{7}, \frac{2}{7}]$.
   (c) $H_1 = [\frac{3}{5}, 0, -\frac{4}{5}]$, $H_2 = [\frac{4}{5}, 0, \frac{3}{5}]$, $H_3 = [0, 1, 0]$.

9. Which of the three transformations in Problem 8 are rotations?

10. Referring to the vectors given in Problem 8, find the transition matrix from (a) the G-basis to the E-basis, (b) E-basis to the H-basis, (c) G-basis to the H-basis, (d) F-basis to the G-basis.

11. Verify that all the transformations in Problem 10 are orthogonal. Which of them are rotations?

12. If $X = [x_1, x_2, x_3]$, $Y = [y_1, y_2, y_3]$ are any two linearly independent vectors of $\mathcal{V}_3(\mathcal{R})$, their vector product is the vector

$$X \times Y = [x_2y_3 - x_3y_2, x_3y_1 - x_1y_3, x_1y_2 - x_2y_1].$$

Prove that X, Y, $X \times Y$ are always positively oriented.

13. Find an equation of the surface

$$x_1^2 + 2x_2^2 + 3x_3^2 - 4x_1x_2 - 4x_2x_3 = 10$$

relative to the coordinate system with basis vectors

$$F_1 = [\frac{2}{3}, \frac{2}{3}, \frac{1}{3}], \qquad F_2 = [\frac{1}{3}, -\frac{2}{3}, \frac{2}{3}], \qquad F_3 = [\frac{2}{3}, -\frac{1}{3}, -\frac{2}{3}].$$

Show that the transformation from the standard basis to the F-basis is a rotation of axes.

| LINEAR TRANSFORMATIONS

5.1 Definition and Basic Properties

Let $\mathscr{V}$ and $\mathscr{W}$ be two vector spaces over the same field $\mathscr{F}$ of scalars.

Definition. *A mapping* $\tau : \mathscr{V} \to \mathscr{W}$ *is called a* linear transformation *from* $\mathscr{V}$ *into* $\mathscr{W}$ *if, for every pair of vectors* V_1, V_2 *in* $\mathscr{V}$ *and every scalar* k,

$$\text{(1)} \qquad \begin{aligned} \tau(V_1 + V_2) &= \tau V_1 + \tau V_2, \\ \tau(kV_1) &= k(\tau V_1). \end{aligned}$$

We note that, putting $k = -1$, equations (1) imply

$$\tau(-V_1) = -\tau V_1,$$

and hence also

$$\tau(V_1 - V_2) = \tau V_1 - \tau V_2.$$

Two linear transformations σ, τ of $\mathscr{V}$ into $\mathscr{W}$ are *equal* if they are equal as mappings, that is, $\sigma = \tau$ if and only if $\sigma V = \tau V$ for all V in $\mathscr{V}$.

Example 1. Any isomorphism of a vector space $\mathscr{V}$ over $\mathscr{F}$ onto a vector space $\mathscr{W}$ over $\mathscr{F}$ (see Section 1.8) is a linear transformation. In particular, if $\mathscr{V}$ has basis

181

$E_1, \ldots, E_n$ and if X is the coordinate vector of V relative to the E-basis, the mapping $\tau : \mathscr{V} \to \mathscr{V}_n(\mathscr{F})$ defined by

$$\tau V = X$$

is a linear transformation of $\mathscr{V}$ onto $\mathscr{V}_n(\mathscr{F})$.

Example 2. Let A be any $m \times n$ matrix over $\mathscr{F}$ and define a mapping $\tau : \mathscr{V}_n(\mathscr{F}) \to \mathscr{V}_m(\mathscr{F})$ as in Section 2.7 by

(2) $$\tau X = AX.$$

Although (2) requires that X and AX are column vectors, we identify these with the corresponding row vectors in $\mathscr{V}_n(\mathscr{F})$ and $\mathscr{V}_m(\mathscr{F})$. That τ satisfies the linearity conditions (1) follows as in Section 2.7.

Example 3. The mapping $\tau : \mathscr{V}_2(\mathscr{F}) \to \mathscr{V}_3(\mathscr{F})$ defined by

$$\tau[x_1, x_2] = [x_1, x_2, 0]$$

is easily shown to be a linear transformation. Verification of equations (1) is left as an exercise.

Example 4. Let $\mathscr{V}$ be any inner product space over $\mathscr{R}$, let $\mathscr{S}$ be a subspace of $\mathscr{V}$ and $\mathscr{S}^\perp$ the orthogonal complement of $\mathscr{S}$. Every vector $V_1 \in \mathscr{V}$ has a unique decomposition $V_1 = S_1 + S_1'$, $S_1 \in \mathscr{S}$ and $S_1' \subset \mathscr{S}^\perp$. The mapping $\tau : \mathscr{V} \to \mathscr{S}$ defined by

$$\tau(V_1) = S_1$$

is a linear transformation called the *orthogonal projection* of $\mathscr{V}$ on $\mathscr{S}$. For if $V_2 = S_2 + S_2'$, $S_2 \in \mathscr{S}$, $S_2' \in \mathscr{S}^\perp$, then $V_1 + V_2 = (S_1 + S_2) + (S_1' + S_2')$ and $kV_1 = kS_1 + kS_1'$. The uniqueness of these decompositions ensure that

$$\tau(V_1 + V_2) = S_1 + S_2 = \tau(V_1) + \tau(V_2),$$
$$\tau(kV_1) = kS_1 = k\tau(V_1),$$

and hence τ is linear.

Theorem 5.1. *Let $\mathscr{V}$ and $\mathscr{W}$ be vector spaces over $\mathscr{F}$ and let $\tau : \mathscr{V} \to \mathscr{W}$ be a linear transformation.*

(a) *If O and O' are the zero vectors of $\mathscr{V}$ and $\mathscr{W}$, then $\tau O = O'$.*
(b) *If $\mathscr{S}$ is any subspace of $\mathscr{V}$ the set*

$$\tau \mathscr{S} = \{\tau V \in \mathscr{W} \mid V \in \mathscr{S}\}$$

is a subspace of $\mathscr{W}$, called the image of $\mathscr{S}$ under τ.

(c) *If $\mathcal{T}$ is any subspace of $\mathcal{W}$ the set*

$$\tau^{-1}\mathcal{T} = \{V \in \mathcal{V} \mid \tau V \in \mathcal{T}\}$$

is a subspace of $\mathcal{V}$ called the inverse image *of $\mathcal{T}$ under τ. This notation does not imply that τ is an invertible mapping.*

Proof.

(a) By (1), $\tau O = \tau(O + O) = \tau O + \tau O$ and hence

$$\tau O = \tau O - \tau O = O'.$$

(b) By (a) $\tau\mathcal{S}$ is nonempty. If $W_1, W_2 \in \tau\mathcal{S}$, then $W_1 = \tau V_1$, $W_2 = \tau V_2$, where $V_1, V_2 \in \mathcal{S}$. Then by (1),

$$W_1 + W_2 = \tau V_1 + \tau V_2 = \tau(V_1 + V_2) \in \tau\mathcal{S}$$

because $V_1 + V_2 \in \mathcal{S}$. Similarly, for every scalar k, $kW_1 = k(\tau V_1) = \tau(kV_1) \in \tau\mathcal{S}$ because $kV_1 \in \mathcal{S}$. Hence, $\tau\mathcal{S}$ is a subspace of $\mathcal{W}$.

(c) By (a) $\tau^{-1}\mathcal{T}$ is nonempty. If $V_1, V_2 \in \tau^{-1}\mathcal{T}$, then $\tau V_1, \tau V_2 \in \mathcal{T}$ and by (1)

$$\tau(V_1 + V_2) = \tau V_1 + \tau V_2 \in \mathcal{T}.$$

Hence, $V_1 + V_2 \in \tau^{-1}\mathcal{T}$. Similarly, $\tau V_1 \in \mathcal{T}$ and, therefore, $k\tau V_1 = \tau(kV_1) \in \mathcal{T}$ and $kV_1 \in \tau^{-1}\mathcal{T}$. Thus, $\tau^{-1}\mathcal{T}$ is a subspace of $\mathcal{V}$.

In particular, if we let $\mathcal{T} = O'$, the zero subspace of $\mathcal{W}$, then $\tau^{-1}O'$ is a subspace of $\mathcal{V}$.

Definition. *If $\tau : \mathcal{V} \to \mathcal{W}$ is a linear transformation and if O' is the zero vector of $\mathcal{W}$, the subspace*

$$\tau^{-1}O' = \{V \in \mathcal{V} \mid \tau V = O'\}$$

is called the null space *(or the* kernel*) of τ.*

Note that if $\tau : \mathcal{V}_n(\mathcal{F}) \to \mathcal{V}_m(\mathcal{F})$ is the linear transformation of Example 2 defined by the matrix A, then the null space of τ is precisely the null space of A as defined in Section 2.7.

Theorem 5.2. *A linear transformation $\tau : \mathcal{V} \to \mathcal{W}$ is a one-to-one mapping of $\mathcal{V}$ into $\mathcal{W}$ if and only if its null space is O.*

Proof. If the null space of τ is O and $\tau V_1 = \tau V_2$, then $\tau(V_1 - V_2) = \tau V_1 - \tau V_2 = O'$. Hence, $V_1 - V_2$ is in the null space and $V_1 = V_2$. Conversely, if τ is one-to-one and $\tau V = O'$, then $\tau V = \tau O$ and $V = O$, since τ is one-to-one.

The transformation of Example 3 is one-to-one and has null space zero. On the other hand, the null space of τ in Example 4 is $\mathscr{S}^{\perp}$ and hence τ is not one-to-one unless $\mathscr{V} = \mathscr{S}$.

Definition. *If $\tau : \mathscr{V} \rightarrow \mathscr{W}$ is a linear transformation from $\mathscr{V}$ onto $\mathscr{W}$ and if, in addition, τ is one-to-one, there is an inverse mapping $\tau^{-1} : \mathscr{W} \rightarrow \mathscr{V}$ and τ is said to be* invertible. *An invertible linear transformation is clearly an isomorphism (see Section 1.8).*

Theorem 5.3. *If $\tau : \mathscr{V} \rightarrow \mathscr{W}$ is an invertible linear transformation, $\tau^{-1} : \mathscr{W} \rightarrow \mathscr{V}$ is also a linear transformation. Moreover, τ^{-1} is invertible and $(\tau^{-1})^{-1} = \tau$.*

Proof. Because τ is onto $\mathscr{W}$, if $W_1, W_2 \in \mathscr{W}$, then $W_1 = \tau V_1$, $W_2 = \tau V_2$ where $V_1, V_2 \in \mathscr{V}$. Hence, $W_1 + W_2 = \tau V_1 + \tau V_2 = \tau(V_1 + V_2)$ and, therefore,

$$(3) \qquad \tau^{-1}(W_1 + W_2) = V_1 + V_2 = \tau^{-1} V_1 + \tau^{-1} V_2.$$

Similarly, for every scalar k, $kW_1 = k\tau V_1 = \tau(kV_1)$ because τ is linear. Hence,

$$(4) \qquad \tau^{-1}(kW_1) = kV_1 = k(\tau^{-1} W_1)$$

and (3) and (4) state that τ^{-1} is a linear transformation. By Theorem 1.0, Corollary 1, τ^{-1} is also invertible and $(\tau^{-1})^{-1} = \tau$.

If $\mathscr{V}$ and $\mathscr{W}$ are finite dimensional, all linear transformations of $\mathscr{V}$ into $\mathscr{W}$ are determined by the following theorem which is an immediate consequence of equations (1).

Theorem 5.4. *Let $\mathscr{V}$ and $\mathscr{W}$ be vector spaces over $\mathscr{F}$. If $E_1, E_2, \ldots, E_n$ is any basis of $\mathscr{V}$ and if $F_1, F_2, \ldots, F_n$ is an arbitrary set of n (not necessarily distinct) vectors of $\mathscr{W}$, there exists a unique linear transformation $\tau : \mathscr{V} \rightarrow \mathscr{W}$ such that $\tau E_i = F_i$, $(i = 1, 2, \ldots, n)$.*

Proof. If $V = \sum_{i=1}^{n} x_i E_i$ is any vector of $\mathscr{V}$ we define τV by

$$(5) \qquad \tau V = \sum_{i=1}^{n} x_i F_i.$$

Equation (5), by putting $x_i = 1$, $x_j = 0$ for $j \neq i$, implies that $\tau E_i = F_i$ and it is a simple exercise to verify that τ satisfies (1) and is hence linear. On the other hand, if $\tau' : \mathscr{V} \rightarrow \mathscr{W}$ is a linear transformation and $\tau' E_i = F_i$ for each i, then by equations (1),

$$\tau' V = \sum_{i=1}^{n} x_i \tau' E_i = \sum_{i=1}^{n} x_i F_i = \tau V.$$

Thus, $\tau' V = \tau V$ for all $V \in \mathscr{V}$ and $\tau' = \tau$.

This theorem states that a linear transformation of $\mathscr{V}$ into $\mathscr{W}$ is completely determined by the images of a set of basis vectors of $\mathscr{V}$ and that an arbitrary choice of these images defines a linear transformation.

5.2 The Algebra of Linear Transformations

Let $\mathscr{V}$ and $\mathscr{W}$ be vector spaces over $\mathscr{F}$ and let $\sigma:\mathscr{V} \to \mathscr{W}$ and $\tau:\mathscr{V} \to \mathscr{W}$ be linear transformations. We define the *sum* $\sigma + \tau$ of these linear transformations to be the mapping from $\mathscr{V}$ into $\mathscr{W}$ defined by

$$(6) \qquad\qquad (\sigma + \tau)V = \sigma V + \tau V.$$

We see that $\sigma + \tau$ is a linear transformation because

$$
\begin{aligned}
(\sigma + \tau)(V_1 + V_2) &= \sigma(V_1 + V_2) + \tau(V_1 + V_2) \quad &\text{by (6)}\\
&= \sigma V_1 + \sigma V_2 + \tau V_1 + \tau V_2 \quad &\text{by (1)}\\
&= \sigma V_1 + \tau V_1 + \sigma V_2 + \tau V_2\\
&= (\sigma + \tau)V_1 + (\sigma + \tau)V_2,
\end{aligned}
$$

and, for any scalar k,

$$
\begin{aligned}
(\sigma + \tau)kV &= \sigma(kV) + \tau(kV)\\
&= k(\sigma V) + k(\tau V)\\
&= k(\sigma V + \tau V)\\
&= k(\sigma + \tau)V.
\end{aligned}
$$

Similarly, if k is any scalar we define the mapping $k\sigma$ by

$$(7) \qquad\qquad (k\sigma)V = k(\sigma V)$$

and check that $k\sigma$ is a linear transformation:

$$
\begin{aligned}
(k\sigma)(V_1 + V_2) &= k[\sigma(V_1 + V_2)] = k(\sigma V_1 + \sigma V_2)\\
&= k(\sigma V_1) + k(\sigma V_2) = (k\sigma)V_1 + (k\sigma)V_2
\end{aligned}
$$

and

$$
\begin{aligned}
(k\sigma)(k'V) &= k[\sigma(k'V)] = k[k'(\sigma V)] = kk'(\sigma V)\\
&= k'[k(\sigma V)] = k'[(k\sigma)V].
\end{aligned}
$$

Theorem 5.5. *If $\mathscr{V}$, $\mathscr{W}$ are vector spaces over $\mathscr{F}$, the set $\mathscr{L}(\mathscr{V}, \mathscr{W})$ of all linear transformations from $\mathscr{V}$ into $\mathscr{W}$ is a vector space over $\mathscr{F}$ when addition and scalar multiplication are defined by (6) and (7).*

Proof. The proof consists in verifying the postulates V1 through V8 (Section 1.4) for a vector space. This is routine and so the full details will be omitted. We have already shown that $\mathscr{L}(\mathscr{V}, \mathscr{W})$ is closed under addition and scalar multiplication. Addition in $\mathscr{L}(\mathscr{V}, \mathscr{W})$ is commutative and associative because addition in $\mathscr{W}$ is. The zero element of $\mathscr{L}(\mathscr{V}, \mathscr{W})$ is the "zero mapping" O defined by

$$OV = O'$$

for all V in $\mathscr{V}$, and is easily seen to be linear. If $\sigma \in \mathscr{L}(\mathscr{V}, \mathscr{W})$, then $-\sigma$ is defined by

$$(-\sigma)V = -(\sigma V).$$

One checks that $-\sigma$ is linear and $\sigma + (-\sigma) = O$. The verification of the remaining postulates is easy and is left to the student.

5.3 Matrix Representations

Now let $E_1, E_2, \ldots, E_n$ be a basis of $\mathscr{V}$, let $F_1, F_2, \ldots, F_m$ be a basis of $\mathscr{W}$, and let $\sigma \in \mathscr{L}(\mathscr{V}, \mathscr{W})$. Since $\sigma E_i \in \mathscr{W}$, we have

$$
\begin{aligned}
\sigma E_1 &= a_{11}F_1 + a_{21}F_2 + \cdots + a_{m1}F_m, \\
\sigma E_2 &= a_{12}F_1 + a_{22}F_2 + \cdots + a_{m2}F_m, \\
&\cdots\cdots\cdots\cdots\cdots\cdots\cdots\cdots\cdots\cdots \\
\sigma E_n &= a_{1n}F_1 + a_{2n}F_2 + \cdots + a_{mn}F_m.
\end{aligned}
$$

(8)

Now if $V = \sum_{i=1}^{n} x_i E_i$ is any vector in $\mathscr{V}$ we have

$$\sigma V = \sum_{i=1}^{n} x_i \sigma E_i = \sum_{i=1}^{m} y_i F_i,$$

where by (8),

(9)
$$y_i = \sum_{j=1}^{n} a_{ij} x_j.$$

Definition. *The matrix $A_\sigma = [a_{ij}]$ of the coefficients in (9) is called the* matrix *of the linear transformation σ relative to the E- and F-bases. It should be noted that A_σ is the transpose of the matrix of the coefficients of the F's in (8).*

We state formally, in matrix notation, the result given by equation (9).

Theorem 5.6. *Let $\mathscr{V}$ and $\mathscr{W}$ be vector spaces over $\mathscr{F}$ with bases $E_1, \ldots, E_n$ and $F_1, \ldots, F_m$ and let $\sigma: \mathscr{V} \to \mathscr{W}$ be a linear transformation. Let $V \in \mathscr{V}$*

and let X be the coordinate (column) vector of V relative to the E-basis and Y the coordinate vector of σV relative to the F-basis. Then

$$(10) \qquad\qquad Y = A_\sigma X,$$

where A_σ is the matrix of σ relative to the E- and F-bases.

Since (10) is simply the matrix form of (9), this theorem has already been proved. Strictly, the matrix A_σ should be designated $A_\sigma^{(E,F)}$ to show its dependence on the bases as well as on σ. When no confusion is likely, we shall drop the superscripts.

Theorem 5.7. *Let $\mathscr{V}$ and $\mathscr{W}$ be vector spaces over $\mathscr{F}$, of dimensions n and m and having bases $E_1, \ldots, E_n$ and $F_1, \ldots, F_m$, respectively. Let $\sigma \in \mathscr{L}(\mathscr{V}, \mathscr{W})$ and let A_σ be the matrix of σ relative to these two bases. Then the mapping*

$$(11) \qquad\qquad \alpha: \sigma \to A_\sigma$$

is an isomorphism of the space $\mathscr{L}(\mathscr{V}, \mathscr{W})$ onto the vector space $\mathscr{M}$ of all $m \times n$ matrices over $\mathscr{F}$.

Proof. By Theorem 5.4 there exists a linear transformation σ for arbitrary choice of the scalars a_{ij} in equations (8). Hence, every $m \times n$ matrix over $\mathscr{F}$ is the matrix of a linear transformation and the mapping α is onto $\mathscr{W}$. If $A_\sigma = A_\tau$, then by equations (8),

$$(\sigma - \tau)E_i = \sigma E_i - \tau E_i = O \qquad (i = 1, 2, \ldots, n).$$

Hence, $(\sigma - \tau)V = O$ for all $V \in \mathscr{V}$ and $\sigma - \tau = O$ or $\sigma = \tau$. Therefore, α is one-to-one. Now if

$$\sigma E_i = \sum_{i=1}^{m} a_{ij} F_i,$$

$$\tau E_i = \sum_{i=1}^{m} b_{ij} F_i,$$

then

$$(\sigma + \tau)E_i = \sigma E_i + \tau E_i$$

$$= \sum_{i=1}^{m} (a_{ij} + b_{ij}) F_i$$

and, therefore,

$$A_{\sigma+\tau} = A_\sigma + A_\tau.$$

Similarly,

$$(k\sigma)E_i = k(\sigma E_i) = \sum_{i=1}^{m} k a_{ij} F_i$$

and, hence,

$$A_{k\sigma} = k A_\sigma.$$

Therefore, $\alpha(\sigma + \tau) = \alpha\sigma + \alpha\tau$, $\alpha(k\sigma) = k(\alpha\sigma)$ and α is an isomorphism.

Corollary. *If $\mathcal{V}$ has dimension n and $\mathcal{W}$ has dimension m, then $\mathcal{L}(\mathcal{V}, \mathcal{W})$ has dimension mn.*

Proof. By Theorem 1.13, Corollary 2, $\mathcal{L}(\mathcal{V}, \mathcal{W})$ has the same dimension as $\mathcal{M}$. Let E_{ij} be the $m \times n$ matrix with 1 in the ith row and jth column and 0 elsewhere. The mn matrices $E_{ij}(i = 1, \ldots, m; j = 1, \ldots, n)$ form a basis of $\mathcal{M}$ because they clearly span $\mathcal{M}$ and are linearly independent by an argument similar to that used in Theorem 1.5. Hence, dim $\mathcal{L}(\mathcal{V}, \mathcal{W})$ = dim $\mathcal{M}$ = mn.

Theorem 5.8. *Let $\mathcal{U}$, $\mathcal{V}$, and $\mathcal{W}$ be three vector spaces over the same field $\mathcal{F}$. If $\tau \in \mathcal{L}(\mathcal{U}, \mathcal{V})$ and $\sigma \in \mathcal{L}(\mathcal{V}, \mathcal{W})$, the product or composite mapping $\sigma\tau \in \mathcal{L}(\mathcal{U}, \mathcal{W})$. Now suppose the three spaces are finite dimensional with bases (E_i), (F_i), and (G_i). If A_σ is the matrix of σ relative to the E- and F-bases and A_τ is the matrix of τ relative to the F- and G-bases, the matrix of $\sigma\tau$ relative to the E- and G-bases is $A_{\sigma\tau} = A_\sigma A_\tau$.*

Proof. First, we must show that the composite mapping $\sigma\tau : \mathcal{U} \to \mathcal{W}$ is linear. This follows from the definition and the linearity of σ and τ:

$$(\sigma\tau)(V_1 + V_2) = \sigma[\tau(V_1 + V_2)] = \sigma(\tau V_1 + \tau V_2)$$
$$= \sigma(\tau V_1) + \sigma(\tau V_2) = (\sigma\tau)V_1 + (\sigma\tau)V_2$$

and

$$(\sigma\tau)(kV) = \sigma[\tau(kV)] = \sigma[k(\tau V)] = k[\sigma(\tau V)] = k[(\sigma\tau)V].$$

Now let $U \in \mathcal{U}$ and let X be the coordinate vector of U relative to the E-basis, Y the coordinate vector of τU relative to the F-basis, and Z the coordinate vector of $(\sigma\tau)U = \sigma(\tau U)$ relative to the G-basis. By Theorem 5.6, $Y = A_\tau X$ and $Z = A_\sigma Y = A_\sigma A_\tau X$. But $Z = A_{\sigma\tau}X$ and, therefore,

(12) $$A_{\sigma\tau}X = A_\sigma A_\tau X.$$

Since U, and hence X, is arbitrary, (12) holds for all X in $\mathcal{V}_n(\mathcal{R})$. If we substitute for X each of the column vectors of the identity matrix I, (12) gives (Theorem 2.5b)

$$A_{\sigma\tau}I = A_\sigma A_\tau I$$

or $A_{\sigma\tau} = A_\sigma A_\tau$, as required.

A special case of interest occurs when $\mathcal{W} = \mathcal{V}$. In this case, if $\sigma \in \mathcal{L}(\mathcal{V}, \mathcal{V})$ we express both V and σV in terms of the same basis $E_1, \ldots, E_n$ of $\mathcal{V}$. Thus, if $\sigma E_i = \sum_{i=1}^{n} a_{ij}E_i$, then $A_\sigma = [a_{ij}]$ is called the matrix of σ relative to the E-basis, and $Y = A_\sigma X$ where X and Y are the coordinate vectors of V and σV relative to the E-basis.

Exercise 5.1

1. Let E_1, E_2, E_3 be the standard basis of $\mathscr{V}_3(\mathscr{R})$, and F_1, F_2 the standard basis of $\mathscr{V}_2(\mathscr{R})$. Write the matrix A_τ of each of the following linear transformations $\tau: \mathscr{V}_3(\mathscr{R}) \to \mathscr{V}_2(\mathscr{R})$ relative to the E- and F-bases. Also write the equations relating the E-coordinates of a vector V with the F-coordinates of τV.

   (a) $\tau E_1 = F_1$,   (b) $\tau E_1 = F_2$,   (c) $\tau E_1 = [1, 2]$,
   $\quad \tau E_2 = F_2$, $\quad \tau E_2 = F_1$, $\quad \tau E_2 = [-5, 1]$,
   $\quad \tau E_3 = O$. $\quad \tau E_3 = F_1 + F_2$. $\quad \tau E_3 = [7, 3]$.

   (d) $\tau E_1 = \tau E_2 = \tau E_3 = O$.   (e) $\tau E_1 = \tau E_2 = \tau E_3 = [4, -3]$.

2. With the same notation as in Problem 1, write the matrix A_σ of each of the following linear transformations $\sigma: \mathscr{V}_2(\mathscr{R}) \to \mathscr{V}_3(\mathscr{R})$, relative to the F- and E-bases.

   (a) $\sigma F_1 = E_1$,   (b) $\sigma F_1 = [1, 2, 3]$,   (c) $\sigma F_1 = O$,
   $\quad \sigma F_2 = E_3$. $\quad \sigma F_2 = [2, -1, 4]$. $\quad \sigma F_2 = E_1$.

   (d) $\sigma F_1 = \sigma F_2 = [1, 2, -6]$.   (e) $\sigma F_1 = \sigma F_2 = O$.

3. Let $\tau: \mathscr{V}_3(\mathscr{R}) \to \mathscr{V}_2(\mathscr{R})$ be defined as in Problem 1(b) and let $\sigma: \mathscr{V}_2(\mathscr{R}) \to \mathscr{V}_3(\mathscr{R})$ be defined as in Problem 2(b). Compute $(\sigma\tau)E_1$, $(\sigma\tau)E_2$, and $(\sigma\tau)E_3$ and write the matrix $A_{\sigma\tau}$ of $\sigma\tau: \mathscr{V}_3(\mathscr{R}) \to \mathscr{V}_3(\mathscr{R})$ relative to the E-basis, that is, $A_{\sigma\tau}^{(E,E)}$. Verify that $A_{\sigma\tau} = A_\sigma A_\tau$.

4. Repeat Problem 3 using τ and σ as defined in:
   (a) Problems 1(c) and 2(b).
   (b) Problems 1(e) and 2(c).
   (c) Problems 1(d) and 2(b).
   (d) Problems 1(a) and 2(c).

5. Let σ and τ be defined as in Problem 3. Compute $(\tau\sigma)F_1$, $(\tau\sigma)F_2$ and write the matrix $A_{\tau\sigma}$ of $\tau\sigma: \mathscr{V}_2(\mathscr{R}) \to \mathscr{V}_2(\mathscr{R})$ relative to the F-basis. Verify that $A_{\tau\sigma} = A_\tau A_\sigma$.

6. Repeat Problem 5 using τ and σ as defined in Problem 4.

7. Let A be an $m \times n$ matrix over $\mathscr{F}$ and let $\tau: \mathscr{V}_n(\mathscr{F}) \to \mathscr{V}_m(\mathscr{F})$ be defined by $\tau X = AX$ (see Example 2, Section 5.1). Prove that:
   (a) τ is onto if and only if rank $A = m$.
   (b) τ is one-to-one if and only if rank $A = n$.
   (c) τ is invertible if and only if $m = n = \text{rank } A$.

8. Let $\mathscr{V}$ be a vector space over $\mathscr{F}$ with basis $E_1, E_2, \ldots, E_n$, and let $\sigma: \mathscr{V} \to \mathscr{V}$ be any linear transformation of $\mathscr{V}$ into itself. Let A_σ be the matrix of σ relative to the E-basis, that is, $A_\sigma^{(E,E)}$. Prove that σ is invertible if and only if A_σ is nonsingular and, in this case, $A_{\sigma^{-1}} = A_\sigma^{-1}$.

9. Let X be any vector of $\mathscr{V}_3(\mathscr{R})$ and let τX be the orthogonal projection of the vector X onto the plane whose equation is $x_1 - 2x_2 + 2x_3 = 0$.
   (a) Prove geometrically or otherwise that τ is a linear transformation.
   (b) Find the matrix of τ relative to the standard basis of $\mathscr{V}_3(\mathscr{R})$.

(c) Choose a basis F_1, F_2, F_3 of $\mathscr{V}_3(\mathscr{R})$ such that F_1 and F_2 lie in the plane $x_1 - 2x_2 + 2x_3 = 0$ and F_3 is perpendicular to this plane.

(d) Write the matrix of τ relative to the F-basis.

5.4 Linear Transformations of a Vector Space into Itself

We now specialize the results of the last two sections to the important particular case in which the range space $\mathscr{W}$ of a linear transformation τ is the same as the domain $\mathscr{V}$. By Theorem 5.5 the set $\mathscr{L}(\mathscr{V}, \mathscr{V})$ of all linear transformations of $\mathscr{V}$ into itself is a vector space. If $\mathscr{V}$ has finite dimension n, then by Theorem 5.7 and its corollary, $\mathscr{L}(\mathscr{V}, \mathscr{V})$ has dimension n^2 and is isomorphic to the space $\mathscr{F}_n$ of all $n \times n$ matrices over $\mathscr{F}$. Given any basis $E_1, \ldots, E_n$, the matrix $A_\tau^{(E,E)}$, in the notation of Section 5.3, is called the matrix of τ relative to the E-basis. It will be written simply A_τ when only one basis is under consideration, or $A_\tau^{(E)}$ when dependence on the basis (E) has to be specified.

If σ, $\tau \in \mathscr{L}(\mathscr{V}, \mathscr{V})$, then $\sigma\tau$ is always defined and $\sigma\tau \in \mathscr{L}(\mathscr{V}, \mathscr{V})$. Thus, in $\mathscr{L}(\mathscr{V}, \mathscr{V})$ we have not only addition and scalar multiplication but also a multiplication defined. In addition to the postulates V1 through V8 for a vector space (Section 1.4), these three operations satisfy the following.

Associative Laws

A1. If $k \in \mathscr{F}$, σ, $\tau \in \mathscr{L}(\mathscr{V}, \mathscr{V})$, then

$$(k\sigma)\tau = k(\sigma\tau) = \sigma(k\tau).$$

A2. For all φ, σ, $\tau \in \mathscr{L}(\mathscr{V}, \mathscr{V})$, $\varphi(\sigma\tau) = (\varphi\sigma)\tau$.

Distributive Laws

For all φ, σ, $\tau \in \mathscr{L}(\mathscr{V}, \mathscr{V})$:

D1. $\varphi(\sigma + \tau) = \varphi\sigma + \varphi\tau$.
D2. $(\sigma + \tau)\varphi = \sigma\varphi + \tau\varphi$.

These laws are easily verified if we recall that two mappings with the same domain and range are equal if and only if they have the same effect on every element in their domain. Thus, for all $V \in \mathscr{V}$,

$$\sigma(k\tau)V = \sigma[(k\tau)V] = \sigma[k(\tau V)] = k[\sigma(\tau V)]$$
$$= k[(\sigma\tau)V] = [k(\sigma\tau)]V,$$

so that $\sigma(k\tau) = k(\sigma\tau)$. Similarly, $(k\sigma)\tau = k(\sigma\tau)$, thus proving A1. The associative law A2 was proved for mappings in Section 1.2. For D1,

$$\varphi(\sigma + \tau)V = \varphi[(\sigma + \tau)V] = \varphi(\sigma V + \tau V) = \varphi(\sigma V) + \varphi(\tau V)$$
$$= (\varphi\sigma)V + (\varphi\tau)V = (\varphi\sigma + \varphi\tau)V.$$

Hence, $\varphi(\sigma + \tau) = \varphi\sigma + \varphi\tau$ and a similar proof gives D2.

Definition. *A* linear associative algebra *over a field $\mathscr{F}$ (briefly an algebra over $\mathscr{F}$) is a set $\mathscr{A}$ of elements such that:*

(1) *$\mathscr{A}$ is a vector space over $\mathscr{F}$.*
(2) *A multiplication is defined on the elements of A such that if σ, $\tau \in \mathscr{A}$, $\sigma\tau \in \mathscr{A}$, and the four laws* A1, A2, D1, D2 *hold.*

It is clear from the preceding that the set $\mathscr{L}(\mathscr{V}, \mathscr{V})$ of all linear transformations of $\mathscr{V}$ into itself is an algebra over $\mathscr{F}$. It is also clear from the laws of matrix algebra that the set of all $n \times n$ matrices over $\mathscr{F}$ is an algebra over $\mathscr{F}$.

Definition. *If $\mathscr{A}$ and $\mathscr{B}$ are two algebras over the same field $\mathscr{F}$, a mapping $\alpha: \mathscr{A} \to \mathscr{B}$ is an (algebra) isomorphism if:*

(1) *α is an isomorphism of the vector space $\mathscr{A}$ onto the vector space $\mathscr{B}$.*
(2) *If A_1, $A_2 \in \mathscr{A}$, then $\alpha(A_1 A_2) = \alpha(A_1)\alpha(A_2)$.*

From Theorems 5.7 and 5.8 we immediately get the following.

Theorem 5.9. *If $\mathscr{V}$ is an n-dimensional vector space over $\mathscr{F}$, the algebra $\mathscr{L}(\mathscr{V}, \mathscr{V})$ of all linear transformations of $\mathscr{V}$ into $\mathscr{V}$ is isomorphic to the algebra $\mathscr{F}_n$ of all $n \times n$ matrices over $\mathscr{F}$.*

Proof. Let $E_1, \ldots, E_n$ be any basis of $\mathscr{V}$. Let $\sigma \in \mathscr{L}(\mathscr{V}, \mathscr{V})$ and let A_σ be the matrix of σ relative to the E-basis. By Theorem 5.7 the mapping $\alpha : \sigma \to A_\sigma$ is an isomorphism of the vector space $\mathscr{L}(\mathscr{V}, \mathscr{V})$ onto the vector space $\mathscr{F}_n$. In Theorem 5.8 we now take $\mathscr{U} = \mathscr{W} = \mathscr{V}$ and $G_i = F_i = E_i$ and we have, for any σ, $\tau \in \mathscr{L}(\mathscr{V}, \mathscr{V})$, $A_{\sigma\tau} = A_\sigma A_\tau$, or $\alpha(\sigma\tau) = \alpha(\sigma)\alpha(\tau)$ and, hence, α is an algebra isomorphism.

We note that the algebra $\mathscr{L}(\mathscr{V}, \mathscr{V})$ contains a multiplicative identity element I defined by $IV = V$ for all V in $\mathscr{V}$. It is clear that I satisfies (1) and is therefore linear. Also, from equations (8) it follows that the matrix A_I of I, relative to *any* basis, is the identity matrix I. We shall therefore use the same symbol I both for the identity transformation and the identity matrix.

Theorem 5.10. *Let* $\sigma: \mathscr{V} \to \mathscr{V}$ *be any linear transformation of a finite dimensional vector space* $\mathscr{V}$ *into itself, and let* A_σ *be the matrix of* σ *relative to a fixed basis* $E_1, \ldots, E_n$. *Then* σ *is invertible if and only if* A_σ *is nonsingular, and if* σ *is invertible, the matrix of* σ^{-1} *relative to the E-basis is* $A_{\sigma^{-1}} = A_\sigma^{-1}$.

Proof. If σ is invertible, then $\sigma\sigma^{-1} = I$ and, by Theorem 5.9, $A_\sigma A_{\sigma^{-1}} = A_I = I$. Hence, A_σ is nonsingular and $A_{\sigma^{-1}} = A_\sigma^{-1}$. Conversely, if A_σ is nonsingular it has an inverse A_σ^{-1} in $\mathscr{F}_n$. By Theorem 5.9, A_σ^{-1} is the matrix of a linear transformation $\tau \in \mathscr{L}(\mathscr{V}, \mathscr{V})$. Since $A_{\sigma\tau} = A_\sigma A_\tau = A_\sigma A_\sigma^{-1} = I$, it follows that $\sigma\tau = I$ and, similarly, $\tau\sigma = I$. Hence, by Theorem 1.0, Corollary 2, in Section 1.2, σ is invertible, $\sigma^{-1} = \tau$, and $A_{\sigma^{-1}} = A_\tau = A_\sigma^{-1}$.

In view of this theorem an invertible linear transformation of $\mathscr{V}$ into $\mathscr{V}$ is also called a nonsingular linear transformation.

The major concern of linear algebra is the study of linear transformations of a vector space into itself or *linear operators* on $\mathscr{V}$ as they are often called. Theorem 5.9 tells us that when $\mathscr{V}$ is finite dimensional, the algebra of linear operators on $\mathscr{V}$ is faithfully reflected in the algebra of $n \times n$ matrices over $\mathscr{F}$. Consequently, the linear operators can be studied by studying matrix algebra. There are two objections to this approach. First, the matrix A_σ of a linear operator σ on $\mathscr{V}$ depends not only on σ but on the particular basis of $\mathscr{V}$ chosen. Thus the matrix approach does not give a "coordinate free" treatment of operator theory. This is not too serious, since in the next section we shall see how a change of basis affects the matrix A_σ. The second objection to the matrix approach is that if $\mathscr{V}$ is not finite dimensional, matrix theory is for the most part no longer applicable, whereas most of the operator theory continues to hold when suitably generalized. In spite of these objections, we shall use primarily the matrix theory approach to the study of linear operators in $\mathscr{V}$ because it seems better suited to an elementary course and because many of the applications to other fields involve matrix properties and methods which therefore need to be emphasized.

As an illustration of the difference (and similarity) between the "matrix approach" and the "operator approach," we now translate the ideas and results of Section 2.7 into operator language.

Let $\mathscr{V}$ be an n-dimensional vector space over $\mathscr{F}$ and let $\tau \in \mathscr{L}(\mathscr{V}, \mathscr{V})$ be any linear operator. Let $\mathscr{N}$ be the null space of τ and let $\text{im } \tau = \{\tau V \mid V \in \mathscr{V}\}$ be the image space of τ. Of course, $\text{im } \tau$ is a subspace of $\mathscr{V}$ by Theorem 5.1.

Definition. *If* $\tau \in \mathscr{L}(\mathscr{V}, \mathscr{V})$, $\dim (\text{im } \tau)$ *is called the rank of* τ *and* $\dim \mathscr{N}$ *is called the nullity of* τ.

Theorem 5.11. *If τ is a linear operator on an n-dimensional vector space $\mathscr{V}$, then*

$$\text{rank } \tau + \text{nullity } \tau = n.$$

Proof.

Method 1. Choose any basis $E_1, \ldots, E_n$ of $\mathscr{V}$. Let $\alpha : \mathscr{V} \to \mathscr{V}_n(\mathscr{F})$ be the isomorphism that maps each $V \in \mathscr{V}$ on its coordinate vector $X = \alpha V$ relative to the E-basis. By Theorem 5.6,

(13)
$$\alpha(\tau V) = A_\tau X = A_\tau(\alpha V).$$

It follows from (13) that α maps the null space of τ onto the null space of A_τ and im τ onto the subspace $\{A_\tau X \mid X \in \mathscr{V}_n(\mathscr{R})\}$ which by Lemma 2.14 is the column space of A_τ. Hence (Theorem 1.13, Corollary 3),

$$\dim (\text{null space of } \tau) = \dim (\text{null space of } A_\tau)$$
and
$$\dim (\text{im } \tau) = \dim (\text{column space of } A_\tau).$$

The theorem now follows from Theorem 2.15.

Method 2. Let $\mathscr{N}$ be the null space of τ and let $s = \dim \mathscr{N}$. Choose a basis $V_1, \ldots, V_s, \ldots, V_n$ of $\mathscr{V}$ such that $V_1, \ldots, V_s$ is a basis for $\mathscr{N}$. Since $\tau V_i = O$ for $i \leq s$, it follows that im τ is spanned by the vectors $\tau V_{s+1}, \ldots, \tau V_n$. But these vectors are linearly independent, for if

$$c_{s+1} \tau V_{s+1} + \cdots + c_n \tau V_n = O,$$
then
$$\tau(c_{s+1} V_{s+1} + \cdots + c_n V_n) = O$$
and
$$c_{s+1} V_{s+1} + \cdots + c_n V_n) \in \mathscr{N}$$

which implies, as in the proof of Theorem 2.15, that $c_{s+1} = \cdots = c_n = 0$. Thus, im τ has dimension $n - s$ and $\dim N + \dim (\text{im } \tau) = n$ as required.

Since Theorem 2.15 clearly follows from Theorem 5.11, our first proof shows that the two theorems are actually equivalent, that is, each implies the other.

Corollary. *If $\mathscr{V}$ is finite dimensional and $\tau \in \mathscr{L}(\mathscr{V}, \mathscr{V})$, then im $\tau = \mathscr{V}$ if and only if the null space of τ is O. Thus, if $\mathscr{V}$ is finite dimensional a linear transformation $\tau : \mathscr{V} \to \mathscr{V}$ is invertible if it is one-to-one.*

We close this section with a few examples.

Example 1. Let $\mathscr{P}$ be the vector space over $\mathscr{R}$ of all polynomials in x with coefficients in $\mathscr{R}$. Define $\delta: \mathscr{P} \to \mathscr{P}$ by

$$\delta p(x) = p'(x),$$

where $p'(x)$ is the derivative of $p(x)$. Since $[p(x) + q(x)]' = p'(x) + q'(x)$ and $[kp(x)]' = kp'(x)$ for all $k \in \mathscr{R}$, δ is a linear transformation. The null space $\mathscr{N}$ of δ is the subspace of $\mathscr{P}$ consisting of all constant polynomials. Thus $\mathscr{N} \neq 0$. On the other hand, im $\tau = \mathscr{P}$ because every polynomial in x is the derivative of a polynomial. This example shows that the corollary to Theorem 5.11 does not hold if $\mathscr{V}$ is not finite dimensional. On the other hand, in the space $\mathscr{P}_n$ of polynomials over $\mathscr{R}$ of degree $\leq n$, im $\delta = \mathscr{P}_{n-1}$ and the corollary holds.

Example 2. Let $\mathscr{V} = \mathscr{V}_2(\mathscr{R})$ and let E_1, E_2 be the standard basis. If $\tau \in \mathscr{L}(\mathscr{V}, \mathscr{V})$ is defined by

$$\tau E_1 = 2E_1,$$

$$\tau E_2 = 3E_2,$$

then

$$\tau[x, y] = [2x, 3y],$$

and the point (x, y) is mapped by τ onto (x', y') where

$$x' = 2x,$$

$$y' = 3y.$$

If (x, y) is on the circle $x^2 + y^2 = 1$, then (x', y') satisfies the equation

$$\frac{x'^2}{4} + \frac{y'^2}{9} = 1$$

and we say that the circle $x^2 + y^2 = 1$ is mapped by τ onto the ellipse

$$\frac{x^2}{4} + \frac{y^2}{9} = 1.$$

Definition. *Let $\mathscr{V}$ be an n-dimensional vector space over $\mathscr{F}$ and let $\tau \in \mathscr{L}(\mathscr{V}, \mathscr{V})$. If there exist n linearly independent vectors $E_1, \ldots, E_n$ in $\mathscr{V}$ and n positive scalars $a_1, \ldots, a_n$ such that*

(27) $$\tau E_i = a_i E_i,$$

then τ is called a magnification.

Since $E_1, \ldots, E_n$ is a basis of $\mathscr{V}$, by (27) the matrix of τ relative to the E-basis is the diagonal matrix

$$A_\tau = \begin{bmatrix} a_1 & 0 & \cdots & 0 \\ 0 & a_2 & \cdots & 0 \\ \cdots & \cdots & \cdots & \cdots \\ 0 & 0 & \cdots & a_n \end{bmatrix}.$$

The transformation in $\mathscr{V}_2(\mathscr{R})$ discussed in Example 2 is a magnification. Geometrically a magnification is an "expansion" (or contraction) of the space $\mathscr{V}$ in each of the directions E_i by a factor a_i.

Example 3. Let $\mathscr{V} = \mathscr{V}_3(\mathscr{R})$ and let E_1, E_2, E_3 be the standard basis. If $\tau \in \mathscr{L}(\mathscr{V}, \mathscr{V})$ is defined by

$$\tau E_1 = E_1,$$

$$\tau E_2 = E_2,$$

$$\tau E_3 = O,$$

then $\tau[x_1, x_2, x_3] = [x_1, x_2, 0]$ and τ is a *projection* of $\mathscr{V}_3(\mathscr{R})$ onto the subspace spanned by E_1, E_2. Also see Example 4, Section 5.1 and Problem 9, Exercise 5.1.

Projections in a more general setting will be discussed in Section 8.10.

Example 4. Let $\mathscr{V} = \mathscr{V}_2(\mathscr{R})$ and let E_1, E_2 be the standard basis. Let τX be the vector obtained by rotating X counter-clockwise through an angle θ. Suppose $\|X\| = r$ and the angle between X and E_1 is α. Then if $X = [x_1, x_2]$, $x_1 = r \cos \alpha$, $x_2 = r \sin \alpha$ and if $\tau X = [x_1', x_2']$,

$$x_1' = r \cos (\alpha + \theta) = r \cos \alpha \cos \theta - r \sin \alpha \sin \theta = x_1 \cos \theta - x_2 \sin \theta,$$

$$x_2' = r \sin (\alpha + \theta) = r \sin \alpha \cos \theta + r \cos \alpha \sin \theta = x_1 \sin \theta + x_2 \cos \theta,$$

or $\tau X = A_\tau X$ where

$$A_\tau = \begin{bmatrix} \cos \theta & -\sin \theta \\ \sin \theta & \cos \theta \end{bmatrix}.$$

Thus, τ is a linear transformation with matrix A_τ relative to the E-basis and

$$\tau E_1 = (\cos \theta) E_1 + (\sin \theta) E_2,$$

$$\tau E_2 = (-\sin \theta) E_1 + (\cos \theta) E_2.$$

Similarly, a rotation of $\mathscr{V}_3(\mathscr{R})$ about a fixed axis is a linear transformation. Rotations in an arbitrary real inner product space will be defined in Section 5.6.

Example 5. Let $\mathscr{V} = \mathscr{V}_3(\mathscr{R})$, let E_1, E_2, E_3 be the standard basis, and let $\tau \in \mathscr{L}(\mathscr{V}, \mathscr{V})$ be defined by $\tau E_1 = E_1$, $\tau E_2 = E_2$, and $\tau E_3 = -E_3$. Then τ is a linear transformation whose matrix relative to the E-basis is

$$A_\tau = \begin{bmatrix} 1 & 0 & 0 \\ 0 & 1 & 0 \\ 0 & 0 & -1 \end{bmatrix}.$$

Clearly, $\tau[x_1, x_2, x_3] = [x_1, x_2, -x_3]$ so that each vector is mapped by τ on its reflection in the $x_1 x_2$-plane. Such a linear transformation is called a *reflection*. Reflections in any real inner product space will be defined in Section 5.6.

5.5 Change of Basis

Let $\mathscr{V}$ be a finite dimensional vector space over $\mathscr{F}$ and let $\tau \in \mathscr{L}(\mathscr{V}, \mathscr{V})$. We shall now prove a theorem that tells us how the matrices of τ relative to different bases of $\mathscr{V}$ are related.

Theorem 5.12. *Let $E_1, \ldots, E_n$ and $F_1, \ldots, F_n$ be two bases of $\mathscr{V}$ and let $\tau \in \mathscr{L}(\mathscr{V}, \mathscr{V})$. If $A_\tau^{(E)}$ is the matrix of τ relative to the E-basis and $A_\tau^{(F)}$ is the matrix of τ relative to the F-basis, then*

(28) $$A_\tau^{(F)} = P^{-1} A_\tau^{(E)} P,$$

where P is the transition matrix from the E- to the F-basis.

Proof. Let $V \in \mathscr{V}$ and let X be the coordinate vector of V relative to the E-basis and X' the coordinate vector of V relative to the F-basis. By Theorem 4.18, $X = PX'$. Now τV has E-coordinate vector $A_\tau^{(E)} X$ and F-coordinate vector $A_\tau^{(F)} X'$ and hence, again by Theorem 4.18,

(29) $$A_\tau^{(E)} X = P A_\tau^{(F)} X' = P A_\tau^{(F)} P^{-1} X.$$

Since (29) holds for all X in $\mathscr{V}_n(\mathscr{F})$, it follows that (see proof of Theorem 5.8) $A_\tau^{(E)} = P A_\tau^{(F)} P^{-1}$ or $A_\tau^{(F)} = P^{-1} A_\tau^{(E)} P$ as required.

Since Theorem 5.12 is very important for subsequent developments, we illustrate it with an example.

Example. Choose $F_1 = [1, 2]$, $F_2 = [3, -1]$ as a basis of $\mathscr{V}_2(\mathscr{R})$ and let $E_1 = [1, 0]$, $E_2 = [0, 1]$ be the standard basis. Since

(30)
$$F_1 = E_1 + 2E_2,$$
$$F_2 = 3E_1 - E_2,$$

the transition matrix from the E- to the F-basis is

$$P = \begin{bmatrix} 1 & 3 \\ 2 & -1 \end{bmatrix}$$

and its inverse is

$$P^{-1} = \begin{bmatrix} \frac{1}{7} & \frac{3}{7} \\ \frac{2}{7} & -\frac{1}{7} \end{bmatrix}.$$

Now let τ be the linear operator in $\mathscr{V}_2(\mathscr{R})$ defined by $\tau F_1 = 2F_1$, $\tau F_2 = 3F_2$ so that the matrix of τ relative to the F-basis is

$$(31) \qquad A_\tau^{(F)} = \begin{bmatrix} 2 & 0 \\ 0 & 3 \end{bmatrix}.$$

From (30), solving for E_1, E_2 we get

$$E_1 = \tfrac{1}{7}F_1 + \tfrac{2}{7}F_2,$$
$$E_2 = \tfrac{3}{7}F_1 - \tfrac{1}{7}F_2$$

and hence

$$\tau E_1 = \tfrac{1}{7}\tau F_1 + \tfrac{2}{7}\tau F_2 = \tfrac{2}{7}F_1 + \tfrac{6}{7}F_2,$$
$$\tau E_2 = \tfrac{3}{7}\tau F_1 - \tfrac{1}{7}\tau F_2 = \tfrac{6}{7}F_1 - \tfrac{3}{7}F_2.$$

Substituting from (30), this gives

$$\tau E_1 = \tfrac{20}{7}E_1 - \tfrac{2}{7}E_2,$$
$$\tau E_2 = -\tfrac{3}{7}E_1 + \tfrac{15}{7} E_2$$

and hence the matrix of τ relative to the E-basis is

$$(32) \qquad A_\tau^{(E)} = \begin{bmatrix} \frac{20}{7} & -\frac{3}{7} \\ -\frac{2}{7} & \frac{15}{7} \end{bmatrix}.$$

We now check that

$$P^{-1}A_\tau^{(F)}P = \begin{bmatrix} \frac{1}{7} & \frac{3}{7} \\ \frac{2}{7} & -\frac{1}{7} \end{bmatrix}\begin{bmatrix} 2 & 0 \\ 0 & 3 \end{bmatrix}\begin{bmatrix} 1 & 3 \\ 2 & -1 \end{bmatrix}$$

$$= \begin{bmatrix} \frac{1}{7} & \frac{3}{7} \\ \frac{2}{7} & -\frac{1}{7} \end{bmatrix}\begin{bmatrix} 2 & 6 \\ 6 & -3 \end{bmatrix} = \begin{bmatrix} \frac{20}{7} & -\frac{3}{7} \\ -\frac{2}{7} & \frac{15}{7} \end{bmatrix} = A_\tau^{(E)},$$

and the statement of Theorem 5.12 is verified in this case. We make a further remark about this example. To say that the matrix of τ relative to the standard basis is given by (32) reveals no obvious information about the geometric nature of the transformation. But to say that the matrix of τ relative to the F-basis is given by (31) tells us at once that τ is a magnification that multiplies vectors in the direction of F_1 by 2 and vectors in the direction of F_2 by 3. This illustrates the fact that although, algebraically speaking, a linear transformation is completely defined when its matrix relative to any given basis is known, the geometric nature of the transformation is not usually revealed automatically by this matrix. The problem of determining the geometric effect on the space of a given linear transformation τ usually involves choosing a basis of the space in such a way that the matrix of τ relative to the chosen basis takes some particularly simple form which will reveal geometric properties of τ to us. Usually what is desired is a matrix in diagonal form, although this is not always possible. The importance of Theorem 5.12 lies partly in the fact that it gives us an algebraic method of investigating the geometric effect of a given linear transformation. A linear transformation τ is completely defined by its matrix A_τ relative to a fixed basis but is equally well defined by its matrix $P^{-1}A_\tau P$ relative to any other basis. Therefore our problem is reduced to finding a nonsingular matrix P such that the matrix $P^{-1}AP$ gives us as much information as possible about the transformation τ. Much of the rest of this book is devoted to the problem of investigating various special types of linear transformations in this way. We shall be particularly interested in determining for exactly what matrices A a nonsingular matrix P can be found such that $P^{-1}AP$ is a diagonal matrix. This problem will be treated in Chapter 6.

Exercise 5.2

1. Let E_1, E_2 be the standard basis of $\mathcal{V}_2(\mathcal{R})$ and consider the linear operators on $\mathcal{V}_2(\mathcal{R})$ defined by:

(i) $\sigma E_1 = E_1 - E_2$,   (ii) $\tau E_1 = O$,   (iii) $\varphi E_1 = -2E_1$,

   $\sigma E_2 = E_1 + E_2$. $\tau E_2 = 2E_1 + E_2$. $\varphi E_2 = 5E_2$.

(a) Write the matrices of σ, τ, φ relative to the E-basis.

(b) Find, directly from the definition, the matrices of σ, τ, and φ relative to the basis $F_1 = [1, -1]$, $F_2 = [1, 1]$.

(c) Find the matrices of σ, τ, φ relative to the F-basis by using Theorem 5.12, and check with your answers to (b).

2. Let σ, τ, φ be linear operators in $\mathcal{V}_2(\mathcal{R})$ with matrices relative to the standard basis:

$$A_\sigma = \begin{bmatrix} 4 & 6 \\ -2 & -3 \end{bmatrix}, \quad A_\tau = \begin{bmatrix} \dfrac{1}{2} & -\dfrac{\sqrt{3}}{2} \\ \dfrac{\sqrt{3}}{2} & \dfrac{1}{2} \end{bmatrix}, \quad A_\varphi = \begin{bmatrix} 7 & 3 \\ -10 & -4 \end{bmatrix}.$$

(a) Show that if $F_1 = [2, -1]$ and $F_2 = [3, -2]$, then $\sigma F_1 = F_1$ and $\sigma F_2 = O$.

(b) Find the matrix of σ relative to the basis F_1, F_2.

(c) Describe the geometric effect of σ.

(d) Show that τ is a rotation of the plane about the origin. What is the angle of rotation? (See Example 4, Section 5.4.)

(e) If $F_1 = [1, -2]$ and $F_2 = [3, -5]$, find φF_1 and φF_2.

(f) Find the matrix of φ relative to the basis F_1, F_2.

(g) What is the geometric effect of φ?

3. Let τ be the linear transformation that rotates every vector of $\mathscr{V}_2(\mathscr{R})$ counterclockwise through an acute angle whose tangent is $\frac{4}{3}$. Let σ be the linear transformation that multiples the first coordinate of every vector of $\mathscr{V}_2(\mathscr{R})$ by 3. Find the matrices of $\sigma\tau$ and $\tau\sigma$ relative to the standard basis.

4. If $\tau: \mathscr{V}_2(\mathscr{R}) \to \mathscr{V}_2(\mathscr{R})$ maps $E_1 = [1, 0]$ and $E_2 = [0, 1]$ onto the vectors $F_1 = [2, 5]$ and $F_2 = [-1, 6]$, respectively, find the matrices of τ and τ^{-1} relative to the basis E_1, E_2 and also relative to the basis F_1, F_2.

5. Suppose $\tau: \mathscr{V}_2(\mathscr{R}) \to \mathscr{V}_2(\mathscr{R})$ maps $F_1 = [2, -1]$ onto $G_1 = [4, 3]$ and $F_2 = [1, 3]$ onto $G_2 = [-1, -2]$. Find the matrix of τ relative to (a) the basis F_1, F_2; (b) the basis G_1, G_2; (c) the basis $E_1 = [1, 0]$, $E_2 = [0, 1]$.

6. Let $\tau: \mathscr{V}_2(\mathscr{R}) \to \mathscr{V}_2(\mathscr{R})$ be the linear transformation whose matrix relative to the standard basis is

$$\begin{bmatrix} -\frac{1}{2} & -\frac{5}{2} \\ -\frac{5}{2} & -\frac{1}{2} \end{bmatrix}.$$

Find the matrix of τ relative to a new basis obtained by rotating the original basis vectors through $45°$, and deduce from this new matrix the geometric effect of τ on the xy-plane.

7. Let $\tau: \mathscr{V}_n(\mathscr{F}) \to \mathscr{V}_n(\mathscr{F})$ be a linear transformation of $\mathscr{V}_n(\mathscr{F})$ and let A be its matrix relative to the standard basis $E_1, \ldots, E_n$. Show that the column vectors of A, are $\tau E_1, \ldots, \tau E_2$.

8. Find the matrix, relative to the basis

$$F_1 = [\tfrac{2}{3}, \tfrac{2}{3}, -\tfrac{1}{3}], \qquad F_2 = [\tfrac{1}{3}, -\tfrac{2}{3}, -\tfrac{2}{3}], \qquad F_3 = [\tfrac{2}{3}, -\tfrac{1}{3}, \tfrac{2}{3}]$$

of $\mathscr{V}_3(\mathscr{F})$, of the linear transformation $\tau: \mathscr{V}_3(\mathscr{R}) \to \mathscr{V}_3(\mathscr{R})$ whose matrix relative to the standard basis is

$$\begin{bmatrix} 2 & 0 & 0 \\ 0 & 4 & 0 \\ 0 & 0 & 3 \end{bmatrix}.$$

Hint. Note that the F-basis is orthonormal relative to the standard inner product and hence the transition matrix from the standard to the F-basis is orthogonal.

9. Let τ be a linear operator on $\mathscr{V}_2(\mathscr{R})$ whose matrix relative to the standard basis is

$$A_\tau = \begin{bmatrix} a & b \\ c & d \end{bmatrix}.$$

(a) Prove that if there are two linearly independent vectors F_1 and F_2 in $\mathscr{V}_2(\mathscr{R})$ and two (real) scalars k_1 and k_2 such that $\tau F_1 = k_1 F_1$ and $\tau F_2 = k_2 F_2$, then

$$P^{-1}A_\tau P = \begin{bmatrix} k_1 & 0 \\ 0 & k_2 \end{bmatrix},$$

where P is the transition matrix from the standard basis to the basis F_1, F_2.

(b) Prove that τ leaves some nonzero vector X invariant (i.e., $\tau X = X$) if and only if

$$\begin{vmatrix} a - 1 & b \\ c & d - 1 \end{vmatrix} = 0.$$

(c) Prove that τ maps some one-dimensional subspace of $\mathscr{V}_n(\mathscr{R})$ into itself if and only if for some real number k

$$\begin{vmatrix} a - k & b \\ c & d - k \end{vmatrix} = 0.$$

10. Find a linear transformation of $\mathscr{V}_2(\mathscr{R})$ into itself that will map the circle with equation $x^2 + y^2 = 1$ onto the ellipse

$$\frac{x^2}{16} + \frac{y^2}{9} = 1.$$

11. Find a linear transformation of $\mathscr{V}_2(\mathscr{R})$ into itself that will map the circle with equation $x^2 + y^2 = 1$ onto the ellipse $13x^2 + 10xy + 13y^2 = 1$.

12. If $E_1, \ldots, E_n$ and $F_1, \ldots, F_n$ are any two bases of a vector space $\mathscr{V}$ over $\mathscr{F}$ and if $\tau \in \mathscr{L}(\mathscr{V}, \mathscr{V})$, prove that $\det A_\tau^{(F)} = \det A_\tau^{(E)}$. Since $\det A_\tau$ therefore depends only on τ and not on the basis, it will be called the determinant of the linear transformation τ.

13. Let τ be any linear transformation of $\mathscr{V}_n(\mathscr{R})$ into itself and let $X_1, X_2, \ldots, X_n$ be n linearly independent vectors of $\mathscr{V}_n(\mathscr{R})$. Prove that the n-volume of the parallelepiped determined by $\tau X_1, \tau X_2, \ldots, \tau X_n$ is equal to the n-volume of the parallelepiped determined by $X_1, X_2, \ldots, X_n$ multiplied by $|\det A_\tau|$ where A_τ is the matrix of τ relative to any basis. Comment on the case when τ is singular.

14. Let τ be any linear transformation of the plane $\mathscr{V}_2(\mathscr{R})$ into itself. Prove that any bounded region A of the plane, which is bounded by continuous curves, is mapped by τ onto a region A' whose area is equal to the area of A multiplied by the absolute value of the determinant of τ.

Hint. Use Problem 13 with $n = 2$ to prove this first for rectangular areas. Then express the area of A as the limit of a sum of rectangular areas by means of an integral.

15. Use Problem 14 to find the area bounded by the ellipse

$$\frac{x^2}{a^2} + \frac{y^2}{b^2} = 1.$$

Hint. Apply a linear transformation that maps the circle $x^2 + y^2 = 1$ onto the given ellipse.

16. Let τ be any linear transformation of $\mathscr{V}_3(\mathscr{R})$ into itself and let R be any bounded region of $\mathscr{V}_3(\mathscr{R})$ bounded by integrable surfaces. Prove (see hint for Problem 14) that R is mapped by τ into a region R' whose volume is equal to the volume of R multiplied by the absolute value of the determinant of τ.

17. Use Problem 16 to find the volume bounded by the ellipsoid

$$\frac{x^2}{a^2} + \frac{y^2}{b^2} + \frac{z^2}{c^2} = 1.$$

18. What is a necessary and sufficient condition that a linear transformation $\tau: \mathscr{V}_3(\mathscr{R}) \to \mathscr{V}_3(\mathscr{R})$ preserve volumes, that is, that τ maps every region R of space into a region having the same volume as R?

19. Prove that every linear transformation of $\mathscr{V}_2(\mathscr{R})$ maps any two regions of equal area onto two regions of equal area.

20. Let $\mathscr{V}$ be an n-dimensional vector space and let $\tau: \mathscr{V} \to \mathscr{V}$ be a linear transformation. A subspace $\mathscr{S}$ of $\mathscr{V}$ is said to be *invariant under* τ if $\tau S \in \mathscr{S}$ for all $S \in \mathscr{S}$. Suppose $\mathscr{V} = \mathscr{S} \oplus \mathscr{T}$ where dim $\mathscr{S} = r$, dim $\mathscr{T} = n - r$, and $\mathscr{S}$ and $\mathscr{T}$ are both invariant under τ. Show that a basis $F_1, \ldots, F_n$ of $\mathscr{V}$ can be chosen such that $A_\tau^{(F)}$ has the form

$$\begin{bmatrix} A_1 & O \\ O & A_2 \end{bmatrix},$$

where A_1 is an $r \times r$ and A_2 an $n - r \times n - r$ matrix.

5.6 Orthogonal Transformations in an Inner Product Space

Throughout this section, the field of scalars will be the field $\mathscr{R}$ of real numbers. Let $\mathscr{V}$ be a vector space of dimension n over $\mathscr{R}$ with an inner product $(\ , \ )$ so that each vector V in $\mathscr{V}$ has a length defined by $\|V\| = (V, V)^{1/2}$. We shall study linear transformations $\tau: \mathscr{V} \to \mathscr{V}$ that *preserve lengths*, that is, such that $\|\tau V\| = \|V\|$ for all V in $\mathscr{V}$.

Theorem 5.13. *A linear transformation $\tau: \mathscr{V} \to \mathscr{V}$ preserves lengths if and only if it preserves inner products, that is, for any two vectors V_1, V_2 in $\mathscr{V}$, $(\tau V_1, \tau V_2) = (V_1, V_2)$.*

Proof. Since $\|V\| = (V, V)^{1/2}$, preservation of inner products certainly implies preservation of lengths. The converse follows from the identity

$$(V_1, V_2) = \tfrac{1}{2}(\|V_1 + V_2\|^2 - \|V_1\|^2 - \|V_2\|^2).$$

Corollary. *If τ preserves lengths, τV_1 is orthogonal to τV_2 if and only if V_1 is orthogonal to V_2, that is, τ preserves orthogonality.*

Definition. *A linear transformation $\tau: \mathscr{V} \to \mathscr{V}$ on a Euclidean vector space $\mathscr{V}$ is called an* orthogonal transformation *on $\mathscr{V}$ if it preserves lengths.*

Theorem 5.14. *Let $\mathscr{V}$ be a Euclidean vector space with orthonormal basis $E_1, E_2, \ldots, E_n$. A linear transformation $\tau: \mathscr{V} \to \mathscr{V}$ is orthogonal if and only if its matrix A_τ relative to the E-basis is orthogonal.*

Proof. Let V, W be any two vectors and let X, Y be their coordinate (column) vectors relative to the E-basis. Since this basis is orthonormal, we have by Theorem 4.16

$$(V, W) = X \cdot Y = X^T Y,$$

$$(\tau V, \tau W) = (A_\tau X) \cdot (A_\tau Y) = (A_\tau X)^T A_\tau Y = X^T A_\tau^T A_\tau Y.$$

Now if A_τ is orthogonal, $A_\tau^T A_\tau = I$, $(V, W) = (\tau V, \tau W)$, and τ is orthogonal. Conversely, if τ is orthogonal,

$$X^T A_\tau^T A_\tau Y = X^T Y$$

for all X and Y in $\mathscr{V}_n(\mathscr{R})$. By substituting for X and Y the ith and jth columns of the identity matrix I, we get $A_\tau^T A_\tau = [\delta_{ij}] = I$ and hence A_τ is orthogonal. Since $E_1, \ldots, E_n$ was an arbitrary orthonormal basis, Theorem 5.14 implies that an orthogonal transformation τ has an orthogonal matrix relative to *every* orthonormal basis. This also follows from

$$A_\tau^{(F)} = P^{-1} A_\tau^{(E)} P,$$

since the transition matrix P from one orthonormal basis (E) to another (F) is orthogonal and hence P^{-1} and $P^{-1} A_\tau^{(E)} P$ are orthogonal because P and $A_\tau^{(E)}$ are.

We recall that $\det A_\tau^{(E)}$ is independent of the basis (E) and that if A_τ is orthogonal, then $\det A_\tau = \pm 1$. Therefore we can distinguish two types of orthogonal transformations τ: those for which $\det A_\tau = 1$ are called *proper* and those for which $\det A_\tau = -1$ are called *improper*.

Definition. *A proper orthogonal transformation $\tau: \mathscr{V} \to \mathscr{V}$ is called a* rotation *of $\mathscr{V}$.*

This amounts to defining a rotation of $\mathscr{V}$ as any linear transformation $\tau:\mathscr{V} \to \mathscr{V}$ that preserves lengths (and hence orthogonality) and orientations. To justify our definition, we shall show that any proper orthogonal transformation in $\mathscr{V}_3(\mathscr{R})$ is a rotation in the ordinary sense.

Let $\tau:\mathscr{V}_3(\mathscr{R}) \to \mathscr{V}_3(\mathscr{R})$ be a proper orthogonal transformation and let A_τ be the matrix of τ relative to the standard basis. Since the standard basis is orthonormal, A_τ is orthogonal and $\det A_\tau = \det A_\tau^T = 1$. Thus we have

$$
\begin{aligned}
\det (A_\tau - I) &= \det (A_\tau - I) \det A_\tau^T \\
&= \det (A_\tau A_\tau^T - A_\tau^T) \\
&= \det (I - A_\tau^T) \\
&= (-1)^3 \det (A_\tau^T - I) \\
&= -\det (A_\tau - I)
\end{aligned}
$$

and, therefore, $\det (A_\tau - I) = 0$ and $A_\tau - I$ is singular. Hence there exists a non-zero vector X such that $(A_\tau - I)X = O$ or

(33)
$$
A_\tau X = X.
$$

Since we are using the standard basis of $\mathscr{V}_3(\mathscr{R})$, $\tau X = A_\tau X$ and (33) states that X is invariant under τ. The same is true of every scalar multiple of X. Let F_3 be the unit vector $\left(\dfrac{1}{\|X\|}\right)X$, let $\mathscr{S}$ be the subspace spanned by F_3 and let $\mathscr{S}^\perp$ be the orthogonal component of $\mathscr{S}$. Choose any orthonormal basis F_1, F_2 of $\mathscr{S}^\perp$, so that F_1, F_2, F_3 is an orthonormal basis of $\mathscr{V}_3(\mathscr{R})$. Clearly, $\mathscr{S}$ is mapped by τ onto itself because $\tau F_3 = F_3$. Since τ is orthogonal it preserves orthogonality of vectors so if $V \in \mathscr{S}^\perp$, then $\tau V \in \mathscr{S}^\perp$. Thus, $\mathscr{S}^\perp$ is also mapped by τ into itself. Hence we must have

$$
\begin{aligned}
\tau F_1 &= aF_1 + bF_2, \\
\tau F_2 &= cF_1 + dF_2, \\
\tau F_3 &= F_3,
\end{aligned}
$$

and the matrix of τ relative to the F-basis is

$$
A_\tau^{(F)} = \begin{bmatrix} a & c & 0 \\ b & d & 0 \\ 0 & 0 & 1 \end{bmatrix},
$$

This matrix is orthogonal, since τ is orthogonal and (F) is orthonormal. Also, $\det A_\tau^{(F)} = 1$ because τ is proper orthogonal. Orthogonality of $A_\tau^{(F)}$ implies $a^2 + c^2 = 1$ so we may choose θ, $0 \le \theta \le \pi$ such that $a = \cos \theta$ and $c = \pm\sin \theta$. Then $a^2 + b^2 = 1$ gives $b = \pm\sin \theta$ and from $b^2 + d^2 = 1$, $d = \pm\cos \theta$. The further orthogonality conditions $ab + cd = 0$, $ac + bd = 0$, together with the fact that $\det A_\tau^{(F)} = ab - bc = 1$, reduce the possibilities to

$$\begin{bmatrix} a & c \\ b & d \end{bmatrix} = \begin{bmatrix} \cos\theta & -\sin\theta \\ \sin\theta & \cos\theta \end{bmatrix}$$

or

$$\begin{bmatrix} a & c \\ b & d \end{bmatrix} = \begin{bmatrix} \cos\theta & \sin\theta \\ -\sin\theta & \cos\theta \end{bmatrix}.$$

Comparison with Example 4, Section 5.4, shows that the first represents a rotation of the plane $\mathscr{S}^\perp$ through an angle θ and the second a rotation of $\mathscr{S}^\perp$ through an angle $-\theta$. Thus the effect of τ is to leave vectors in the line $\mathscr{S}$ invariant and to rotate vectors in the plane $\mathscr{S}^\perp$ through a fixed angle θ. Now let X be any vector of $\mathscr{V}_3(\mathscr{R})$ so that $X = S + T$, $S \in \mathscr{S}$, and $T \in \mathscr{S}^\perp$. Then $\tau X = \tau S + \tau T = S + \tau T$ and it is clear that τ rotates the whole space $\mathscr{V}_3(\mathscr{R})$ about the line $\mathscr{S}$ as axis.

Example 1. Let $\tau : \mathscr{V}_3(\mathscr{R}) \to \mathscr{V}_3(\mathscr{R})$ be the linear transformation $\tau X = AX$, where

$$A = \begin{bmatrix} \frac{6}{7} & \frac{2}{7} & \frac{3}{7} \\ \frac{2}{7} & \frac{3}{7} & -\frac{6}{7} \\ -\frac{3}{7} & \frac{6}{7} & \frac{2}{7} \end{bmatrix}.$$

Prove that τ is a rotation, find the axis and the angle of rotation.

Solution. One checks that A, the matrix of τ relative to the standard basis, is proper orthogonal. Hence, τ is a rotation. To find the axis we look for a nonzero vector X invariant under τ. If $AX = X$, then $(A - I)X = O$, where

$$A - I = \begin{bmatrix} -\frac{1}{7} & \frac{2}{7} & \frac{3}{7} \\ \frac{2}{7} & -\frac{4}{7} & -\frac{6}{7} \\ -\frac{3}{7} & \frac{6}{7} & -\frac{6}{7} \end{bmatrix}.$$

By our previous discussion we know that $A - I$ is singular. The system $(A - I)X = O$ is equivalent to

$$-x_1 + 2x_2 + 3x_3 = 0,$$

$$3x_1 - 6x_2 + 6x_3 = 0,$$

of which a solution is $x_1 = 2$, $x_2 = 1$, $x_3 = 0$. The axis of rotation is therefore the line

$$x_1 - 2x_2 = 0,$$

$$x_3 = 0$$

in the $x_1 x_2$-plane, or the line determined by the vector $[2, 1, 0]$. To find the angle of rotation, choose any vector orthogonal to $X = [2, 1, 0]$, say, $Y = [0, 0, 1]$. Then $\tau Y = A Y = [\frac{3}{7}, -\frac{6}{7}, \frac{2}{7}]$. The required angle θ is the angle between Y and τY, so

$$\cos \theta = \frac{Y \cdot (\tau Y)}{\|Y\| \, \|\tau Y\|} = \frac{2}{7}.$$

Hence, relative to an orthonormal basis (F) of which the third vector is $\left[\dfrac{2}{\sqrt{5}}, \dfrac{1}{\sqrt{5}}, 0 \right]$, on the axis of rotation, the matrix of τ will be

$$A_\tau^{(F)} = \begin{bmatrix} \dfrac{2}{7} & -\dfrac{3\sqrt{5}}{7} & 0 \\ \dfrac{3\sqrt{5}}{7} & \dfrac{2}{7} & 0 \\ 0 & 0 & 1 \end{bmatrix}.$$

Definition. *Let $\mathscr{V}$ be an n-dimensional Euclidean vector space and let $E_1, E_2, \ldots, E_n$ be an orthonormal basis. A linear transformation $\tau : \mathscr{V} \to \mathscr{V}$ defined by $\tau E_j = -E_j$ for some fixed j and $\tau E_i = E_i$ for all $i \neq j$ is called a reflection.*

Taking $j = 1$ in this definition, the matrix of τ relative to the E-basis is

$$Q = \begin{bmatrix} -1 & 0 & \cdots & 0 \\ 0 & 1 & \cdots & 0 \\ \multicolumn{4}{c}{\dotfill} \\ 0 & 0 & \cdots & 1 \end{bmatrix}.$$

This is clearly an improper orthogonal matrix and, therefore, a reflection is an improper orthogonal transformation.

If, on the other hand, P is any improper orthogonal matrix we can write $P = QR$ where R is obtained from P by multiplying the first row by -1. Thus, R is a proper orthogonal matrix. It follows that every improper orthogonal transformation $\tau : \mathscr{V} \to \mathscr{V}$ is the product of a rotation and a reflection.

Example 2. Find the matrix relative to the standard basis of the reflection τ of $\mathscr{V}_3(\mathscr{R})$ in the plane $x_1 + 2x_2 - 2x_3 = 0$.

Solution. A unit vector perpendicular to the plane is $F_1 = [\frac{1}{3}, \frac{2}{3}, -\frac{2}{3}]$. Hence, $\tau F_1 = -F_1$ and τ leaves invariant all vectors in the given plane. Choose F_2 and F_3

mutually orthogonal unit vectors in the plane. One such choice is

$$F_2 = [-\tfrac{2}{3}, \tfrac{2}{3}, \tfrac{1}{3}],$$
$$F_3 = [\tfrac{2}{3}, \tfrac{1}{3}, \tfrac{2}{3}].$$

Then $\tau F_1 = -F_1$, $\tau F_2 = F_2$, and $\tau F_3 = F_3$ so that

$$A_\tau^{(F)} = \begin{bmatrix} -1 & 0 & 0 \\ 0 & 1 & 0 \\ 0 & 0 & 1 \end{bmatrix}.$$

The transition matrix from the standard basis to the F-basis is

$$P = \begin{bmatrix} \tfrac{1}{3} & -\tfrac{2}{3} & \tfrac{2}{3} \\ \tfrac{2}{3} & \tfrac{2}{3} & \tfrac{1}{3} \\ -\tfrac{2}{3} & \tfrac{1}{3} & \tfrac{2}{3} \end{bmatrix},$$

and the transition matrix from the F-basis to the standard basis is therefore P^{-1}, which is equal to P^T since P is orthogonal. Hence the matrix of τ relative to the standard basis is

$$(P^T)^{-1}A_\tau^{(E)}P^T = P A_\tau^{(F)}P^T$$

$$= \begin{bmatrix} \tfrac{1}{3} & -\tfrac{2}{3} & \tfrac{2}{3} \\ \tfrac{2}{3} & \tfrac{2}{3} & \tfrac{1}{3} \\ -\tfrac{2}{3} & \tfrac{1}{3} & \tfrac{2}{3} \end{bmatrix} \begin{bmatrix} -\tfrac{1}{3} & -\tfrac{2}{3} & \tfrac{2}{3} \\ -\tfrac{2}{3} & \tfrac{2}{3} & \tfrac{1}{3} \\ \tfrac{2}{3} & \tfrac{1}{3} & \tfrac{2}{3} \end{bmatrix} = \begin{bmatrix} \tfrac{7}{9} & -\tfrac{4}{9} & \tfrac{4}{9} \\ -\tfrac{4}{9} & \tfrac{1}{9} & \tfrac{8}{9} \\ \tfrac{4}{9} & \tfrac{8}{9} & \tfrac{1}{9} \end{bmatrix}$$

Exercise 5.3

1. Show that the linear transformation of $\mathscr{V}_3(\mathscr{R})$, whose matrix relative to the standard basis is

$$\begin{bmatrix} \tfrac{2}{3} & \tfrac{2}{3} & \tfrac{1}{3} \\ -\tfrac{2}{3} & \tfrac{1}{3} & \tfrac{2}{3} \\ \tfrac{1}{3} & -\tfrac{2}{3} & \tfrac{2}{3} \end{bmatrix},$$

is a rotation. Find the axis of rotation and the angle through which the plane perpendicular to this axis is rotated.

2. Prove that a rotation in a Euclidean space of dimension n must leave a nonzero vector invariant if n is odd.

3. Find the matrix relative to the basis $F_1 = [\tfrac{2}{3}, -\tfrac{2}{3}, \tfrac{1}{3}]$, $F_2 = [\tfrac{2}{3}, \tfrac{1}{3}, -\tfrac{2}{3}]$, $F_3 = [\tfrac{1}{3}, \tfrac{2}{3}, \tfrac{2}{3}]$ of (a) a counterclockwise rotation of $\mathscr{V}_3(\mathscr{R})$ through $45°$ about the x-axis; (b) a reflection of $\mathscr{V}_3(\mathscr{R})$ in the xy-plane.

4. Find the matrix relative to the standard basis of the following operators in $\mathscr{V}_3(\mathscr{R})$: (a) a reflection in the x_2x_3 plane; (b) a reflection in the plane $x_1 - x_2 = 0$; (c) a

reflection in the plane $2x_1 + x_2 - x_3 = 0$; (d) a counterclockwise rotation through $45°$ about the x-axis; (e) a counterclockwise rotation through $120°$ about the line $x_1 = x_2 = x_3$.

5. If A is an improper orthogonal matrix, prove that det $(A + I) = 0$.

6. Prove that the product of two reflections is a rotation.

7. If τ is the reflection in $V_3(\mathcal{R})$ described in Problem 4(b) and σ is the reflection in Problem 4(c), find the matrix of the product transformation $\tau\sigma$, show that $\tau\sigma$ is a rotation, and find the axis about which this rotation takes place. *Hint.* Each reflection leaves all vectors in a plane invariant.

8. Let σ be a reflection in a plane p_1 and τ be a reflection in a plane p_2. Show that the transformation $\sigma\tau$ leaves the line of intersection of p_1 and p_2 invariant and must therefore represent a rotation about this line.

5.7 Invariant Subspaces and Direct Sum Decompositions

If a vector space $\mathscr{V}$ of dimension n has subspaces $\mathscr{S}$ and $\mathscr{T}$ such that $\mathscr{V} = \mathscr{S} + \mathscr{T}$ and $\mathscr{S} \cap \mathscr{T} = O$, we saw in Section 4.3 that dim $\mathscr{S}$ + dim $\mathscr{T}$ = n and that every vector V in $\mathscr{V}$ has a unique representation in the form $V = S + T$, $S \in \mathscr{S}$, $T \in \mathscr{T}$. In these circumstances, we called $\mathscr{V}$ the direct sum of $\mathscr{S}$ and $\mathscr{T}$ and wrote $\mathscr{V} = \mathscr{S} \oplus \mathscr{T}$. We now extend these ideas to define direct sums of more than two subspaces.

Definition. *A vector space* $\mathscr{V}$ *is said to be the* direct sum *of subspaces* $\mathscr{S}_1, \mathscr{S}_2, \ldots, \mathscr{S}_r$ *if* $\mathscr{V}$ *is the sum of the subspaces* $\mathscr{S}_i$ *and if the intersection of each of the subspaces* $\mathscr{S}_i$ *with the sum of the* $r - 1$ *remaining subspaces* $\mathscr{S}_j$ $(j \neq i)$ *is O.*

If $\mathscr{V}$ is the direct sum of $\mathscr{S}_1, \mathscr{S}_2, \ldots, \mathscr{S}_r$ we write

$$\mathscr{V} = \mathscr{S}_1 \oplus \mathscr{S}_2 \oplus \cdots \oplus \mathscr{S}_r.$$

It follows from the definition that if $\mathscr{V} = \mathscr{S}_1 \oplus \mathscr{T}$ and $\mathscr{T} = \mathscr{S}_2 \oplus \cdots \oplus \mathscr{S}_r$, then $\mathscr{V} = \mathscr{S}_1 \oplus \mathscr{S}_2 \oplus \cdots \oplus \mathscr{S}_r$ so that insertion of brackets is unnecessary and $\oplus$ is an associative operation.

Theorem 5.15. *If* $\mathscr{V} = \mathscr{S}_1 \oplus \mathscr{S}_2 \oplus \cdots \oplus \mathscr{S}_r$, *every vector* V *in* $\mathscr{V}$ *has a unique representation in the form*

$$V = S_1 + S_2 + \cdots + S_r,$$

where $S_i \in \mathscr{S}_i$ $(i = 1, 2, \ldots, r)$.

Proof. Since $\mathscr{V}$ is the sum of $\mathscr{S}_1, \mathscr{S}_2, \ldots, \mathscr{S}_r$, V certainly has such a representation. If there were two such, we would have

$$V = S_1 + S_2 + \cdots + S_r = S_1' + S_2' + \cdots + S_r',$$

where $S_i' \in S_i$, and therefore

$$S_1 - S_1' = (S_2' - S_2) + \cdots + (S_r' - S_r).$$

Since $S_i' - S_i \in \mathscr{S}_i$, it follows that $S_1 - S_1'$ belongs both to $\mathscr{S}_1$ and to $\mathscr{S}_2 + \mathscr{S}_3 + \cdots + \mathscr{S}_r$. Hence, by the definition of direct sum, $S_1 - S_1' = O$ and $S_1 = S_1'$. Similarly, $S_i' = S_i$ for $i = 2, 3, \ldots, r$.

Theorem 5.16. *If* $\mathscr{V} = \mathscr{S}_1 \oplus \mathscr{S}_2 \oplus \cdots \oplus \mathscr{S}_r$, *where* $\mathscr{S}_i$ *has dimension* n_i *and* $\mathscr{V}$ *has dimension* n, *then* $n = n_1 + n_2 + \cdots + n_r$ *and there exists a basis of* $\mathscr{V}$ *of which the first* n_1 *vectors constitute a basis of* S_1, *the next* n_2 *vectors a basis of* $S_2, \ldots,$ *and the last* n_r *vectors a basis of* S_r.

Proof. This follows at once from Theorems 1.8 and 1.10.

Definition. *Let* $\tau : \mathscr{V} \to \mathscr{V}$ *be a linear transformation of* $\mathscr{V}$. *A vector* V *is invariant under* τ *if* $\tau V = V$. *A subspace* $\mathscr{S}$ *of* $\mathscr{V}$ *is invariant under* τ *if for every vector* $S \in \mathscr{S}$, $\tau S \in \mathscr{S}$.

Note that invariance of $\mathscr{S}$ requires only $\tau S \in \mathscr{S}$, not $\tau S = S$, for all S in $\mathscr{S}$. For example, the null space of τ is always invariant under τ.

Theorem 5.17. *Let* $\mathscr{V}$ *be a space of dimension* n *that is the direct sum of subspaces* $\mathscr{S}_i$ *of dimension* n_i ($i = 1, 2, \ldots, r$). *If* τ *is a linear transformation of* $\mathscr{V}$ *that leaves each of the subspaces* $\mathscr{S}_i$ *invariant, then the matrix of* τ *relative to a suitably chosen basis of* $\mathscr{V}$ *has the form*

$$(34) \qquad A_\tau = \begin{bmatrix} A_1 & O & \cdots & O \\ O & A_2 & \cdots & O \\ \multicolumn{4}{c}{\cdots\cdots\cdots\cdots\cdots} \\ O & O & \cdots & A_r \end{bmatrix},$$

where A_i *is a square matrix of order* n_i ($i = 1, 2, \ldots, r$) *and all elements of* A_τ *not in the diagonal blocks* A_i *are zero.*

Proof. Choose a basis $E_1, E_2, \ldots, E_n$ of $\mathscr{V}$, as in Theorem 5.16, so that $E_1, \ldots, E_{n_1}$ is a basis of $\mathscr{S}_1$, $E_{n_1+1}, \ldots, E_{n_1+n_2}$ is a basis of $\mathscr{S}_2$, etc. Because τ leaves $\mathscr{S}_1$ invariant, $\tau E_1, \ldots, \tau E_{n_1}$ are linear combinations of $E_1, \ldots, E_{n_1}$. Similarly,

$\tau E_{n_1+1}, \ldots, \tau E_{n_1+n_2}$ are linear combinations of $E_{n_1+1}, \ldots, E_{n_1+n_2}$, etc. Hence, $A_\tau^{(E)}$ has the form (34).

If each of the invariant subspaces $\mathscr{S}_i$ is one dimensional, each $n_i = 1$ and (34) is a diagonal matrix. In this case τ is said to be diagonalizable. This means that $\mathscr{V}$ is the direct sum of n one-dimensional subspaces each of which is invariant under τ. A magnification, for example, is diagonalizable. In the next chapter we shall study diagonalizability from the matrix point of view.

Exercise 5.4

1. Let a vector space $\mathscr{V}$ be the direct sum of subspaces $\mathscr{S}_1, \ldots, \mathscr{S}_r$. For any $V \in \mathscr{V}$, let $V = S_1 + \cdots + S_r$ where $S_i \in \mathscr{S}_i$. For each i $(i = 1, 2, \ldots, r)$, define $\pi_i \colon \mathscr{V} \to \mathscr{V}$ by $\pi_i V = S_i$.
   (a) Prove that π_i (called a projection of $\mathscr{V}$ onto $\mathscr{S}_i$) is a linear transformation of $\mathscr{V}$ into $\mathscr{V}$ and im $\pi_i = \mathscr{S}_i$.
   (b) Prove that $\pi_i^2 = \pi_i$ and $\pi_i \pi_j = O$ if $i \neq j$.
   (c) Prove that $\pi_1 + \pi_2 + \cdots + \pi_r = I$.

2. Let $\mathscr{V}$ be a vector space and let $\pi_i \colon \mathscr{V} \to \mathscr{V}$, $(i = 1, 2, \ldots, r)$ be r linear transformations which satisfy (b) and (c) of Problem 1. Let $\mathscr{S}_i = $ im π_i and prove that

$$\mathscr{V} = \mathscr{S}_1 \oplus \mathscr{S}_2 \oplus \cdots \oplus \mathscr{S}_r.$$

CHAPTER SIX

SIMILARITY AND DIAGONALI- ZATION THEOREMS

6.1 Similar Matrices

Let A be any $n \times n$ matrix with elements in $\mathscr{F}$. By Theorem 5.12 and the subsequent discussion in Sections 5.5 and 5.6, it is clear that considerable interest attaches to the problem of finding, when this is possible, a nonsingular matrix P such that $P^{-1}AP$ is a diagonal matrix. In this chapter we shall determine when such a matrix P exists and how it may be found. We also discuss important special cases and applications.

We shall assume throughout that the field of scalars is the complex field $\mathscr{C}$. Although some of our theorems concern matrices with real elements, it is still advantageous to consider these as matrices over $\mathscr{C}$ which they certainly are, since $\mathscr{R} \subset \mathscr{C}$. Thus, all matrices considered are assumed to have elements in $\mathscr{C}$.

Definition. *Let A and B be $n \times n$ matrices. We say that B is* similar *to A if there exists a nonsingular matrix P such that $B = P^{-1}AP$.*

Theorem 6.1. *Similarity of matrices is an equivalence relation.*

Proof. Using $P = I$, we see that A is similar to itself, and similarity is therefore reflexive. If $B = P^{-1}AP$, then $A = PBP^{-1} = Q^{-1}BQ$, where $Q = P^{-1}$. Hence, similarity is symmetric. Finally, if $B = P^{-1}AP$ and $C = R^{-1}BR$, then $C = R^{-1}P^{-1}APR = S^{-1}AS$, where $S = PR$. This proves the transitivity.

In view of the symmetry of the similarity relationship we may speak without ambiguity of "similar matrices A and B."

Theorem 6.2. *If A is similar to B, then* det A = det B.

Proof. If $B = P^{-1}AP$,

$$\det B = \det P^{-1} \det A \det P = \det P^{-1} \det P \det A = \det A,$$

since $\det P^{-1} \det P = \det P^{-1}P = 1$.

Theorem 6.3. *Let $\mathscr{C}_n$ be the algebra of all $n \times n$ matrices over $\mathscr{C}$, and let P be a fixed nonsingular $n \times n$ matrix. The mapping $\alpha : \mathscr{C}_n \to \mathscr{C}_n$ defined by $\alpha A = P^{-1}AP$ is an isomorphism.*

Proof. We have

$$\alpha(A + B) = P^{-1}(A + B)P = P^{-1}AP + P^{-1}BP = \alpha A + \alpha B,$$

$$\alpha(kA) = P^{-1}(kA)P = k(P^{-1}AP) = k(\alpha A),$$

$$\alpha(AB) = P^{-1}ABP = (P^{-1}AP)(P^{-1}BP) = (\alpha A)(\alpha B).$$

Moreover, because $P^{-1}AP = P^{-1}BP$ implies $A = B$, α is one-to-one and because for any matrix B in $\mathscr{C}_n$, $B = \alpha(PBP^{-1})$ clearly α is onto $\mathscr{C}_n$. Hence, α is an algebra isomorphism of $\mathscr{C}_n$ onto itself.

Definition. *If P is nonsingular the matrix $P^{-1}AP$ is called the* transform *of A by P and the mapping $\alpha : A \to P^{-1}AP$ is called a* similarity transformation *by P.*

6.2 Eigenvalues and Eigenvectors of a Matrix

Let $\mathscr{V}$ be an n-dimensional vector space over $\mathscr{C}$ with basis $E_1, \ldots, E_n$ and let $\tau : \mathscr{V} \to \mathscr{V}$ be a linear transformation with matrix A_τ relative to the E-basis. A first step in investigating the geometric properties of τ is to find the one-dimensional subspaces of $\mathscr{V}$ left invariant by τ. Such a subspace must be spanned by a single nonzero vector V such that $\tau V = kV$ for some scalar k. Now if X is the coordinate vector of V relative to the E-basis, then $A_\tau X$ is the coordinate vector of τV and the equation $\tau V = kV$ implies $A_\tau X = kX$ or

(1) $$(A_\tau - kI)X = O.$$

Thus the existence of a one-dimensional subspace invariant under τ is equivalent to the existence of a nonzero solution X for equation (1). Such a solution exists if and only if the matrix $A - kI$ is singular and hence if and only if

(2) $$\det (A_\tau - kI) = 0.$$

Therefore we are sufficiently interested in values of k that satisfy (2) and nonzero vectors X of $\mathscr{V}_n(\mathscr{C})$ that satisfy (1) to justify giving them names.

Definition. *Let $A = [a_{ij}]$ be any $n \times n$ matrix with elements in $\mathscr{C}$. The determinant*

$$\det (A - xI) = \begin{vmatrix} a_{11} - x & a_{12} & \cdots & a_{1n} \\ a_{21} & a_{22} - x & \cdots & a_{2n} \\ \cdots\cdots\cdots\cdots\cdots\cdots\cdots\cdots\cdots \\ a_{n1} & a_{n2} & \cdots & a_{nn} - x \end{vmatrix}$$

is a polynomial of degree n in x, with coefficients in $\mathscr{C}$, called the characteristic polynomial *of A. The equation* $\det (A - xI) = 0$ *is called the* characteristic equation *of A. The roots of the characteristic equation are called the* eigenvalues (*or characteristic roots*) *of A.*

A polynomial of degree n with coefficients in $\mathscr{C}$ always has exactly n roots in $\mathscr{C}$ if we follow the usual convention of counting a root of multiplicity r as r equal roots. Hence an $n \times n$ matrix with elements in $\mathscr{C}$ always has n eigenvalues in $\mathscr{C}$. An $n \times n$ matrix with real elements need not, of course, have n real eigenvalues, but will have n eigenvalues in $\mathscr{C}$.

Example 1. Find the eigenvalues of the matrix

$$A = \begin{bmatrix} -17 & 18 & -6 \\ -18 & 19 & -6 \\ -9 & 9 & -2 \end{bmatrix}.$$

Solution.

$$\det (A - xI) = \begin{vmatrix} -17 - x & 18 & -6 \\ -18 & 19 - x & -6 \\ -9 & 9 & -2 - x \end{vmatrix}$$

$$= -x^3 + 3x - 2 = -(x + 2)(x - 1)^2.$$

Therefore the eigenvalues are $-2, 1, 1$, the 1 being counted twice because of the square factor $(x - 1)^2$ in the characteristic polynomial.

If k is an eigenvalue of A, then det $(A - kI) = 0$ and $A - kI$ is singular. Hence there exists a nonzero vector X such that $(A - kI)X = O$ or $AX = kX$. Here, as usual, we identify the column vector X with the corresponding row vector in $\mathscr{V}_n(\mathscr{C})$.

Definition. *If k is an eigenvalue of an $n \times n$ matrix A, a nonzero vector X in $\mathscr{V}_n(\mathscr{C})$ such that $AX = kX$ is called an eigenvector of A corresponding to the eigenvalue k.*

Theorem 6.4. *Similar matrices have the same characteristic polynomial and hence the same eigenvalues. If X is an eigenvector of A corresponding to the eigenvalue k, then $P^{-1}X$ is an eigenvector of $P^{-1}AP$ corresponding to the eigenvalue k.*

Proof. By the distributive law for matrix multiplication,

$$P^{-1}(A - xI)P = P^{-1}AP - P^{-1}(xI)P = P^{-1}AP - xI.$$

Hence, by Theorem 6.2, det $(P^{-1}AP - xI) =$ det $(A - xI)$ and the first result follows.

Now if k is an eigenvalue of A and X is a corresponding eigenvector, $AX = kX$, and hence

$$(P^{-1}AP)(P^{-1}X) = P^{-1}AX = P^{-1}(kX) = k(P^{-1}X)$$

and $P^{-1}X$ is an eigenvector of $P^{-1}AP$ corresponding to k.

Corollary. *If A is similar to a diagonal matrix D, the diagonal elements of D are the eigenvalues of A.*

Proof. By Theorem 6.4, A and D have the same eigenvalues. But the eigenvalues of the diagonal matrix D are clearly the diagonal elements of D.

The next theorem tells us when a matrix is similar to a diagonal matrix and how the transforming matrix P can be found.

Theorem 6.5. *A necessary and sufficient condition that an $n \times n$ matrix A, with elements in $\mathscr{C}$, be similar to a diagonal matrix is that A have n linearly independent eigenvectors in $\mathscr{V}_n(\mathscr{C})$. Moreover,*

$$P^{-1}AP = \begin{bmatrix} k_1 & 0 & \cdots & 0 \\ 0 & k_2 & \cdots & 0 \\ \cdots\cdots\cdots\cdots\cdots \\ 0 & 0 & \cdots & k_n \end{bmatrix}$$

if and only if the jth column of P is an eigenvector of A corresponding to the eigenvalue k_j of A, $(j = 1, 2, \ldots, n)$ and these n column vectors are linearly independent.

Proof. If A is similar to a diagonal matrix D with diagonal elements $k_1, \ldots, k_n$, there is a nonsingular matrix P such that $P^{-1}AP = D$ or

$$(3) \qquad\qquad AP = PD.$$

Moreover, $k_1, k_2, \ldots, k_n$ are the eigenvalues of A by the corollary to Theorem 6.4. If we let $P_1, P_2, \ldots, P_n$ be the column vectors of P we can equate corresponding columns on each side of (3) to get (Theorem 2.5):

$$(4) \qquad\qquad AP_i = k_iP_i \qquad (i = 1, 2, \ldots, n).$$

But (4) states that P_i is an eigenvector of A corresponding to the eigenvalue k_i and, since P is nonsingular, the n eigenvectors $P_1, \ldots, P_n$ are linearly independent.

Conversely, if $P_1, \ldots, P_n$ are linearly independent eigenvectors of A corresponding to eigenvalues $k_1, \ldots, k_n$ equations (4) and therefore (3) hold where P is the matrix with columns $P_1, \ldots, P_n$. Since its columns are linearly independent, P is nonsingular and hence (3) implies $P^{-1}AP = D$. This proves both statements in the theorem.

Definition. *A matrix A is said to be* diagonalizable *if it is similar to a diagonal matrix. A matrix P such that $P^{-1}AP$ is diagonal is said to* diagonalize A *or* transform A to *diagonal form.*

Example 2. Find a matrix P that will diagonalize the matrix

$$A = \begin{bmatrix} 1 & 1 \\ -1 & 1 \end{bmatrix}$$

and find the diagonal form $P^{-1}AP$ of A.

Solution. The characteristic equation of A is

$$\begin{vmatrix} 1 - x & 1 \\ -1 & 1 - x \end{vmatrix} = 0$$

which simplifies to $x^2 - 2x + 2 = 0$. The eigenvalues are the roots of this equation, namely, $1 + i$ and $1 - i$. The corresponding eigenvectors are found by solving the systems

$$\begin{bmatrix} -i & 1 \\ -1 & -i \end{bmatrix} \begin{bmatrix} x_1 \\ x_2 \end{bmatrix} = \begin{bmatrix} 0 \\ 0 \end{bmatrix} \quad \text{and} \quad \begin{bmatrix} i & 1 \\ -1 & i \end{bmatrix} \begin{bmatrix} y_1 \\ y_2 \end{bmatrix} = \begin{bmatrix} 0 \\ 0 \end{bmatrix}$$

to be $X = [1, i]$ and $Y = [1, -i]$. These are the columns of the matrix P so we take

$$P = \begin{bmatrix} 1 & 1 \\ i & -i \end{bmatrix}$$

and Theorem 6.5 assures us that

(5) $$P^{-1}AP = \begin{bmatrix} 1+i & 0 \\ 0 & 1-i \end{bmatrix}.$$

The student should verify that

$$P^{-1} = \begin{bmatrix} \dfrac{1}{2} & -\dfrac{i}{2} \\[2ex] \dfrac{1}{2} & \dfrac{i}{2} \end{bmatrix}$$

and that (5) does indeed hold.

Example 3. Find a matrix P that will diagonalize the matrix A of Example 1.

Solution. In Example 1 we found the eigenvalues of A to be $-2, 1, 1$. An eigenvector corresponding to -2 is found by solving the system (with matrix $A + 2I$)

$$\begin{bmatrix} -15 & 18 & -6 \\ -18 & 21 & -6 \\ -9 & 9 & 0 \end{bmatrix} \begin{bmatrix} x_1 \\ x_2 \\ x_3 \end{bmatrix} = \begin{bmatrix} 0 \\ 0 \\ 0 \end{bmatrix}.$$

The matrix of coefficients has rank 2 and a nonzero solution is found, by solving any two of the equations, to be $X_1 = [2, 2, 1]$. This solution is unique except for a scalar factor.

To find eigenvectors corresponding to the eigenvalue 1 we must solve the system (with matrix $A - I$):

$$\begin{bmatrix} -18 & 18 & -6 \\ -18 & 18 & -6 \\ -9 & 9 & -3 \end{bmatrix} \begin{bmatrix} x_1 \\ x_2 \\ x_3 \end{bmatrix} = \begin{bmatrix} 0 \\ 0 \\ 0 \end{bmatrix}.$$

This time the matrix has rank 1 and we need only solve the single equation

$$-3x_1 + 3x_2 - x_3 = 0.$$

Therefore we can find *two* linearly independent solutions, for example, $X_2 = [1, 1, 0]$ and $X_2 = [-1, 0, 3]$. Since our three eigenvectors X_1, X_2, X_3 are linearly independent, we choose

$$P = \begin{bmatrix} 2 & 1 & -1 \\ 2 & 1 & 0 \\ 1 & 0 & 3 \end{bmatrix}$$

and, by Theorem 6.5,

$$P^{-1}AP = \begin{bmatrix} -2 & 0 & 0 \\ 0 & 1 & 0 \\ 0 & 0 & 1 \end{bmatrix}.$$

Example 4. Prove that the matrix

$$A = \begin{bmatrix} 1 & 2 \\ 0 & 1 \end{bmatrix}$$

is not diagonalizable.

Solution. Since $\det (A - xI) = (1 - x)^2$, the eigenvalues are both equal to 1. Hence, every eigenvector $X = [x_1, x_2]$ must satisfy

$$\begin{bmatrix} 0 & 2 \\ 0 & 0 \end{bmatrix} \begin{bmatrix} x_1 \\ x_2 \end{bmatrix} = \begin{bmatrix} 0 \\ 0 \end{bmatrix}$$

or $2x_2 = 0$. Thus, every eigenvector has the form $[x_1, 0]$. It is therefore impossible to find two linearly independent eigenvectors and, by Theorem 6.5, A is not diagonalizable.

Theorem 6.6. *If an $n \times n$ matrix A has n distinct eigenvalues, A is diagonalizable.*

Proof. Let $k_1, k_2, \ldots, k_n$ be the eigenvalues of A and let X_i be an eigenvector corresponding to k_i. If $X_1, X_2, \ldots, X_n$ are linearly dependent we can choose r so that $1 \leq r < n$ and $X_1, X_2, \ldots, X_r$ are linearly independent but $X_1, X_2, \ldots, X_{r+1}$ are linearly dependent. Hence we can choose $c_1, c_2, \ldots, c_{r+1}$, not all zero, such that

(6) $$c_1X_1 + c_2X_2 + \cdots + c_{r+1}X_{r+1} = O.$$

Multiplying (6) on the left by A and using $AX_i = k_iX_i$, we get

(7) $$c_1k_1X_1 + c_2k_2X_2 + \cdots + c_{r+1}k_{r+1}X_{r+1} = O.$$

Now multiply (6) by k_{r+1} and subtract from (7) to get

$$c_1(k_1 - k_{r+1})X_1 + c_2(k_2 - k_{r+1})X_2 + \cdots + c_r(k_r - k_{r+1})X_r = O.$$

But since $X_1, X_2, \ldots, X_r$ are linearly independent and $k_{r+1} \neq k_i$ for $i = 1, 2, \ldots, r$, the last equation implies $c_1 = c_2 = \cdots = c_r = 0$ and then, since $X_{r+1} \neq O$, (6) implies $c_{r+1} = 0$ which contradicts our assumption. Hence the eigenvectors $X_1, \ldots,$ X_n are linearly independent and, by Theorem 6.5, A is diagonalizable.

Exercise 6.1

1. Find the eigenvalues and eigenvectors of the following matrices:

   (a) $\begin{bmatrix} 2 & 4 \\ 3 & 13 \end{bmatrix}$.

   (b) $\begin{bmatrix} 2 & -3 \\ -3 & 1 \end{bmatrix}$.

   (c) $\begin{bmatrix} 3 & -2 \\ 2 & 1 \end{bmatrix}$.

   (d) $\begin{bmatrix} 3 & 2 & 4 \\ 2 & 0 & 2 \\ 4 & 2 & 3 \end{bmatrix}$.

   (e) $\begin{bmatrix} \frac{2}{3} & \frac{2}{3} & \frac{1}{3} \\ -\frac{2}{3} & \frac{1}{3} & \frac{2}{3} \\ \frac{1}{3} & -\frac{2}{3} & \frac{2}{3} \end{bmatrix}$.

2. What are the eigenvalues and eigenvectors of the identity matrix?

3. For each of the following matrices, find the diagonal form and a diagonalizing matrix P.

   (a) $\begin{bmatrix} -\frac{23}{7} & \frac{12}{7} \\ -\frac{40}{7} & \frac{23}{7} \end{bmatrix}$.

   (b) $\begin{bmatrix} 20 & 18 \\ -27 & -25 \end{bmatrix}$.

   (c) $\begin{bmatrix} 3 & 4 \\ -4 & 3 \end{bmatrix}$.

   (d) $\begin{bmatrix} \cos \varphi & -\sin \varphi \\ \sin \varphi & \cos \varphi \end{bmatrix}$.

   (e) $\begin{bmatrix} \dfrac{1}{2} & -\dfrac{\sqrt{3}}{2} \\ \dfrac{\sqrt{3}}{2} & \dfrac{1}{2} \end{bmatrix}$.

   (f) $\begin{bmatrix} 4 & 2 & -2 \\ -5 & 3 & 2 \\ -2 & 4 & 1 \end{bmatrix}$.

   (g) $\begin{bmatrix} 1 & 0 & 0 & -5 \\ 0 & -1 & 0 & 6 \\ 0 & 0 & 2 & 0 \\ 0 & 0 & 0 & 2 \end{bmatrix}$.

4. If $\mathscr{V}$ is a Euclidean space of dimension n where n is odd and $\tau: \mathscr{V} \to \mathscr{V}$ is a rotation, show that the matrix A_τ of τ relative to any orthonormal basis has an eigenvalue equal to 1.

5. If $\mathscr{V}$ is a Euclidean space of dimension n and $\tau: \mathscr{V} \to \mathscr{V}$ is a reflection, prove that the matrix A_τ of τ relative to any orthonormal basis has $n - 1$ eigenvalues equal to 1 and one eigenvalue equal to -1.

6. Prove that the following matrices are not diagonalizable:

   (a) $\begin{bmatrix} 2 & 1 & 0 \\ 0 & 2 & 1 \\ 0 & 0 & 2 \end{bmatrix}$.

   (b) $\begin{bmatrix} 1 & 2 & -4 \\ 0 & -1 & 6 \\ 0 & -1 & 4 \end{bmatrix}$.

7. Let D be a diagonal matrix and let D_1 be a diagonal matrix obtained from D by permuting the diagonal elements. Prove that D_1 is similar to D. *Hint.* By Theorem

3.1 we can reduce the problem to the case of interchanging two diagonal elements. Now use Theorem 2.22.

8. If $A = \begin{bmatrix} a & b \\ c & d \end{bmatrix}$ and $Q = \begin{bmatrix} 0 & 1 \\ 1 & 0 \end{bmatrix}$, find $Q^{-1}AQ$.

9. If A is any matrix of order n, the *secondary diagonal* of A is the diagonal from the lower left-hand to the upper right-hand corner of A. Let Q be the matrix of order n with all its secondary diagonal elements equal to one and all other elements equal to zero. Show that the effect of transforming A by Q is to rotate the elements of A about its central point through an angle of $180°$; that is, to shift the element in the ith row and jth column of A to the $(n - i + 1)$th row and $(n - j + 1)$th column, $(i, j = 1, 2, \ldots, n)$. Verify this by actual multiplication for the general third-order matrix, $A = [a_{ij}]$.

10. Show that transforming A by the matrix Q defined in Problem 9 has the effect of reflecting A first in its secondary diagonal and then in its main diagonal.

11. Deduce from Problem 10 that if A is symmetric about its secondary diagonal, then $Q^{-1}AQ = A^T$.

12. If A^S is the matrix obtained by reflecting A in its secondary diagonal, show that $Q^{-1}AQ = (A^S)^T = (A^T)^S$, and hence that $\det A^S = \det A$.

13. If A is a matrix of order n and I is the identity matrix of order n, show that the eigenvalues of

$$B = \begin{bmatrix} O & I \\ A & O \end{bmatrix}$$

are $\pm \sqrt{\lambda_1}, \ldots, \pm \sqrt{\lambda_n}$, where $\lambda_1, \ldots, \lambda_n$ are the eigenvalues of A. *Hint.* Reduce the characteristic determinant of B to a determinant of order n by adding x times the $(n - i)$th column to the $(i + 1)$th column $(i = 0, 1, \ldots, n - 1)$ and applying Theorem 3.13, Corollary 1, n times.

6.3 The Triangularization Theorem

Although not every matrix is diagonalizable, we can show that every matrix with elements in $\mathscr{C}$ is similar to a triangular matrix, that is, a matrix with only zero elements below the main diagonal.

Theorem 6.7. *Let A be an $n \times n$ matrix with elements in $\mathscr{C}$ and eigenvalues $k_1, k_2, \ldots, k_n$. There exists a nonsingular matrix P with elements in $\mathscr{C}$ such that*

$$P^{-1}AP = \begin{bmatrix} k_1 & b_{12} & \cdots & b_{1n} \\ 0 & k_2 & \cdots & b_{2n} \\ \multicolumn{4}{c}{\dotfill} \\ 0 & 0 & \cdots & k_n \end{bmatrix}.$$

Proof. Let S_1 be an eigenvector of A corresponding to k_1. Let S be any nonsingular matrix of which S_1 is the first column. By Theorem 2.5 the first column of AS is $AS_1 = k_1 S_1$ and, therefore, the first column of $S^{-1}AS$ is $S^{-1}(k_1 S_1) = k_1 (S^{-1} S_1)$, which is the first column of $k_1 (S^{-1} S)$, namely,

$$\begin{bmatrix} k_1 \\ 0 \\ \cdot \\ \cdot \\ \cdot \\ 0 \end{bmatrix}.$$

Hence we have

$$S^{-1}AS = \begin{bmatrix} k_1 & B_1 \\ O & A_1 \end{bmatrix},$$

where A_1 has order $n - 1$. Since

$$\det (S^{-1}AS - xI) = (k_1 - x) \det (A_1 - xI),$$

and since, by Theorem 6.4, $S^{-1}AS$ and A have the same eigenvalues, it follows that the eigenvalues of A_1 are $k_2, k_3, \ldots, k_n$. The theorem is therefore proved for the case $n = 2$. We assume it true for matrices of order $n - 1$ and proceed by induction. Since A_1 is of order $n - 1$, there exists, by our induction assumption, a nonsingular matrix Q such that

$$Q^{-1}A_1 Q = \begin{bmatrix} k_2 & b_{23} & \cdots & b_{2n} \\ 0 & k_3 & \cdots & b_{3n} \\ \multicolumn{4}{c}{\dotfill} \\ 0 & 0 & \cdots & k_n \end{bmatrix}.$$

Now let

$$R = \begin{bmatrix} 1 & O \\ O & Q \end{bmatrix},$$

and we have

$$(SR)^{-1}A(SR) = R^{-1}S^{-1}ASR = \begin{bmatrix} 1 & O \\ O & Q^{-1} \end{bmatrix} \begin{bmatrix} k_1 & B_1 \\ O & A_1 \end{bmatrix} \begin{bmatrix} 1 & O \\ O & Q \end{bmatrix}$$

$$= \begin{bmatrix} k_1 & B_1 Q \\ O & Q^{-1}A_1 Q \end{bmatrix}$$

$$= \begin{bmatrix} k_1 & b_{12} & \cdots & b_{1n} \\ 0 & k_2 & \cdots & b_{2n} \\ \multicolumn{4}{c}{\dotfill} \\ 0 & 0 & \cdots & k_n \end{bmatrix}.$$

Therefore the matrix $P = SR$ transforms A to triangular form as required.

If

$$f(x) = a_0 x^m + a_1 x^{m-1} + \cdots + a_m$$

is any polynomial in x and A is any square matrix, we denote by $f(A)$ the matrix

$$a_0 A^m + a_1 A^{m-1} + \cdots + a_m I.$$

For example, if $f(x) = x^2 - 2x + 3$ and

$$A = \begin{bmatrix} 1 & -1 \\ 2 & 3 \end{bmatrix},$$

$$f(A) = \begin{bmatrix} 1 & -1 \\ 2 & 3 \end{bmatrix}^2 - 2 \begin{bmatrix} 1 & -1 \\ 2 & 3 \end{bmatrix} + \begin{bmatrix} 3 & 0 \\ 0 & 3 \end{bmatrix} = \begin{bmatrix} 0 & -2 \\ 4 & 4 \end{bmatrix}.$$

Theorem 6.8. *If $k_1, k_2, \ldots, k_n$ are the eigenvalues of A and $f(x)$ is any polynomial, then the eigenvalues of $f(A)$ are $f(k_1), f(k_2), \ldots, f(k_n)$.*

Proof. If B and C are triangular matrices with zeros below the diagonal and diagonal elements $b_1, b_2, \ldots, b_n$ and $c_1, c_2, \ldots, c_n$, it is easy to verify that kB, $B + C$, and BC are all triangular with zeros below the diagonal and diagonal elements, respectively, $kb_1, kb_2, \ldots, kb_n$, $b_1 + c_1, b_2 + c_2, \ldots, b_n + c_n$, and $b_1 c_1, b_2 c_2, \ldots, b_n c_n$. It follows that if $T = P^{-1}AP$ is the triangular matrix similar to A with diagonal elements $k_1, k_2, \ldots, k_n$, then $f(T)$ is triangular with diagonal elements $f(k_1), f(k_2), \ldots, f(k_n)$. However, by Theorem 6.3, $f(T) = f(P^{-1}AP) = P^{-1}f(A)P$. Thus, $f(T)$ is a triangular matrix similar to $f(A)$ and its diagonal elements $f(k_1), f(k_2), \ldots, f(k_n)$ are therefore the eigenvalues of $f(A)$.

We conclude this section with the statements of two important theorems that we shall not prove.

Theorem 6.9. *(Cayley-Hamilton theorem) If $f(x)$ is the characteristic polynomial of a matrix A, then $f(A) = O$.*

For the proof of this theorem, the student is referred to [7]. However, in the special case when A is diagonalizable, the result follows at once from Theorem 6.8 which tells us that all the eigenvalues of $f(A)$ must be zero. Hence, if A, and therefore $f(A)$ also, is diagonalizable we have $P^{-1}f(A)P = O$ and thus $f(A) = O$.

Theorem 6.10. *Every matrix A with elements in $\mathscr{C}$ is similar to a triangular matrix of the form*

$$J = \begin{bmatrix} F_1 & O & \cdots & O \\ O & F_2 & \cdots & O \\ \multicolumn{4}{c}{\dotfill} \\ O & O & \cdots & F_r \end{bmatrix},$$

where each F_i has the form

$$F_i = \begin{bmatrix} k_i & 1 & 0 & \cdots & 0 \\ 0 & k_i & 1 & \cdots & 0 \\ & & \cdots & & \\ 0 & 0 & 0 & \cdots & k_i \end{bmatrix},$$

and $k_1, k_2, \ldots, k_r$ are the eigenvalues of A but are not necessarily distinct.

The matrix J is called the *Jordan canonical form* of A. Two matrices are similar if and only if they have the same Jordan form except possibly for the order of the matrices F_i, and A is diagonalizable if and only if each F_i has order 1. For proof of these results, the student is referred to [5].

Exercise 6.2

1. Find a nonsingular matrix P that will transform the matrix in Problem 6(b), Exercise 6.1, to triangular form.

2. Check that the matrix

$$A = \begin{bmatrix} -20 & 23 & -8 \\ -26 & 30 & -10 \\ -21 & 24 & -7 \end{bmatrix}$$

   has eigenvalues 2 and -1 and use this information to find a matrix P such that $P^{-1}AP$ is triangular.

3. Prove that an $n \times n$ matrix A is singular if and only if it has an eigenvalue 0.

4. If A is nonsingular, prove that the eigenvalues of A^{-1} are the reciprocals of the eigenvalues of A.

5. A nonzero matrix A is said to be nilpotent if, for some positive integer r, $A^r = 0$. Show by successive multiplication that a triangular matrix in which all the diagonal elements are zero is nilpotent.

6. Show that a nonzero matrix is nilpotent if and only if all its eigenvalues are equal to zero.

7. Prove that a nonzero nilpotent matrix cannot be similar to a diagonal matrix.

8. If $A^2 = A$ prove that all eigenvalues of A are equal to 1 or 0.

9. Prove that if A is similar to a diagonal matrix, A^T is similar to A.

10. Prove that every matrix A is similar to its transpose. *Hint.* Note that each of the matrices F_i in the Jordan normal form J of A is symmetric about its secondary diagonal. Hence, by Problem 11, Exercise 6.1, $Q_i^{-1}F_iQ_i = F_i^T$, where Q_i is a

matrix of the type defined in Problem 9, Exercise 6.1. Let

$$Q = \begin{bmatrix} Q_1 & O & \cdots & O \\ O & Q_2 & \cdots & O \\ \multicolumn{4}{c}{\dotfill} \\ O & O & \cdots & Q_r \end{bmatrix}$$

and show that $Q^{-1}JQ = J^T$. Now show that J^T is similar to A^T and the result follows.

11. By modifying the proof of Theorem 6.7, prove that if A is an $n \times n$ matrix with real elements and n real eigenvalues, there exists an orthogonal matrix P such that $P^{-1}AP$ is triangular. *Hint.* Let S_1 be a *unit* eigenvector corresponding to k_1 and let S be an *orthogonal* matrix with S_1 as its first column. Continue as in the proof of Theorem 6.7.

6.4 Application to Systems of Linear Differential Equations

Let $y_1, y_2, \ldots, y_n$ be functions of an independent variable x that satisfy the system of linear differential equations

$$(8) \qquad \frac{dy_i}{dx} = \sum_{j=1}^{n} a_{ij}y_j + b_i(x) \qquad (i = 1, 2, \ldots, n),$$

where $b_i(x)$ are integrable functions of x. The coefficients a_{ij} are, in general, functions of x, but we shall consider only the important special case in which a_{ij} are constants. In order to write (8) in matrix form we let $A = [a_{ij}]$,

$$Y = \begin{bmatrix} y_1 \\ y_2 \\ \cdot \\ \cdot \\ \cdot \\ y_n \end{bmatrix}, \quad \text{and} \quad B = \begin{bmatrix} b_1 \\ b_2 \\ \cdot \\ \cdot \\ \cdot \\ b_n \end{bmatrix}.$$

We call Y and B vector functions of x and define the derivative of Y by

$$\frac{dY}{dx} = \begin{bmatrix} y_1' \\ y_2' \\ \cdot \\ \cdot \\ \cdot \\ y_n' \end{bmatrix},$$

where $y'_i = \dfrac{dy_i}{dx}$. Equations (8) can now be written

(9)
$$\frac{dY}{dx} = AY + B.$$

To solve (9) we make a change of variables and let

$$y_i = \sum_{j=1}^{n} p_{ij} z_j$$

or, in matrix notation, let $Y = PZ$, where $P = [p_{ij}]$ is a matrix of order n with constant elements. Therefore we have

$$\frac{dY}{dx} = P \frac{dZ}{dx},$$

and equation (9) becomes

$$P \frac{dZ}{dx} = APZ + B$$

or

(10)
$$\frac{dZ}{dx} = (P^{-1}AP)Z + P^{-1}B.$$

Now suppose that the matrix A has n linearly independent eigenvectors and, therefore, P can be chosen so that $P^{-1}AP$ is a diagonal matrix. If the eigenvalues of A are $k_1, k_2, \ldots, k_n$, equation (10) is then equivalent to

$$\frac{dz_i}{dx} = k_i z_i + h_i(x) \qquad\qquad (i = 1, 2, \ldots, n),$$

where $h_i(x)$ is a linear combination, with constant coefficients, of $b_1(x), \ldots, b_n(x)$. These are standard linear first order differential equations in z_i. Their solutions are

$$z_i = c_i e^{k_i x} + e^{k_i x} \int e^{-k_i x} h_i(x) \, dx,$$

where the c_i are constants.

The above solution of equations (8) is theoretically extremely simple. To apply it, however, to a numerical example, it is necessary first to find the eigenvalues of A and then to find the linearly independent eigenvectors of A that constitute the column vectors of P. If n is not greater than 3 this could be done by actually finding the characteristic equation of A and solving for the eigenvalues using, if necessary, some method of approximate solution. For larger values of n, iterative procedures for approximating the eigenvalues and solving for the corresponding eigenvectors are available and a computer can be programmed to carry these out. Descriptions of these can be found in [3] or [10].

The solution of equations (8) when A is not similar to a diagonal matrix can be made to depend on Theorem 6.7. If we choose P so that $P^{-1}AP$ is triangular (with zeros *above* the main diagonal), equation (10) is equivalent to the system

$$\frac{dz_1}{dx} = k_1 z_1 + h_1(x),$$

$$\frac{dz_2}{dx} = b_{21} z_1 + k_2 z_2 + h_2(x),$$

(11) $\qquad\qquad \cdot\ \cdot$

$$\frac{dz_n}{dx} = b_{n1} z_1 + b_{n2} z_2 + \cdots + k_n z_n + h_n(x).$$

The first equation of (11) is linear in z_1 and hence immediately solvable. Its solution can be substituted in the second equation, which then becomes a linear equation in z_2. In general, the solutions of the first $r - 1$ equations for $z_1, z_2, \ldots, z_{r-1}$ can be substituted in the rth equation, which can then be solved for z_r.

If the original system is homogeneous, we have $b_i(x) = h_i(x) = 0$ ($i = 1, 2, \ldots, n$). In this case we can see what form the solutions of (11) will take without writing them down explicitly. From the first equation

$$z_1 = c_1 e^{k_1 x},$$

and the second equation becomes

$$\frac{dz_2}{dx} = k_2 z_2 + b_{21} c_1 e^{k_1 x}.$$

This is a linear first-order equation in z_2 whose solution has the form

$$z_2 = c_2 e^{k_1 x} + c_3 e^{k_2 x} \qquad \text{if } k_2 \neq k_1,$$
$$z_2 = (c_2 x + c_3) e^{k_1 x} \qquad \text{if } k_2 = k_1,$$

and, in each case, $c_2 = 0$ if $b_{21} = 0$. Continuing thus, if $k_1 = k_2 = k_3$ the third equation will yield a solution of the form

$$z_3 = (c_4 x^2 + c_5 x + c_6) e^{k_1 x}$$

and, in general, r equal eigenvalues of A give rise to a solution of the form $e^{kx} p(x)$, where $p(x)$ is a polynomial of degree at most $r - 1$. We can therefore conclude that if $k_1, \ldots, k_m$ are the distinct eigenvalues of A, the solutions of the homogeneous system

(12) $\qquad\qquad\qquad\qquad \dfrac{dy_i}{dx} = \displaystyle\sum_{j=1}^{n} a_{ij} y_j$

are linear combinations of $e^{k_1 x}, \ldots, e^{k_m x}$ with coefficients that are either constants or polynomials in x whose degrees are bounded by the maximum multiplicity with which any one of the eigenvalues occurs. In applications, it is often important to know whether the y_i's tend to 0 as $x \to \infty$. This is the problem of stability of the solutions. Our result shows that a necessary and sufficient condition that all the solutions of (12) tend to 0 as $x \to \infty$ is that all the eigenvalues of A have negative real parts.

As an application of the methods described above, consider an electric circuit containing a resistance R_1, capacitance C_1, and inductance L_1, connected in series. If this circuit is coupled with a similar circuit designated by R_2, C_2, and L_2, and if the mutual inductance between the circuits is M, then the currents I_1 and I_2 in the two circuits satisfy the differential equations

$$L_1 \frac{d^2 I_1}{dt^2} + M \frac{d^2 I_2}{dt^2} + R_1 \frac{dI_1}{dt} + \frac{I_1}{C_1} = 0,$$

$$M \frac{d^2 I_1}{dt^2} + L_2 \frac{d^2 I_2}{dt^2} + R_2 \frac{dI_2}{dt} + \frac{I_2}{C_2} = 0.$$

If we let $\dfrac{dI_1}{dt} = y_3$ and $\dfrac{dI_2}{dt} = y_4$, these equations are equivalent to the system

$$\frac{dI_1}{dt} = y_3,$$

$$\frac{dI_2}{dt} = y_4,$$

$$L_1 \frac{dy_3}{dt} + M \frac{dy_4}{dt} = -\frac{1}{C_1} I_1 - R_1 y_3,$$

$$M \frac{dy_3}{dt} + L_2 \frac{dy_4}{dt} = -\frac{1}{C_2} I_2 - R_2 y_4.$$

Denoting by Y the column vector $[I_1, I_2, y_3, y_4]^T$, these equations can now be written

$$A \frac{dY}{dt} = BY,$$

where

$$A = \begin{bmatrix} 1 & 0 & 0 & 0 \\ 0 & 1 & 0 & 0 \\ 0 & 0 & L_1 & M \\ 0 & 0 & M & L_2 \end{bmatrix}, \quad B = \begin{bmatrix} 0 & 0 & 1 & 0 \\ 0 & 0 & 0 & 1 \\ -\dfrac{1}{C_1} & 0 & -R_1 & 0 \\ 0 & -\dfrac{1}{C_1} & 0 & -R_2 \end{bmatrix}.$$

We let $D = L_1L_2 - M^2$. If $D \neq 0$, A is nonsingular and

$$\frac{dY}{dt} = A^{-1}BY,$$

where

$$A^{-1}B = \begin{bmatrix} 0 & 0 & 1 & 0 \\ 0 & 0 & 0 & 1 \\ -\dfrac{L_2}{C_1D} & \dfrac{M}{C_2D} & -\dfrac{R_1L_2}{D} & \dfrac{MR_2}{D} \\ \dfrac{M}{C_1D} & -\dfrac{L_1}{C_2D} & \dfrac{R_1M}{D} & -\dfrac{L_1R_2}{D} \end{bmatrix}.$$

If we assume that the resistances R_1 and R_2 are negligible and put $R_1 = R_2 = 0$, the matrix $A^{-1}B$ takes the form that was considered in Problem 13, Exercise 6.1. Its eigenvalues are therefore $\pm\sqrt{\lambda}$ and $\pm\sqrt{\lambda'}$, where λ and λ' are the roots of

$$\begin{vmatrix} -\dfrac{L_2}{C_1D} - x & \dfrac{M}{C_2D} \\ \dfrac{M}{C_1D} & -\dfrac{L_1}{C_2D} - x \end{vmatrix} = 0.$$

This reduces to the quadratic equation

$$x^2 + \frac{L_1C_1 + L_2C_2}{DC_1C_2}x + \frac{1}{DC_1C_2} = 0,$$

whose discriminant is always positive, being equal to

$$\frac{(L_1C_1 - L_2C_2)^2 + 4M^2C_1C_2}{D^2C_1^2C_2^2}.$$

It follows that λ and λ' are real and unequal. Moreover, since the coefficient of x and the constant term agree in sign with D, a necessary and sufficient condition that λ and λ' both be negative is that $D > 0$ or $L_1L_2 > M^2$. Assuming that this condition is satisfied, we let $\lambda = -\omega^2$ and $\lambda' = -\omega'^2$, and the eigenvalues of $A^{-1}B$ are the pure imaginary numbers $\pm i\omega$ and $\pm i\omega'$. Since these are distinct, $A^{-1}B$ is similar to a diagonal matrix, and the solutions for I_1 and I_2 are therefore linear combinations of $e^{\pm i\omega t}$ and $e^{\pm i\omega't}$ or of $\sin \omega t$, $\cos \omega t$, $\sin \omega't$, and $\cos \omega't$. Thus the condition $L_1L_2 > M^2$ ensures periodic solutions. Two frequencies occur in the solutions, namely, $\omega/2\pi$ and $\omega'/2\pi$. It is clear that if the original differential equations are to be satisfied, both I_1 and I_2 must contain terms with frequency $\omega/2\pi$ and also terms with frequency $\omega'/2\pi$. In other words, both frequencies occur in both circuits. Once the frequencies are known, the solutions for I_1 and I_2 could be determined by actually finding the linear transformation that diagonalizes $A^{-1}B$. However, it is probably easier to substitute expressions of the form $a_1 \cos \omega t + b_1 \sin \omega t + c_1 \cos \omega' t + d_1 \sin \omega' t$ for I_1 and I_2 and determine the relations between the unknown constants so that the differential equations are satisfied. There will be two arbitrary constants in the general solution for I_1 and two in the solution for I_2.

For additional information on the use of matrix methods in the theory of differential equations and their solution, the student is referred to [2] or [8].

Exercise 6.3

1. Solve the following systems of differential equations:

(a) $\dfrac{dy_1}{dx} = y_2,$

$\dfrac{dy_2}{dx} = y_3,$

$\dfrac{dy_3}{dx} = 3y_1 + y_2 - 3y_3.$

(b) $\dfrac{dy_1}{dx} = y_2,$

$\dfrac{dy_2}{dx} = y_3,$

$\dfrac{dy_3}{dx} = y_3 - 4y_2 + 4y_1.$

(c) $\dfrac{dy_1}{dx} = 3y_1 - y_2,$

$\dfrac{dy_2}{dx} = y_1 + 3y_2.$

(d) $\dfrac{dy_1}{dx} = y_1 + y_2 - y_3 + e^x$

$\dfrac{dy_2}{dx} = y_1 - y_2 + y_3 + 2$

$\dfrac{dy_3}{dx} = -y_1 + y_2 + y_3 + 2e^{-x}$

2. Using the result of Problem 1, Exercise 6.2, solve the system

$$\frac{dy_1}{dx} = y_1 + 2y_2 - 4y_3,$$

$$\frac{dy_2}{dx} = 6y_3 - y_2,$$

$$\frac{dy_3}{dx} = 4y_3 - y_2.$$

3. Prove that the substitution $y_2 = dy/dx$, $y_2 = d^2y/dx^2, \ldots, y_n = d^{n-1}y/dx^{n-1}$ will reduce the linear nth-order equation

$$\frac{d^n y}{dx^n} + a_1 \frac{d^{n-1} y}{dx^{n-1}} + \cdots + a_{n-1} \frac{dy}{dx} + a_n y = 0$$

to a linear system of the form (12). Write down the matrix of the resulting system and its characteristic equation.

4. Use the method indicated in Problem 3 to solve

$$\frac{d^4y}{dx^4} - y = 0.$$

6.5 Real Symmetric Matrices

We recall that a matrix A is symmetric if $A^T = A$. Real symmetric matrices constitute an important class of diagonalizable matrices and this particular diagonalization process has many applications. We shall prove the basic theorems in this section and consider applications in Chapter 7.

Theorem 6.11. *The eigenvalues of a real symmetric matrix are real.*

Proof. Although this follows trivially from a later result (see Theorem 8.8), we include an independent proof here. Let A be a real symmetric matrix with a complex eigenvalue $r + is$. We shall prove $s = 0$. Since the matrix $A - (r + is)I$ is singular, so also is the matrix

$$\begin{aligned} B &= [A - (r + is)I][A - (r - is)I] \\ &= A^2 - 2rA + (r^2 + s^2)I \\ &= (A - rI)^2 + s^2I. \end{aligned}$$

Since B is real and singular, there exists a real nonzero column vector X such that $BX = O$ and hence $X^TBX = 0$. Therefore,

$$\begin{aligned} 0 = X^TBX &= X^T(A - rI)^2X + s^2X^TX \\ &= X^T(A - rI)^T(A - rI)X + s^2X^TX \end{aligned}$$

because $A - rI$ is symmetric. Now if we let $Y = (A - rI)X$ we have

(13) $$Y^TY + s^2X^TX = 0.$$

But X and Y are both real and $X \neq 0$. Hence, $Y^TY \geq 0$ and $X^TX > 0$. Thus (13) implies $s = 0$ and the eigenvalue $r + is = r$ and is real.

Theorem 6.12. *If A is a real symmetric matrix, there exists an orthogonal matrix P such that $P^{-1}AP$ is a diagonal matrix.*

Proof. The proof is a simple modification of that of Theorem 6.7. Let $k_1, k_2, \ldots,$ k_n be the eigenvalues of A. Since k_1 is real, there is a real unit eigenvector S_1 of A

corresponding to k_1. By Theorem 4.15, there is an orthogonal matrix S with S_1 as its first column. The first column of AS is $AS_1 = k_1 S_1$ and hence the first column of $S^{-1} A S_1$ is $k_1 S^{-1} S_1$, which is the first column of $k_1 S^{-1} S$ or

$$[k_1, 0, \ldots, 0]^T.$$

But $S^{-1} A S$ is symmetric because, since S is orthogonal,

$$(S^{-1} A S)^T = (S^T A S)^T = S^T A^T S = S^{-1} A S.$$

Therefore,

$$S^{-1} A S = \begin{bmatrix} k_1 & O \\ O & A_1 \end{bmatrix},$$

where A_1 is symmetric, of order $n - 1$, and has eigenvalues $k_2, k_3, \ldots, k_n$. The proof is now completed by induction as in the proof of Theorem 6.7. If Q is an orthogonal matrix of order $n - 1$ that diagonalizes A_1, then SR, where

$$R = \begin{bmatrix} 1 & O \\ O & Q \end{bmatrix},$$

is also orthogonal and diagonalizes A.

Corollary. *A real symmetric matrix of order n has n mutually orthogonal[1] eigenvectors in $\mathscr{V}_n(\mathscr{R})$.*

Proof. By Theorem 6.5 the column vectors of P are eigenvectors of A. These column vectors are mutually orthogonal unit vectors because P is orthogonal.

Conversely, it is clear that if we can find n mutually orthogonal unit eigenvectors of A, then the matrix P of which these are the column vectors is orthogonal and will transform A to diagonal form. The problem of diagonalizing a real symmetric matrix is therefore reduced to that of finding n mutually orthogonal eigenvectors of A. The two following theorems show how this may be done.

Theorem 6.13. *If two eigenvectors S_1 and S_2 of a real symmetric matrix A correspond to different eigenvalues of A, then S_1 and S_2 are orthogonal.*

Proof. Let k_1, k_2 be distinct eigenvalues of A and let S_1, S_2 be corresponding eigenvectors. Then $AS_1 = k_1 S_1$ and $S_2^T A S_1 = k_1 S_2^T S_1$. However, since A is symmetric, $S_2^T A S_1 = (A S_2)^T S_1 = k_2 S_2^T S_1$. Hence, $k_1 S_2^T S_1 = k_2 S_2^T S_1$ and, since $k_1 \neq k_2$, $S_2^T S_1 = S_2 \cdot S_1 = 0$ and S_1 and S_2 are orthogonal.

---

[1] Whenever $\mathscr{V}_n(\mathscr{R})$ is endowed with an inner product, it is assumed, unless otherwise stated, to be the standard inner product. Thus, orthogonality and unit vector mean orthogonality and unit vector relative to the ordinary dot product.

Theorem 6.14. *If k occurs exactly p times as an eigenvalue of a real symmetric matrix A, then A has p but not more than p mutually orthogonal eigenvectors corresponding to k.*

Proof. By Theorem 6.12 there exists a matrix P such that $P^{-1}AP$ is a diagonal matrix in which k occurs exactly p times in the main diagonal. Hence, $P^{-1}AP - kI$ has rank $n - p$. Since $P^{-1}AP - kI = P^{-1}(A - kI)P$ and P and P^{-1} are nonsingular, $A - kI$ also has rank $n - p$. Hence, by Theorem 2.12, the solution space of the equations

$$(A - kI)X = 0$$

has dimension $n - (n - p) = p$, and there are therefore p but not more than p mutually orthogonal unit vectors in this space. These are p mutually orthogonal eigenvectors of A. They are not, of course, uniquely determined.

Example. Find an orthogonal matrix P that will diagonalize the symmetric matrix

$$A = \begin{bmatrix} 0 & 1 & 1 \\ 1 & 0 & -1 \\ 1 & -1 & 0 \end{bmatrix}.$$

Solution. The characteristic equation is

$$\det (A - xI) = -x^3 + 3x - 2$$
$$= -(x - 1)^2(x + 2) = 0,$$

and hence the eigenvalues are $1, 1, -2$. By Theorem 6.14, we can find two mutually orthogonal eigenvectors corresponding to the eigenvalue 1 by solving

$$(A - I)X = \begin{bmatrix} -1 & 1 & 1 \\ 1 & -1 & -1 \\ 1 & -1 & -1 \end{bmatrix} \begin{bmatrix} x_1 \\ x_2 \\ x_3 \end{bmatrix} = \begin{bmatrix} 0 \\ 0 \\ 0 \end{bmatrix}$$

or $-x_1 + x_2 + x_3 = 0$. Orthogonal solutions are $S_1 = [1, 0, 1]$ and $S_2 = [1, 2, -1]$. An eigenvector corresponding to the eigenvalue -2 is found by solving

$$2x_1 + x_2 + x_3 = 0,$$
$$x_1 + 2x - x_3 = 0$$

to be $S_3 = [-1, 1, 1]$ which (as Theorem 6.13 requires) is orthogonal to both S_1 and S_2. The required matrix P is therefore a matrix whose columns are unit vectors which are scalar multiples of S_1, S_2 and S_3, namely,

$$P = \begin{bmatrix} \dfrac{1}{\sqrt{2}} & \dfrac{1}{\sqrt{6}} & -\dfrac{1}{\sqrt{3}} \\[2ex] 0 & \dfrac{2}{\sqrt{6}} & \dfrac{1}{\sqrt{3}} \\[2ex] \dfrac{1}{\sqrt{2}} & -\dfrac{1}{\sqrt{6}} & \dfrac{1}{\sqrt{3}} \end{bmatrix}.$$

The order of the columns of P determines the order in which the eigenvalues of A appear in the diagonal form of A.

We recall from Section 4.8 that a real $n \times n$ symmetric matrix $A = [a_{ij}]$ determines a quadratic form

(13) $$X^T A X = \sum_{i=1}^{n} \sum_{j=1}^{n} a_{ij} x_i x_j$$

and that A is said to be positive definite if, for all nonzero $X \in \mathcal{V}_n(\mathcal{R})$, $X^T A X > 0$. The two theorems that follow characterize this property in matrix terms.

Theorem 6.15. *A real symmetric matrix is positive (nonnegative) definite if and only if all its eigenvalues are positive (nonnegative).*

Proof. Let A be an $n \times n$ real symmetric matrix with eigenvalues $k_1, k_2, \ldots, k_n$. Let P be an orthogonal matrix that diagonalizes A, so that

$$P^{-1}AP = D = \begin{bmatrix} k_1 & 0 & \cdots & 0 \\ 0 & k_2 & \cdots & 0 \\ \multicolumn{4}{c}{\dotfill} \\ 0 & 0 & \cdots & k_n \end{bmatrix}.$$

Let $X^T A X$ be the quadratic form (13) and define $Y = [y_1, y_2, \ldots, y_n]^T$ by $Y = P^T X$. Since $P^T = P^{-1}$, we have also $X = PY$ and

$$X^T A X = (PY)^T A P Y = Y^T P^T A P Y = Y^T D Y$$

and, therefore,

(14) $$X^T A X = k_1 y_1^2 + k_2 y_2^2 + \cdots + k_n y_n^2.$$

Now if $k_1, \ldots, k_n$ are all positive, (14) ensures that $X^T A X \geq 0$ for all X in $\mathcal{V}_n(\mathcal{R})$. Moreover, if $X^T A X = 0$, then $Y = O$ and $X = PY = O$. Hence, A is positive definite. Similarly, if each $k_i \geq 0$, then by (14), $X^T A X \geq 0$ for all X in

$\mathscr{V}_n(\mathscr{R})$ and A is nonnegative definite. Conversely, if $k_1, \ldots, k_n$ are not all positive, suppose, to be definite, that $k_1 \leq 0$. By Theorem 1.1, we can choose real values for $x_1, x_2, \ldots, x_n$ not all zero, such that $y_2 = y_3 = \cdots = y_n = 0$ because this involves solving $n - 1$ linear homogeneous equations in $x_1, x_2, \ldots, x_n$. From $X = PY$, it then follows that $y_1 \neq 0$. For this choice of $x_1, \ldots, x_n$, (14) becomes

$$X^T A X = k_1 y_1^2$$

and, since $k_1 \leq 0$, $X^T A X$ is negative or zero for a nonzero vector X. Hence, A is not positive definite. Moreover, if $k_1 < 0$, then $X^T A X < 0$ when X is chosen as indicated. Thus, if the eigenvalues are not all ≥ 0, A is not nonnegative definite.

Corollary. *A positive definite real symmetric matrix A is nonsingular.*

Proof. Because all the eigenvalues of A are positive, the diagonal matrix similar to A is nonsingular and so therefore is A.

Theorem 6.16. *A real symmetric matrix A is positive definite if and only if there exists a nonsingular matrix Q such that $A = Q^T Q$.*

Proof. If $A = Q^T Q$ where Q is nonsingular we have for all X in $\mathscr{V}_n(\mathscr{R})$,

$$X^T A X = X^T Q^T Q X = (QX)^T Q X = Y \cdot Y \geq 0,$$

where $Y = QX$. Since Q is nonsingular, $Y = O$ if and only if $X = O$ and hence $X^T A X = 0$ if and only if $X = O$ and therefore A is positive definite.

Conversely, if A is positive definite, by Theorem 6.15, we can find an orthogonal matrix P such that

$$P^{-1}AP = P^T A P = D = \begin{bmatrix} k_1 & 0 & \cdots & 0 \\ 0 & k_2 & \cdots & 0 \\ \cdots & \cdots & \cdots & \cdots \\ 0 & 0 & \cdots & k_n \end{bmatrix},$$

where each $k_i > 0$. Let D_1 be the diagonal matrix with diagonal elements $\sqrt{k_1}, \sqrt{k_2}, \ldots, \sqrt{k_n}$ so that $D_1^2 = D$ and $D_1^T = D_1$. Then

$$A = PDP^{-1} = PD_1^2 P^T = PD_1 D_1^T P^T = (PD_1)(PD_1)^T$$

and $A = Q^T Q$ where $Q = (PD_1)^T$. Clearly, Q is nonsingular since P and D_1 are.

Theorem 6.17. *Every real nonsingular matrix A can be written as a product $A = PS$, where S is a positive definite symmetric matrix and P is orthogonal.*

Proof. Since A is nonsingular, $A^T A$ is positive definite symmetric by Theorem 6.16. Choose an orthogonal matrix Q such that

$$Q^T A^T A Q = D = \begin{bmatrix} k_1 & 0 & \cdots & 0 \\ 0 & k_2 & \cdots & 0 \\ \cdots & \cdots & \cdots & \cdots \\ 0 & 0 & \cdots & k_n \end{bmatrix},$$

where $k_1, \ldots, k_n$ are the positive real eigenvalues of $A^T A$. Let $p_i = \sqrt{k_i}$ $(i = 1, 2, \ldots, n)$ and let D_1 be the diagonal matrix, with $p_1, \ldots, p_n$ as diagonal elements, so that $D_1^2 = D$. Now let $S = Q D_1 Q^T$. Clearly, $S^T = S$, so S is symmetric. Moreover, S is positive definite because it is similar to D_1 which has positive eigenvalues. Also, $S^2 = Q D_1 Q^T Q D_1 Q^T = Q D Q^T = A^T A$. Now let $P = A S^{-1}$ and

$$P^T P = (S^{-1})^T A^T A S^{-1} = (S^{-1})^T S = S^{-1} S = I$$

because $A^T A = S^2$ and S, and therefore S^{-1} is symmetric. Hence, P is orthogonal, S is positive definite symmetric, and $A = PS$.

The decomposition $A = SP$ obtained in Theorem 6.17 is called the *polar factorization* of A. Its implication for linear transformations will be discussed in the next section.

Exercise 6.4

1. Find orthogonal matrices that will diagonalize each of the following symmetric matrices:

   (a) $\begin{bmatrix} 4 & 3 \\ 3 & -4 \end{bmatrix}$.

   (b) $\begin{bmatrix} \dfrac{23}{25} & -\dfrac{36}{25} \\ -\dfrac{36}{25} & \dfrac{2}{25} \end{bmatrix}$.

   (c) $\begin{bmatrix} \dfrac{5}{4} & -\dfrac{3\sqrt{3}}{4} \\ -\dfrac{3\sqrt{3}}{4} & -\dfrac{1}{4} \end{bmatrix}$.

   (d) $\begin{bmatrix} 1 & 2 & 0 \\ 2 & 2 & 2 \\ 0 & 2 & 3 \end{bmatrix}$.

   (e) $\begin{bmatrix} 10 & 0 & 2 \\ 0 & 6 & 0 \\ 2 & 0 & 7 \end{bmatrix}$.

   (f) $\begin{bmatrix} 0 & 1 & 1 \\ 1 & 0 & 1 \\ 1 & 1 & 0 \end{bmatrix}$.

2. Using the proof of Theorem 6.7 as a guide, express each of the following as a product of an orthogonal matrix and a positive definite symmetric matrix:

   (a) $\begin{bmatrix} \frac{27}{10} & -\frac{29}{10} \\ \frac{11}{10} & \frac{3}{10} \end{bmatrix}$.

   (b) $\begin{bmatrix} 4\sqrt{2} & 4\sqrt{2} \\ -\sqrt{2} & \sqrt{2} \end{bmatrix}$.

   (c) $\begin{bmatrix} 1 & -1 & -1 \\ 1 & 1 & -1 \\ 1 & 1 & 1 \end{bmatrix}$.

3. Show that if A is symmetric and P orthogonal, then $P^{-1}AP$ is symmetric and use this to deduce Theorem 6.12 from Theorem 6.7 and Problem 11, Exercise 6.2.

4. Prove that every positive definite symmetric matrix A has a unique positive definite "square root" B such that $B^2 = A$.

5. Use the result of Problem 4 to prove the uniqueness of the polar factorization $A = PS$, when A is nonsingular. *Hint.* Let $A = PS = P_1 S_1$ and show $A^T A = S^2 = S_1^2$. Now use Problem 4.

6. Prove that every nonsingular matrix is similar to the product of an orthogonal matrix and a diagonal matrix with positive diagonal elements.

7. Prove that every real nonsingular matrix A is equal to a product $S_1 P_1$, where S_1 is positive definite symmetric and P_1 is orthogonal.

8. Prove that A and A^T have the same eigenvalues. If k_1 and k_2 are distinct eigenvalues of A, prove that any eigenvector of A corresponding to k_1 is orthogonal to any eigenvector of A^T corresponding to k_2. Deduce Theorem 6.13 from this.

9. If A is both real symmetric and orthogonal, prove that all its eigenvalues are $+1$ or -1.

6.6 Application to Linear Operators

Let $\mathscr{V}$ be an n-dimensional vector space over the field $\mathscr{F}$. We return now to consideration of the algebra $\mathscr{L}(\mathscr{V}, \mathscr{V})$ of linear operators in $\mathscr{V}$. We shall use the basic isomorphism $\alpha : \mathscr{L}(\mathscr{V}, \mathscr{V}) \to \mathscr{F}_n$ (see Theorem 5.9) to show that our theorems about matrices are essentially theorems about linear operators.

Definition. *If $\tau : \mathscr{V} \to \mathscr{V}$ is a linear operator and I is the identity operator on $\mathscr{V}$, a scalar k such that $\tau - kI$ is a singular operator (i.e., has nonzero null space) is called an eigenvalue of τ. Any nonzero vector V in the null space of $\tau - kI$ is called an eigenvector of τ corresponding to k or a k-eigenvector of τ.*

Theorem 6.18. *Let $E_1, \ldots, E_n$ be any basis of $\mathscr{V}$ and let $\tau \in \mathscr{L}(\mathscr{V}, \mathscr{V})$. A scalar k is an eigenvalue of τ if and only if k is an eigenvalue of $A_\tau^{(E)}$. A vector V in $\mathscr{V}$ is an eigenvector of τ corresponding to k if and only if its coordinate vector X relative to the E-basis is an eigenvector of $A_\tau^{(E)}$ corresponding to k.*

Proof. We agree to write A_τ for $A_\tau^{(E)}$ and note that the identity matrix I is the matrix relative to the E-basis (or any basis!) of the identity operator I. With this understanding, we have by Theorem 5.7

(15) $$A_{\tau - kI} = A_\tau - kI$$

and, by Theorem 5.10, $\tau - kI$ is singular if and only if $A_\tau - kI$ is singular. Thus, k is an eigenvalue of τ if and only if it is an eigenvalue of A_τ.

Now let X be the E-coordinate vector of V. By Theorem 5.6 and equation (15) the E-coordinate vector of $(\tau - kI)V$ is

$$A_{\tau-kI}X = (A_\tau - kI)X,$$

and hence $(\tau - kI)V = O$ if and only if $(A_\tau - kI)X = O$. Thus, V is a k-eigenvector of τ if and only if X is a k-eigenvector of A_τ. If $F_1, \ldots, F_n$ is a second basis and P is the transition matrix from the E- to the F-basis, it follows that $P^{-1}X$ is an eigenvector corresponding to k of $A_\tau^{(F)} = P^{-1}A_\tau^{(E)}P$. (See Theorem 6.4.)

Definition. *An operator $\tau \in \mathscr{L}(\mathscr{V}, \mathscr{V})$ is said to be diagonalizable if τ has n linearly independent eigenvectors where $n = \dim \mathscr{V}$.*

Theorem 6.19. *An operator $\tau \in \mathscr{L}(\mathscr{V}, \mathscr{V})$ is diagonalizable if and only if its matrix A_τ relative to any basis is diagonalizable.*

Proof. Since vectors in $\mathscr{V}$ are linearly independent if and only if their coordinate vectors are linearly independent (Theorem 1.13), this theorem follows from Theorem 6.18 and Theorem 6.5.

If $\mathscr{F}$ is the real field and $\mathscr{V}$ is endowed with an inner product, then by Theorem 4.16 the vectors $V_1, V_2, \ldots, V_r$ in $\mathscr{V}$ are mutually orthogonal if and only if their coordinate vectors $X_1, X_2, \ldots, X_r$ relative to an orthonormal basis are orthogonal (relative to the dot product) in $\mathscr{V}_n(\mathscr{R})$.

Definition. *If $\mathscr{V}$ is Euclidean with inner product (,) an operator $\tau \in \mathscr{L}(\mathscr{V}, \mathscr{V})$ is called* symmetric *(or* self-adjoint*) if $(\tau V_1, V_2) = (V_1, \tau V_2)$, for all V_1, V_2 in $\mathscr{V}$.*

Theorem 6.20. *If $\mathscr{V}$ is Euclidean, an operator $\tau \in \mathscr{L}(\mathscr{V}, \mathscr{V})$ is symmetric if and only if its matrix relative to an orthonormal basis is symmetric.*

Proof. Let X_1, X_2 be the coordinate vectors of V_1, V_2 relative to an orthonormal basis $E_1, E_2, \ldots, E_n$, and let A_τ be the matrix of τ relative to the same basis. By Theorem 4.16

$$(\tau V_1, V_2) = (A_\tau X_1) \cdot X_2 = (A_\tau X_1)^T X_2 = X_1^T A_\tau^T X_2,$$

whereas

$$(V_1, \tau V_2) = X_1 \cdot (A_\tau X_2) = X_1^T A_\tau X_2$$

Hence, τ is symmetric if and only if $X_1^T A_\tau^T X_2 = X_1^T A_\tau X_2$ for all X_1, X_2 in $\mathscr{V}_n(\mathscr{R})$. By substituting for X_1 and X_2 the ith and jth columns of the identity matrix $(i, j = 1, 2, \ldots, n)$, we see that this implies $A_\tau^T = A_\tau$.

Corollary 1. *A symmetric linear operator in an n-dimensional Euclidean space is diagonalizable, and has real eigenvalues and n mutually orthogonal eigenvectors.*

Definition. *A symmetric linear operator on an n-dimensional Euclidean vector space is said to be* positive *if all its eigenvalues are positive.*

Corollary 2. *A positive symmetric operator on an n-dimensional Euclidean vector space is a magnification in n mutually orthogonal directions.*

Proof. Let τ be a positive operator on $\mathscr{V}$. By Corollary 1, we can choose an orthonormal basis $E_1, \ldots, E_n$ consisting of eigenvectors of τ and hence for $i = 1, 2, \ldots, n$,

$$\tau E_i = k_i E_i \qquad k_i > 0$$

and τ is a magnification in the directions E_i.

Theorem 6.21. *Let $\mathscr{V}$ be an n-dimensional Euclidean vector space. Every nonsingular linear operator $\tau \in \mathscr{L}(\mathscr{V}, \mathscr{V})$ can be written in the form $\tau = \rho\sigma$ where ρ is orthogonal (i.e., length-preserving) and σ is a positive symmetric operator.*

Proof. Let A_τ be the matrix of τ relative to a fixed orthonormal basis $E_1, \ldots, E_n$. By Theorem 6.17, $A_\tau = PS$, where P is orthogonal and S is positive definite symmetric. By Theorem 5.9, $P = A_\rho^{(E)}$ and $S = A_\sigma^{(E)}$, where ρ, $\sigma \in \mathscr{L}(\mathscr{V}, \mathscr{V})$. Since $A_\rho^{(E)}$ is orthogonal and the E-basis is orthonormal, ρ is orthogonal by Theorem 5.14. Similarly, since $A_\sigma^{(E)}$ is positive definite symmetric, σ is symmetric and positive by Theorem 6.20. Since $A_\tau^{(E)} = A_\rho^{(E)} A_\sigma^{(E)} = A_{\rho\sigma}^{(E)}$, we have $\tau = \rho\sigma$. Thus, every nonsingular linear operator $\tau : \mathscr{V} \to \mathscr{V}$ is equivalent either to a magnification σ followed by a rotation (if ρ is proper) or to a magnification followed by a rotation and a reflection (if ρ is improper).

CHAPTER SEVEN

REDUCTION OF QUADRATIC FORMS AND APPLICATIONS

7.1 Equivalence of Quadratic Forms

We want to investigate the effect on the real quadratic form

$$X^T A X = \sum_{i=1}^{n} \sum_{j=1}^{n} a_{ij} x_i x_j,$$

with symmetric matrix A, of a change of variables $X = PY$, or

$$x_i = \sum_{j=1}^{n} p_{ij} y_j \qquad (i = 1, 2, \ldots, n).$$

Since $X^T A X = (PY)^T A P Y = Y^T P^T A P Y$, the effect is to transform $X^T A X$ into $Y^T B Y$, where $B = P^T A P$. We note that $B^T = P^T A^T P = P^T A P = B$, so that B is also symmetric. This suggests the following definition of equivalent forms.

Definition. *Two quadratic forms $X^T A X$ and $Y^T B Y$ (or $X^T B X$) are said to be equivalent if there exists a nonsingular matrix P such that $B = P^T A P$. More specifically, the forms are* orthogonally *equivalent if P can be chosen to be orthogonal,* real-equivalent *if P can be chosen with real elements, and* complex-equivalent, *or simply equivalent, if P has elements in $\mathscr{C}$.*

237

Clearly, orthogonal equivalence implies real-equivalence and real-equivalence implies complex-equivalence but not conversely. It is also easy to verify that each of these equivalences is an equivalence relation in the sense of Section 2.4. Moreover, equivalence of $X^T A X$ and $Y^T B Y$ means that $X^T A X$ can be transformed into $Y^T B Y$ by a transformation $X = PY$, where P is nonsingular.

Theorem 7.1. *Two real quadratic forms $X^T A X$ and $X^T B X$ are orthogonally equivalent if and only if A and B have the same eigenvalues and these occur with the same multiplicities.*

Proof. If A and B have eigenvalues $k_1, k_2, \ldots, k_n$ and D is a diagonal matrix with $k_1, \ldots, k_n$ as diagonal elements, then there exist orthogonal matrices P and Q such that

$$P^T A P = Q^T B Q = D.$$

Hence, $B = (Q^T)^{-1} P^T A P Q^{-1} = (PQ^{-1})^T A P Q^{-1}$, and since PQ^{-1} is orthogonal, $X^T B X$ is orthogonally equivalent to $X^T A X$. Conversely, if the two forms are orthogonally equivalent, B is similar to A (because $P^T = P^{-1}$) and A and B have the same eigenvalues with the same multiplicities.

Corollary. *The real quadratic form $X^T A X$ is orthogonally equivalent to the form*

$$X^T D X = k_1 x_1^2 + k_2 x_2^2 + \cdots + k_n x_n^2,$$

where $k_1, k_2, \ldots, k_n$ are the eigenvalues of A.

Theorem 7.2. *Every real quadratic form $X^T A X$ in n variables $x_1, x_2, \ldots, x_n$ is real-equivalent to the form*

$$(1) \qquad x_1^2 + \cdots + x_p^2 - x_{p+1}^2 - \cdots - x_r^2,$$

where r is the rank of A and p is the number of positive eigenvalues of A.

Proof. If A has rank r its diagonal form has rank r and hence A has exactly r nonzero eigenvalues $k_1, k_2, \ldots, k_r$. Suppose that $k_1, \ldots, k_p$ are positive and $k_{p+1}, \ldots, k_r$ are negative. Let D be the $n \times n$ diagonal matrix with diagonal elements $k_1, k_2, \ldots, k_r, 0, \ldots, 0$ and let Q be the $n \times n$ (real) diagonal matrix with diagonal elements

$$\frac{1}{\sqrt{k_1}}, \ldots, \frac{1}{\sqrt{k_p}}, \frac{1}{\sqrt{-k_{p+1}}}, \ldots, \frac{1}{\sqrt{-k_r}}, 1, \ldots, 1.$$

Let P be an orthogonal matrix such that $P^T A P = D$. Then

$$B = (PQ)^T A P Q = Q^T D Q = \begin{bmatrix} I_p & O & O \\ O & -I_{r-p} & O \\ O & O & O_{n-r} \end{bmatrix}.$$

Naturally, the zero matrix O_{n-r} in B is missing if $r = n$, I_p is missing if $p = 0$, and $-I_{r-p}$ if $r = p$. Since P and Q are real and nonsingular, so is PQ and hence $X^T A X$ is real-equivalent to

$$X^T B X = x_1^2 + \cdots + x_p^2 - x_{p+1}^2 - \cdots - x_r^2.$$

Theorem 7.3. *The two forms*

$$X^T A X = x_1^2 + \cdots + x_p^2 - x_{p+1}^2 - \cdots - x_r^2,$$
$$Y^T B Y = y_1^2 + \cdots + y_q^2 - y_{q+1}^2 - \cdots - y_s^2$$

are real-equivalent if and only if $r = s$ and $p = q$. It is assumed here that A, B are $n \times n$ matrices with $n \geq r$ and $n \geq s$, $X^T = [x_1, x_2, \ldots, x_n]$, $Y^T = [y_1, y_2, \ldots, y_n]$.

Proof. If $r = s$ and $p = q$, then $A = B$ and the forms are clearly real-equivalent. Conversely, if the forms are real-equivalent, then A and B have the same rank and therefore $s = r$. It still must be proved that $q = p$. Suppose $B = P^T A P$, where $P = [c_{ij}]$ is a real nonsingular $n \times n$ matrix. Then

(2) $$Y^T B Y = (PY)^T A(PY) = X^T A X,$$

where $X = PY$. Now, if $q > p$ we can choose real values of $y_1, y_2, \ldots, y_n$, not all zero, to satisfy the $n - q + p$ equations:

$$x_1 = c_{11} y_1 + c_{12} y_2 + \cdots + c_{1n} y_n = 0,$$
$$\cdot \quad \cdot \quad \cdot \quad \cdot \quad \cdot \quad \cdot \quad \cdot \quad \cdot \quad \cdot \quad \cdot \quad \cdot \quad \cdot$$
$$x_p = c_{p1} y_1 + c_{p2} y_2 + \cdots + c_{pn} y_n = 0,$$
$$y_{q+1} = 0$$
$$\cdot$$
$$\cdot$$
$$\cdot$$
$$y_n = 0.$$

Substituting this choice of the y's in (2), we get

$$y_1^2 + y_2^2 + \cdots + y_q^2 = -x_{p+1}^2 - \cdots - x_r^2,$$

which is impossible since X and Y are real and $y_1, y_r, \ldots, y_q$ are not all zero. Hence, $q > p$ is impossible and a similar argument shows $p \not> q$, so that $p = q$.

The quadratic form (1) is called the *real canonical form of $X^T A X$*. As a result of Theorems 7.2 and 7.3, we have the following.

Corollary 1. *Two real quadratic forms are real-equivalent if and only if they have the same real canonical form.*

Corollary 2. *Two real quadratic forms are real-equivalent if and only if their matrices have the same rank r and the same number p of positive eigenvalues. These two numbers r and p are therefore invariant under real nonsingular transformations of the form $X = PY$.*

Theorem 7.4. *Every real quadratic form $X^T A X$ is complex-equivalent to the form $x_1^2 + x^2 + \cdots + x_r^2$ where r is the rank of A.*

Proof. Let p be the number of positive eigenvalues of A and choose a real nonsingular matrix P such that $X^T P^T A P X$ is the real canonical form of $X^T A X$, so that

$$P^T A P = \begin{bmatrix} I_p & & \\ & -I_{r-p} & \\ & & O \end{bmatrix}.$$

Now let Q be the nonsingular diagonal matrix in $\mathscr{C}$ with 1's in the first p and last $n - r$ diagonal places and i in each of the $r - p$ places in which -1's occur in $P^T A P$. Clearly,

$$(PQ)^T A (PQ) = Q^T (P^T A P) Q = \begin{bmatrix} I_r & O \\ O & O \end{bmatrix},$$

PQ is nonsingular, and $X = (PQ)Y$ transforms $X^T A X$ into $y_1^2 + y_2^2 + \cdots + y_r^2$.

Corollary. *Two real quadratic forms in $x_1, x_2, \ldots, x_n$ are complex-equivalent if and only if they have the same rank.*

Theorem 7.5. *A real quadratic form $X^T A X$ is a product of two linear factors with coefficients in $\mathscr{C}$ if and only if rank $A \le 2$.*

Proof. We assume $A \ne 0$ since otherwise the theorem is trivial. Suppose, first, that $X^T A X$ factors and

(3) $\qquad X^T A X = (a_1 x_1 + a_2 x_2 + \cdots + a_n x_n)(b_1 x_1 + b_2 x_2 + \cdots + b_n x_n).$

If the two factors are linearly independent, we can assume, by a permutation of the x_i if necessary, that

$$\begin{vmatrix} a_1 & a_2 \\ b_1 & b_2 \end{vmatrix} \neq 0.$$

Now let

$$y_1 = a_1 x_1 + a_2 x_2 + \cdots + a_n x_n,$$

$$y_2 = b_1 x_1 + b_2 x_2 + \cdots + b_n x_n,$$

$$y_i = x_i \qquad (i = 3, 4, \ldots, n),$$

and we have $Y = QX$ with Q nonsingular. The transformation $Y = QX$, or $X = PY$ where $P = Q^{-1}$, gives $X^T A X = y_1 y_2$. But $y_1 y_2$ is a quadratic form of rank 2 and, by the corollary to Theorem 7.4, $X^T A X$ also has rank 2.

If the two factors in (3) are not linearly independent, we can assume $a_1 \neq 0$ and

$$X^T A X = k(a_1 x_1 + a_2 x_2 + \cdots + a_n x_n)^2 \qquad k \neq 0.$$

In this case let

$$y_1 = a_1 x_1 + a_2 x_2 + \cdots + a_n x_n,$$

$$y_i = x_i \qquad (i = 2, 3, \ldots, n),$$

and we see as above that $X^T A X$ is complex-equivalent to the form $k y_1^2$ which has rank 1. Thus, if $X^T A X$ factors it has rank ≤ 2.

Conversely, if $X^T A X$ has rank ≤ 2 but $A \neq 0$, by Theorem 7.4 it is complex-equivalent either to y_1^2 or to $y_1^2 + y_2^2$, where $X = PY$ and P is nonsingular. Since both y_1^2 and $y_1^2 + y_2^2$ are products of linear factors in $\mathscr{C}$, the substitution $Y = P^{-1}X$ gives a factorization of $X^T A X$.

Example 1. Find a transformation $X = PY$ that will transform $x^2 + 2y^2 + 3z^2 + 4xy + 4yz$ to real canonical form.

Solution. The matrix of the form is

$$A = \begin{bmatrix} 1 & 2 & 0 \\ 2 & 2 & 2 \\ 0 & 2 & 3 \end{bmatrix}$$

and has eigenvalues 2, 5, and -1. Corresponding unit eigenvectors are $[\frac{2}{3}, \frac{1}{3}, -\frac{2}{3}]$, $[\frac{1}{3}, \frac{2}{3}, \frac{2}{3}]$, and $[\frac{2}{3}, -\frac{2}{3}, \frac{1}{3}]$. Hence the orthogonal matrix

$$Q = \begin{bmatrix} \frac{2}{3} & \frac{1}{3} & \frac{2}{3} \\ \frac{1}{3} & \frac{2}{3} & -\frac{2}{3} \\ -\frac{2}{3} & \frac{2}{3} & \frac{1}{3} \end{bmatrix}$$

transforms A to diagonal form

$$Q^T A Q = Q^{-1} A Q = \begin{bmatrix} 2 & 0 & 0 \\ 0 & 5 & 0 \\ 0 & 0 & -1 \end{bmatrix}.$$

Now let

$$R = \begin{bmatrix} \dfrac{1}{\sqrt{2}} & 0 & 0 \\ 0 & \dfrac{1}{\sqrt{5}} & 0 \\ 0 & 0 & 1 \end{bmatrix}$$

and then

$$(QR)^T A(QR) = R^T Q^T A Q R = \begin{bmatrix} 1 & 0 & 0 \\ 0 & 1 & 0 \\ 0 & 0 & -1 \end{bmatrix}.$$

Therefore we let

$$P = QR = \begin{bmatrix} \dfrac{2}{3\sqrt{2}} & \dfrac{1}{3\sqrt{5}} & \dfrac{2}{3} \\ \dfrac{1}{3\sqrt{2}} & \dfrac{2}{3\sqrt{5}} & -\dfrac{2}{3} \\ -\dfrac{2}{3\sqrt{2}} & \dfrac{2}{3\sqrt{5}} & \dfrac{1}{3} \end{bmatrix}$$

and $X = PY$ will reduce the form to $Y^T P^T A P Y$ or $y_1^2 + y_2^2 - y_3^2$.

Example 2. Show that the quadratic form

$$5x_1^2 + 4x_3^2 + 4x_1 x_2 - 12x_1 x_3 - 8x_2 x_3$$

is a product of linear factors and find these factors.

Solution. The eigenvalues of the matrix A of the form are found to be 12, -3, and 0 so that rank $A = 2$ and, by Theorem 7.5, $X^T A X$ factors. The orthogonal matrix that diagonalizes A is found to be the matrix Q of Example 1. Therefore the transformation $X = QY$ transforms $X^T A X$ into $12y_1^2 - 3y_2^2 = 3(2y_1 + y_2)(2y_1 - y_2)$. Since $Y = Q^T X$, we have

$$y_1 = \tfrac{1}{3}(2x_1 + x_2 - 2x_3),$$

$$y_2 = \tfrac{1}{3}(x_1 + 2x_2 + 2x_3)$$

and

$$(2y_1 + y_2) = \tfrac{1}{3}(5x_1 + 4x_2 - 2x_3),$$

$$2y_1 - y_2 = (x_1 - 2x_3).$$

Therefore

$$X^T A X = (5x_1 + 4x_2 - 2x_3)(x_1 - 2x_3).$$

Exercise 7.1

1. Write the matrix and find the rank of each of the following quadratic forms:

   (a) $x_1^2 - 2x_1x_2 + 2x_2^2$.

   (b) $x_1x_2 - x_2^2$.

   (c) $9x_1^2 - 12x_1x_2 + 4x_2^2$.

   (d) $4x_1^2 + x_2^2 - 8x_3^2 + 4x_1x_2 - 4x_1x_3 + 8x_2x_3$

   (e) $x_1^2 - x_3^2 + 4x_1x_2 + 4x_2x_3$.

   (f) $x_1x_2 - x_3x_4$.

2. Which of the quadratic forms in Problem 1 are products of linear factors?

3. Find the real canonical form of each of the quadratic forms in Problem 1. For 1(e) and 1(f) find the transformation $X = PY$ that effects the reduction to real canonical form.

4. Show that the following are products of linear factors and find these factors:

   (a) $x_1^2 + x_2^2 + 2x_3^2 - 2x_1x_2 - 6\sqrt{2}\,x_1x_3 + 6\sqrt{2}\,x_2x_3$.

   (b) $x_1^2 - x_2^2 - 5x_3^2 - 4x_1x_3 - 6x_2x_3$.

5. Write down the matrix of the quadratic form

   $$(ax_1 + bx_2 + cx_3)^2$$

   and check that it has rank 1. Do the same for the form

   $$(ax_1 + bx_2 + cx_3 + dx_4)^2.$$

6. If A is the matrix

   $$\begin{bmatrix} 2ad & ae + bd & af + cd \\ ae + bd & 2be & bf + ce \\ af + cd & bf + ce & 2cf \end{bmatrix},$$

   show that det $A = 0$. *Hint.* Show that $\tfrac{1}{2}A$ is the matrix of the quadratic form

   $$(ax_1 + bx_2 + cx_3)(dx_1 + ex_2 + fx_3).$$

7. A real quadratic form $X^T A X$ defines a function $f : \mathscr{V}_n(\mathscr{R}) \to \mathscr{R}$ such that $f(X) = X^T A X$. The subset im f of $\mathscr{R}$ is called the *range of values* of the quadratic form.

   (a) Prove that real-equivalent forms have the same range of values.

   (b) Prove that the range of values of a nonnegative definite form is the set of nonnegative real numbers.

(c) A real quadratic form is said to be *indefinite* if its range of values contains both positive and negative real numbers. Prove that the range of values of an indefinite form is the set $\mathscr{R}$ of all real numbers.

7.2 Classification of Quadric Surfaces

The general equation of the second degree in x, y, and z has the form

(4) $\quad ax^2 + by^2 + cz^2 + 2hxy + 2gxz + 2fyz + 2lx + 2my + 2nz + d = 0,$

in which the coefficients are assumed to be real numbers. We use the term *quadric surface*, or simply *quadric*, for the graph of any equation of this form. Our object here is to give, by means of suitable rotations and translations of axes, a complete classification of these surfaces which will include, of course, the quadric surfaces discussed in Chapter 0.

We assume that (4) is the equation of the quadric relative to rectangular coordinate axes OX, OY, and OZ. Let E_1, E_2, and E_3 be unit vectors along OX, OY, and OZ, respectively, that is, the standard orthonormal basis of $\mathscr{V}_3(\mathscr{R})$.

The second-degree terms of equation (4) constitute a quadratic form in x, y, z whose matrix is

$$H = \begin{bmatrix} a & h & g \\ h & b & f \\ g & f & c \end{bmatrix}.$$

By the corollary to Theorem 6.12, H has three mutually orthogonal unit eigenvectors which may be chosen to be positively oriented. Let these be

$$F_1 = [\lambda_1, \mu_1, \nu_1] = \lambda_1 E_1 + \mu_1 E_2 + \nu_1 E_3,$$
$$F_2 = [\lambda_2, \mu_2, \nu_2] = \lambda_2 E_1 + \mu_2 E_2 + \nu_2 E_3,$$
$$F_3 = [\lambda_3, \mu_3, \nu_3] = \lambda_3 E_1 + \mu_3 E_2 + \nu_3 E_3,$$

and choose new coordinate axes OX', OY', OZ' along F_1, F_2, F_3, respectively.

The transition matrix from the E- to the F-basis is

$$P = \begin{bmatrix} \lambda_1 & \lambda_2 & \lambda_3 \\ \mu_1 & \mu_2 & \mu_3 \\ \nu_1 & \nu_2 & \nu_3 \end{bmatrix}$$

and, since both bases are orthonormal and similarly oriented, P is proper orthogonal and the coordinate transformation is a rotation of axes.

Now if (x, y, z) and (x', y', z') are the E- and F-coordinates of the same point and if $X = [x, y, z]^T$, $X' = [x', y', z']^T$ are the corresponding column vectors, we have $X = PX'$ or

$$x = \lambda_1 x' + \lambda_2 y' + \lambda_3 z',$$
$$y = \mu_1 x' + \mu_2 y' + \mu_2 z',$$
$$z = \nu_1 x' + \nu_2 y' + \nu_3 z'$$

as the equations connecting the two coordinate systems. The quadratic terms in (4) which constitute the quadratic form $X^T H X$ are therefore transformed into

$$X'^T P^T H P X' = X'^T D X' = k_1 x'^2 + k_2 y'^2 + k_3 z'^2,$$

where k_1, k_2, k_3 are the eigenvalues of H and D is the diagonal form of H. Hence the equation of the quadric (4) in the F-coordinate system has the form

(5) $$k_1 x'^2 + k_2 y'^2 + k_3 z'^2 + 2l' x' + 2m' y' + 2n' z' + d = 0.$$

We now classify the different quadric surfaces that can be represented by equation (5) by comparing (5), after suitable translation of axes, with the standard forms discussed in Chapter 0.

Case I (k_1, k_2, k_3 all different from zero). In this case the translation defined by

$$x'' = x' + l'/k_1,$$
$$y'' = y' + m'/k_2,$$
$$z'' = z' + n'/k_3$$

reduces equation (5) to the form

(6) $$k_1 x^2 + k_2 y^2 + k_3 z^2 + d' = 0,$$

where $d' = d - \dfrac{l'^2}{k_1} - \dfrac{m'^2}{k_2} - \dfrac{n'^2}{k_3}$ and, for convenience, the double primes have been omitted.

Case I(a) (k_1, k_2, k_3 all have the same sign).

1. If d' agrees in sign with k_1, the surface contains no real points but will be called an imaginary ellipsoid.
2. If d' differs in sign from k_1, the surface is a real ellipsoid.
3. If $d' = 0$, the surface is an imaginary quadric cone containing one real point $(0, 0, 0)$.

Case I(b) (k_1, k_2, k_3 not all of the same sign).

4. If the sign of d' agrees with the sign of two of the k_i's, the surface is a hyperboloid of two sheets.

5. If the sign of d' differs from that of two of the k_i's, the surface is a hyperboloid of one sheet.

6. If $d' = 0$, the surface is a real quadric cone.

Case II (one of k_1, k_2, k_3 is zero, two different from zero). In this case, assuming, for example, that $k_3 = 0$, the translation

$$x'' = x' + l'/k_1,$$

$$y'' = y' + m'/k_2,$$

$$z'' = z'$$

reduces equation (5) to the form

$$k_1 x^2 + k_2 y^2 + 2n'z + d' = 0.$$

Case II(a). If $n' \neq 0$, the translation $z' = z + d'/2n'$ gives $k_1 x^2 + k_2 y^2 + 2n'z = 0$.

7. If k_1, k_2 have like signs, the surface is an elliptic paraboloid.

8. If k_1, k_2 have opposite signs, the surface is a hyperbolic paraboloid.

Case II(b). If $n' = 0$, the equation has the form $k_1 x^2 + k_2 y^2 + d' = 0$.

9. If k_1, k_2 have the same sign and d' the opposite sign, the surface is a real elliptic cylinder.

10. If k_1, k_2, and d' all have the same sign, the surface is an imaginary elliptic cylinder.

11. If k_1, k_2 have opposite signs and $d' \neq 0$, the surface is a hyperbolic cylinder.

12. If k_1, k_2 have the same sign and $d' = 0$, the surface is two imaginary planes intersecting in a real line, namely, the z-axis.

13. If k_1 and k_2 have opposite signs and $d' = 0$, the surface is two real intersecting planes.

Case III (if two of k_1, k_2, k_3 are zero). In this case, assuming $k_1 \neq 0$, $k_2 = k_3 = 0$, equation (5) takes the form $k_1 x^2 + 2m'y + 2n'z + d' = 0$ after the translation $x'' = x' + l'/k_1$, $y'' = y'$, $z'' = z'$.

Case III(a) (m' and n' not both zero). If m', n' are both different from zero, a suitable rotation about the x-axis will reduce the equation to

$$k_1 x^2 + m''y + d' = 0 \qquad m'' \neq 0,$$

and a translation will remove the constant term.

14. Hence, if m', n' are not both zero, the surface is a parabolic cylinder.

Case III(b) ($m' = n' = 0$).

15. If k_1 and d' have like signs, then the surface is two imaginary parallel planes.
16. If k_1 and d' differ in sign, the surface is two real parallel planes.
17. If $d' = 0$, the surface is two real coincident planes.

This completes the first classification of the 17 different types of quadric surfaces. Of these, numbers 1, 2, 4, 5, 7, and 8 are the six *true quadrics*. Five of these are real and one imaginary. All the rest are called *singular* quadrics. The reason for this distinction will appear in the next section where we shall give another classification of quadrics according to rank.

Example 1. By rotation and translation of axes, reduce the equation

$$2x^2 + 2y^2 + 5z^2 - 4xy + 2yz - 2xz - 10x - 6y - 2z = 0$$

to standard form and determine what type of quadric it represents.

Solution. The matrix of the relevant quadratic form is

$$H = \begin{bmatrix} 2 & -2 & -1 \\ -2 & 2 & 1 \\ -1 & 1 & 5 \end{bmatrix}$$

and $\det(H - xI) = -x^3 + 9x^2 - 18x$. We equate to 0 and solve to find eigenvalues 6, 3, and 0. Corresponding eigenvectors are $[-1, 1, 2]$, $[1, -1, 1]$, and $[1, 1, 0]$. The transition matrix P of the necessary rotation is therefore

$$P = \begin{bmatrix} -\dfrac{1}{\sqrt{6}} & \dfrac{1}{\sqrt{3}} & \dfrac{1}{\sqrt{2}} \\[2mm] \dfrac{1}{\sqrt{6}} & -\dfrac{1}{\sqrt{3}} & \dfrac{1}{\sqrt{2}} \\[2mm] \dfrac{2}{\sqrt{6}} & \dfrac{1}{\sqrt{3}} & 0 \end{bmatrix},$$

where the columns are chosen to be unit eigenvectors such that $\det P = 1$. The transformation

$$x = -\frac{x'}{\sqrt{6}} + \frac{y'}{\sqrt{3}} + \frac{z'}{\sqrt{2}},$$

$$y = \frac{x'}{\sqrt{6}} - \frac{y'}{\sqrt{3}} + \frac{z'}{\sqrt{2}},$$

$$z = \frac{2x'}{\sqrt{6}} + \frac{y'}{\sqrt{3}}$$

then reduces the given equation to

$$6x'^2 + 3y'^2 - 2\sqrt{3}\,y' - 8\sqrt{2}\,z' = 0,$$

where only the first degree terms need be computed because we know how the rotation affects the quadratic terms. The equation may now be written

$$6x'^2 + 3\left(y'^2 - \frac{2}{\sqrt{3}}y' + \frac{1}{3}\right) - 8\sqrt{2}\left(z' + \frac{1}{8\sqrt{2}}\right) = 0$$

and a translation to the new origin $\left(0, \dfrac{1}{\sqrt{3}}, -\dfrac{1}{\sqrt{2}}\right)$ reduces the equation to the form

$$6x^2 + 3y^2 - 8\sqrt{2}\,z = 0.$$

The surface is therefore an elliptic paraboloid.

Example 2. By a suitable rotation of axes, find what curve is represented by the equation

$$-x^2 + 4xy + 2y^2 = 6.$$

Solution. The left-hand side is a quadratic form with matrix

$$A = \begin{bmatrix} -1 & 2 \\ 2 & 2 \end{bmatrix}.$$

The characteristic equation of A is

$$\det(A - xI) = x^2 - x - 6 = 0,$$

and its eigenvalues are 3 and -2. Corresponding unit eigenvectors are $\left[\dfrac{1}{\sqrt{5}}, \dfrac{2}{\sqrt{5}}\right]$ and $\left[-\dfrac{2}{\sqrt{5}}, \dfrac{1}{\sqrt{5}}\right]$. Hence the necessary rotation of axes has transition matrix

$$P = \begin{bmatrix} \dfrac{1}{\sqrt{5}} & -\dfrac{2}{\sqrt{5}} \\ \dfrac{2}{\sqrt{5}} & \dfrac{1}{\sqrt{5}} \end{bmatrix}$$

(chosen with $\det P = 1$), and the transformation $X = PX'$ reduces the equation to

$$3x'^2 - 2y'^2 = 6.$$

The graph is therefore a hyperbola.

Exercise 7.2

1. Reduce the following equations to standard form by a suitable rotation of axes followed if necessary by a translation. Hence, determine the type of quadric that each equation represents.
   (a) $2xy + 2yz + 2xz = 1$.
   (b) $2x^2 + z^2 + 2xy + 3xz + yz - 2 = 0$.
   (c) $x^2 - yz + 2x - 4y + 3z = 12$.
   (d) $6x^2 - 2y^2 + z^2 + 12xz - 36x - 4y - 16z + 62 = 0$.
   (e) $x^2 + 4y^2 + 4z^2 - 4xy - 4xz + 8yz + 6x - 6y - 3 = 0$.

2. Use the orthogonal reduction of quadratic forms to find rotations of axes that will reduce the following equations to standard equations of conics. Hence, determine what type of conic each equation represents.
   (a) $x^2 - 4xy + 4y^2 + 2x - 2y + 6 = 0$.
   (b) $2x^2 - 4xy + 5y^2 = 36$.
   (c) $4x^2 + 6xy - 4y^2 + 8x = 20$.

3. Classify all curves, real or imaginary, which can be represented by the general equation

$$ax^2 + 2hxy + by^2 + 2gx + 2fy + c = 0,$$

according to the nature of the eigenvalues of the matrix

$$\begin{bmatrix} a & h \\ h & b \end{bmatrix}.$$

4. Find a necessary and sufficient condition on the coefficients of the equation in Problem 3 that this equation represent two (real or imaginary) straight lines. *Hint.* Show that the left-hand side of this equation factors if and only if the quadratic form $ax^2 + 2hxy + by^2 + 2gxt + 2fyt + ct^2$ factors, and then use Theorem 7.5.

7.3 Classification of Quadrics by Rank

Consider the general quadratic equation (4) and denote its left-hand side by $\varphi(x, y, z)$. In order to exploit to the full our matrix notation for quadratic forms, we introduce a quadratic form in four variables associated with $\varphi(x, y, z)$. This is the quadratic form

(7)
$$\psi(x, y, z, u) = u^2 \varphi\left(\frac{x}{u}, \frac{y}{u}, \frac{z}{u}\right)$$
$$= ax^2 + by^2 + cz^2 + du^2 + 2hxy$$
$$+ 2gxz + 2fyz + 2lxu + 2myu + 2nzu$$

whose matrix is

$$D = \begin{bmatrix} a & b & g & l \\ h & b & f & m \\ g & f & c & n \\ l & m & n & d \end{bmatrix}.$$

The determinant det D is called the *discriminant* of the quadric and the rank of D is called the rank of the quadric. We shall see in Section 7.4 that both det D and rank D are invariant under rotations and translations of coordinate axes.

It is easy to see that $\varphi(x, y, z)$ factors into (real or imaginary) linear factors if and only if $\psi(x, y, z, u)$ does. For a factorization of ψ gives a factorization of φ on putting $u = 1$ whereas a factorization of φ gives a factorization of ψ via equation (7). By Theorem 7.5, therefore, φ is a product of linear factors if and only if rank $D \leq 2$ and hence a quadric consists of two planes (real, imaginary, or coincident) if and only if its rank ≤ 2. It is not difficult to show that the two planes are distinct if the rank is 2 but coincident if the rank is 1.

In order to distinguish the quadrics of rank 3 from those of rank 4, we apply to $\varphi(x, y, z, u)$ the transformation

(8)
$$\begin{aligned} x &= \lambda_1 x' + \lambda_2 y' + \lambda_3 z', \\ y &= \mu_1 x' + \mu_2 y' + \mu_3 z', \\ z &= \nu_1 x' + \nu_2 y' + \nu_3 z', \\ u &= u' \end{aligned}$$

whose matrix is $Q = \begin{bmatrix} P & O \\ O & 1 \end{bmatrix}$. Here, P is the matrix of Section 7.2 whose columns are unit eigenvectors of H. Writing X and X' for the vectors $[x, y, z, u]^T$ and $[x', y', z', u']^T$, the form $\varphi(x, y, z, u) = X^T D X$ is transformed by (8) into $X'^T D' X'$, where

$$D' = Q^T D Q = \begin{bmatrix} k_1 & 0 & 0 & l' \\ 0 & k_2 & 0 & m' \\ 0 & 0 & k_3 & n' \\ l' & m' & n' & d \end{bmatrix}$$

and k_1, k_2, k_3 are, as before, the eigenvalues of H. Since Q is nonsingular, the rank of D' is equal to that of D. If k_1, k_2, k_3 are all different from 0, D' is equivalent under elementary row and column transformations to the diagonal matrix

$$D'' = \begin{bmatrix} k_1 & 0 & 0 & 0 \\ 0 & k_2 & 0 & 0 \\ 0 & 0 & k_3 & 0 \\ 0 & 0 & 0 & d' \end{bmatrix},$$

in which $d' = d - l'^2/k_1 - m'^2/k_2 - n'^2/k_3$. Hence, D'' is the matrix associated with equation (6). Since its rank is the same as that of D, we conclude that the two ellipsoids and the two hyperboloids have rank 4 but that the quadric cones ($d' = 0$) have rank 3.

Turning to Case II, if k_1 and k_2 are different from 0 but $k_3 = 0$, D' is *r-c-*equivalent to

$$\begin{bmatrix} k_1 & 0 & 0 & 0 \\ 0 & k_2 & 0 & 0 \\ 0 & 0 & 0 & n' \\ 0 & 0 & n' & d' \end{bmatrix},$$

which has rank 4 if $n' \neq 0$ but rank 3 if $n' = 0$ and $d' \neq 0$. Therefore the two paraboloids have rank 4 and the elliptic and hyperbolic cylinders rank 3.

Continuing in this way we can easily complete the following classification of the 17 quadrics by rank.

1. Quadrics of rank 4: real and imaginary ellipsoids, the two hyperboloids, and the two paraboloids.
2. Quadrics of rank 3: real and imaginary cones; the elliptic, hyperbolic, and parabolic cylinders.
3. Quadrics of rank 2: two intersecting planes (real or imaginary); two parallel planes (real or imaginary).
4. Quadric of rank 1: two coincident planes.

A quadric of rank 4 is called a *true* or *nonsingular* quadric. A quadric of rank less than 4 is called a *singular* quadric, and one of rank less than 3, a *degenerate* quadric.

Example. Without rotating axes, find the nature of the quadric represented by the equation

$$x^2 + 4y^2 + z^2 - 4xy - 2xz + 4yz - 8x + 4z + 5 = 0.$$

Solution. We have

$$H = \begin{bmatrix} 1 & -2 & -1 \\ -2 & 4 & 2 \\ -1 & 2 & 1 \end{bmatrix}, \qquad D = \begin{bmatrix} 1 & -2 & -1 & -4 \\ -2 & 4 & 2 & 0 \\ -1 & 2 & 1 & 2 \\ -4 & 0 & 2 & 5 \end{bmatrix}.$$

The eigenvalues of H are easily found to be 0, 0, and 6. Applying elementary row transformations, we find that D is row-equivalent to

$$D' = \begin{bmatrix} 1 & -2 & -1 & -4 \\ 0 & 0 & 0 & -8 \\ 0 & 0 & 0 & -2 \\ 0 & -8 & -2 & -11 \end{bmatrix}.$$

Since D' is clearly singular, $\det D = 0$, and further reduction by elementary row transformations easily shows that rank $D =$ rank $D' = 3$. Referring to Case III Section 7.2, we see that the only possibility is a parabolic cylinder, since this is the only rank 3 quadric for which two of the eigenvalues of H are zero. The graph of the equation is therefore a parabolic cylinder.

7.4 Invariants

In this section, we shall show that certain functions of the coefficients of equation (4) are invariant under rotation and translation of axes.

Theorem 7.6. *Under a rotation of axes the matrices H and D associated with equation (4) are replaced by similar matrices.*

Proof. Let P be the transition matrix of the rotation. We have already seen that H is then replaced by $P^T H P = P^{-1} H P$ which is similar to H.

Now suppose the rotation, with transition matrix P, transforms (4) into

$$(9) \qquad \Phi(x', y', z') = 0,$$

so that

$$\Phi(x', y', z') = \varphi(x, y, z),$$

where $X = PX'$. The quadratic forms associated with (4) and (9) are

$$(10) \qquad u^2 \varphi\left(\frac{x}{u}, \frac{y}{u}, \frac{z}{u}\right)$$

and

$$(11) \qquad u'^2 \Phi\left(\frac{x'}{u'}, \frac{y'}{u'}, \frac{z'}{u'}\right),$$

and we see that (10) is transformed into (11) by equations (8). The transition matrix of (8) is

$$Q = \begin{bmatrix} P & O \\ O & 1 \end{bmatrix},$$

which is orthogonal because P is. Hence the matrix D is replaced by $Q^T DQ = Q^{-1}DQ$ which is similar to D.

Theorem 7.7. *The following functions of the coefficients of equation* (4) *are invariant under rotation of axes.*

1.
$$\det H = \begin{vmatrix} a & h & g \\ h & b & f \\ g & f & c \end{vmatrix}.$$

2. $j = ab + bc + ca - h^2 - f^2 - g^2$.

3. $k = a + b + c$.

4.
$$\det D = \begin{vmatrix} a & h & g & l \\ h & b & f & m \\ g & f & c & n \\ l & m & n & d \end{vmatrix}.$$

5.
$$q = \begin{vmatrix} a & h & g \\ h & b & f \\ g & f & c \end{vmatrix} + \begin{vmatrix} a & h & l \\ h & b & m \\ l & m & d \end{vmatrix} + \begin{vmatrix} a & g & l \\ g & c & n \\ l & n & d \end{vmatrix} + \begin{vmatrix} b & f & m \\ f & c & n \\ m & n & d \end{vmatrix}.$$

6. $r = ab + bc + ca + ad + bd + cd - f^2 - g^2 - h^2 - l^2 - m^2 - n^2$.

7. $s = a + b + c + d$.

Proof. Since a rotation of axes replaces D and H by similar matrices, the characteristic polynomials $\det (D - xI)$ and $\det (H - xI)$ are invariant under rotation of axes. The proof of the theorem, therefore, consists in noting that $\det H, -j$, and k are, respectively, the constant term, the coefficient of x, and the coefficient of x^2 in $\det (H - xI)$ and that $\det D, -q, r, -s$ are, respectively, the constant term and the coefficients of x, x^2, and x^3 in $\det (D - xI)$.

Theorem 7.8. *The functions* $\det D, \det H, j, k$ *are invariant under translation of axes.*

Proof. Since the translation

$$x = x' + \alpha,$$

(12)
$$y = y' + \beta,$$

$$z = z' + \gamma,$$

when applied to equation (4), leaves the coefficients of all second-degree terms in x, y, z unchanged, it follows that the matrix H is unchanged. Hence, det $H, j,$ and k are invariant under translation. To prove that det D is also invariant, we must investigate the effect of this translation on the quadratic form (10). Under the translation defined by (12), equation (4) becomes

$$\varphi(x' + \alpha, y' + \beta, z' + \gamma) = 0,$$

and the associated quadratic form is

$$u'^2 \varphi\left(\frac{x' + \alpha u'}{u'}, \frac{y' + \beta u'}{u'}, \frac{z' + \gamma u'}{u'}\right).$$

Comparing this with $u^2\varphi(x/u, y/u, z/u)$, we see that under the translation (12) the associated quadratic form undergoes the transformation

$$x = x' + \alpha u',$$
$$y = y' + \beta u',$$
$$z = z' + \gamma u',$$
$$u = u',$$

whose matrix is

$$C = \begin{bmatrix} 1 & 0 & 0 & \alpha \\ 0 & 1 & 0 & \beta \\ 0 & 0 & 1 & \gamma \\ 0 & 0 & 0 & 1 \end{bmatrix}.$$

Under this translation, therefore, D is replaced by $C^T D C = D'$, and since det $C = $ det $C^T = 1$, we have det $D' = $ det D.

Example 1. Determine the nature of the surface

$$2x^2 + 2y^2 - z^2 - 8xy + 4xz - 4yz - 8x + 4y - 4z + 2 = 0$$

without actually performing a rotation of axes.

Solution. We have

$$H = \begin{bmatrix} 2 & -4 & 2 \\ -4 & 2 & -2 \\ 2 & -2 & -1 \end{bmatrix} \quad \text{and} \quad D = \begin{bmatrix} 2 & -4 & 2 & -4 \\ -4 & 2 & -2 & 2 \\ 2 & -2 & -1 & -2 \\ -4 & 2 & -2 & 2 \end{bmatrix},$$

and easily check that det $D = 0$ so the quadric is singular. The eigenvalues of H are found to be -2, -2, and 7, the roots of

$$\det (H - xI) = -x^3 + 3x^2 + 24x + 28 = 0.$$

It immediately follows [Case I(6) in Section 7.2] that the surface is a real quadric cone whose equation, after a suitable rotation of axes, will be

$$2x^2 + 2y^2 - 7z^2 = 0.$$

It is not necessary to compute the transition matrix P of the rotation.

Example 2. Find det D and the eigenvalues of H and deduce from these the nature of the quadric

$$3x^2 + 2xy + 2xz - 2yz + 4x + 5 = 0.$$

Solution. We have

$$D = \begin{bmatrix} 3 & 1 & 1 & 2 \\ 1 & 0 & -1 & 0 \\ 1 & -1 & 0 & 0 \\ 2 & 0 & 0 & 5 \end{bmatrix} \quad \text{and} \quad H = \begin{bmatrix} 3 & 1 & 1 \\ 1 & 0 & -1 \\ 1 & -1 & 0 \end{bmatrix},$$

from which det $D = -21$,

$$\det (H - xI) = -x^3 + 3x^2 + 3x - 5,$$

the eigenvalues of H are 1, $1 + \sqrt{6}$ and $1 - \sqrt{6}$, and det $H = -5$. Because the eigenvalues of H are nonzero, a translation following the appropriate rotation of axes will reduce the equation to

$$x^2 + (1 + \sqrt{6})y^2 + (1 - \sqrt{6})z^2 + c = 0.$$

Since det D is invariant under rotation and translation, the discriminant of this equation is equal to det D and hence $-5c = -21$ or $c = 21/5$. The quadric is therefore a hyperboloid of two sheets with equation, in the final coordinate system,

$$5x^2 + 5(1 + \sqrt{6})y^2 + 5(1 - \sqrt{6})z^2 + 21 = 0.$$

Exercise 7.3

1. For each of the following quadrics, find the rank, and evaluate det D and the eigenvalues of H. Deduce from this information, without carrying out coordinate

transformations, the nature of the quadric represented and a standard equation for it.

(a) $3x^2 + 10y^2 + 3z^2 - 2xz + 4y + 4z = 9$.

(b) $2x^2 - y^2 + 2z^2 + xy + 4xz + yz - 3x + 3y - 3z = 2$.

(c) $3x^2 + y^2 + z^2 + 8xy - 8xz - 4yz + 16x + 4y - 8z + 4 = 0$.

(d) $6x^2 - 2y^2 - 6xy - 6yz + 2xz = 35$.

(e) $15x^2 - 5y^2 + 3z^2 - 22xy + 14xz - 14yz + 6x - 10y + 2z = 5$.

(f) $x^2 - 2y^2 - z^2 - 4yz + 2xz + 2y - 2z = 0$.

2. Given the general equation of a conic

$$ax^2 + 2hxy + by^2 + 2gx + 2fy + c = 0,$$

let

$$D_1 = \begin{bmatrix} a & h & g \\ h & b & f \\ g & f & c \end{bmatrix}, \qquad H_1 = \begin{bmatrix} a & h \\ h & b \end{bmatrix}.$$

(a) Show that the conic is degenerate (i.e., its graph is two real or imaginary straight lines) if and only if det $D_1 = 0$.

(b) If det $D_1 \neq 0$, prove that the graph of the equation is an ellipse (real or imaginary) if det $H_1 > 0$, a hyperbola if det $H_1 < 0$, and a parabola if det $H_1 = 0$.

(c) If det $D_1 = 0$, show that the graph is two imaginary lines intersecting in a real point if det $H_1 > 0$, two intersecting real lines if det $H_1 < 0$, and two parallel or coincident lines if det $H_1 = 0$.

3. Prove that a nonsingular quadric is a paraboloid if and only if det $H = 0$.

4. Prove that equation (4) represents an elliptic paraboloid if and only if det $D < 0$, det $H = 0$, and $j > 0$.

5. Derive necessary and sufficient conditions that equation (4) represent a hyperbolic paraboloid.

6. Prove that equation (4) represents a real ellipsoid if and only if its coefficients satisfy either

(a) $\qquad\qquad$ det $D < 0$, $\qquad$ det $H > 0$, $\qquad$ $j > 0$, and $k > 0$

or

(b) $\qquad\qquad$ det $D < 0$, $\qquad$ det $H < 0$, $\qquad$ $j > 0$, and $k < 0$.

7. Prove that for a hyperboloid of one sheet det $D > 0$ but for a hyperboloid of two sheets det $D < 0$.

8. Prove that if det $D > 0$, det $H < 0$, and $k > 0$, equation (4) represents a hyperboloid of one sheet.

7.5 A Problem in Dynamics

We shall now prove a theorem on the simultaneous reduction of two quadratic forms and apply it to a problem in dynamics.

Theorem 7.9. *Let $X^T A X$ and $X^T B X$ be two real quadratic forms in n variables and let $X^T A X$ be positive definite. There exists a real nonsingular matrix P such that the transformation $X = PY$ reduces $X^T A X$ to the form*

$$(13) \qquad Y^T Y = y_1^2 + y_2^2 + \cdots + y_n^2$$

and $X^T B X$ to the form

$$(14) \qquad Y^T D Y = k_1 y_1^2 + k_2 y_2^2 + \cdots + k_n y_n^2,$$

where $k_1, \ldots, k_n$ are the roots of the determinantal equation

$$(15) \qquad \det (B - xA) = 0.$$

Moreover, $k_1, \ldots, k_n$ are real and are positive if and only if $X^T B X$ is positive definite.

Proof. Since $X^T A X$ is positive definite, by Theorem 7.2 it is real-equivalent to the form (13). Let R be the real nonsingular matrix such that $X = RY$ reduces $X^T A X$ to (13) so that $R^T A R = I$. Under the same transformation, $X^T B X$ becomes $Y^T R^T B R Y$. Now since $R^T B R$ is real symmetric, we can choose an orthogonal matrix Q such that Q diagonalizes $R^T B R$. We let $Q^T R^T B R Q = D$ and let $k_1, k_2, \ldots, k_n$ be the diagonal elements of D and hence the eigenvalues of $R^T B R$. Now let $P = RQ$ and we have

$$P^T B P = Q^T R^T B R Q = D,$$
$$P^T A P = Q^T R^T A R Q = Q^T I Q = I,$$

and hence the transformation $X = PY$ reduces $X^T A X$ to the form (13) and $X^T B X$ to the form (14).

It remains to be shown that $k_1, \ldots, k_n$ are the roots of (15). By definition, $k_1, \ldots, k_n$ are the eigenvalues of $R^T B R$ and hence the roots of

$$\det (R^T B R - xI) = 0.$$

Because $R^T A R = I$, this equation can be written

$$\det [R^T B R - x(R^T A R)] = 0$$

or

$$\det [R^T (B - xA)R] = 0,$$

and since R is nonsingular, this is equivalent (by Theorem 3.12) to (15). Finally, since $X^T B X$ is real-equivalent to $Y^T D Y$, the matrices B and D must (Theorem 7.3, Corollary 2) have the same number of positive eigenvalues. Hence, $k_1, \ldots, k_n$ are all positive if and only if $X^T B X$ is positive definite.

We shall apply Theorem 7.9 to the solution of the problem of small oscillations of a conservative dynamical system about a position of stable equilibrium. We assume that the position of the vibrating system is specified at any time by n generalized coordinates $q_1, q_2, \ldots, q_n$ whose derivatives with respect to time are denoted by $\dot{q}_1, \dot{q}_2, \ldots, \dot{q}_n$. We assume also that the equilibrium position is defined by $q_1 = q_2 = \cdots = q_n = 0$. The kinetic energy of the system is then a quadratic form

$$T = \sum_{i=1}^{n} \sum_{j=1}^{n} a_{ij} \dot{q}_i \dot{q}_j$$

in the velocities $\dot{q}_i$. Although, in general, the coefficients a_{ij} depend on the coordinates, if the oscillations are small an approximation to the motion can be obtained by assuming a_{ij} to be constant. If we make this assumption, Lagrange's equations of motion (see [11]), take the form

$$(16) \qquad \frac{d}{dt}\left(\frac{\partial T}{\partial \dot{q}_i}\right) = -\frac{\partial V}{\partial q_i},$$

where V is the potential energy of the system. Let V be expanded in a Taylor's series about the point $q_i = 0$ $(i = 1, 2, \ldots, n)$. We may assume that the constant term is 0, since it will not affect equations (16). The first-degree terms in q_i also are absent, since V has a minimum value at a point of stable equilibrium and, therefore, $\partial V / \partial q_i = 0$ $(i = 1, 2, \ldots, n)$ at the point $q_1 = q_2 = \cdots = q_n = 0$. Hence, by neglecting cubic and higher terms, since the q_i are small, we can approximate V by a quadratic form

$$V = \sum_{i=1}^{n} \sum_{j=1}^{n} b_{ij} q_i q_j$$

in the coordinates q_i. The coefficients b_{ij} are constants.

Now if we introduce new coordinates $q_1', \ldots, q_n'$ defined by

$$(17) \qquad q_i = \sum_{j=1}^{n} p_{ij} q_j',$$

where p_{ij} are constants and $P = [p_{ij}]$ is nonsingular, it follows, on differentiation with respect to time, that

$$\dot{q}_i = \sum_{j=1}^{n} p_{ij} \dot{q}_j',$$

and hence the velocities $\dot{q}_i$ are subjected to the same transformation as the coordinates q_i. By the nature of kinetic energy the quadratic form T is positive definite. Hence, by Theorem 7.9, the transformation (17) can be chosen so that

$$T = \dot{q}_1'^2 + \dot{q}_2'^2 + \cdots + \dot{q}_n'^2$$

and

$$V = k_1 q_1'^2 + k_2 q_2'^2 + \cdots + k_n q_n'^2.$$

Moreover, since V has a minimum value 0 at the point $q_1' = q_2' = \cdots = q_n' = 0$, it follows that V is positive definite and $k_i > 0$, $(i = 1, 2, \ldots, n)$. Therefore we can write $k_i = \omega_i^2$, where $\omega_1, \ldots, \omega_n$ are real and positive.

Substituting in (16), the equations of motion now become

$$\frac{d^2 q_i'}{dt^2} = -\omega_i^2 q_i' \qquad\qquad (i = 1, 2, \ldots, n),$$

and their solutions are

(18) $$q_i' = a_i \cos{(\omega_i t + b_i)} \qquad\qquad (i = 1, 2, \ldots, n).$$

The coordinates $q_1', \ldots, q_n'$ are called *normal coordinates* for the vibrating system. Equations (18) imply that each normal coordinate is subject to a simple harmonic vibration.

CHAPTER EIGHT

COMPLEX INNER PRODUCT SPACES

8.1 The Standard Inner Product in $V_n(C)$

Although we assume a knowledge of complex numbers, we recall some basic facts and terminology. Let $x = s + it$ be any complex number. The real numbers s and t are called, respectively, the real part and the imaginary part of x and are denoted by $R(x)$ and $I(x)$. The conjugate of x is $s - it$ and will be denoted by $\bar{x}$. We note that $x + \bar{x} = 2s = 2R(x)$ and $x\bar{x} = s^2 + t^2$ are both real. Clearly, $x = \bar{x}$ if and only if x is real. The absolute value of x is defined by

$$|x| = \sqrt{x\bar{x}} = \sqrt{s^2 + t^2}.$$

If x is real this is the ordinary absolute value of a real number. For all $x, y \in \mathscr{C}$ we have $|xy| = |x|\,|y|$ and $|x + y| \leq |x| + |y|$.

If $x, y \in \mathscr{C}$ it is easy to verify that $\bar{\bar{x}} = x$ and

(1)
$$\overline{x + y} = \bar{x} + \bar{y},$$
$$\overline{xy} = \bar{x}\bar{y},$$

so that the mapping $x \to \bar{x}$ is an isomorphism of $\mathscr{C}$ onto itself. If $A = [a_{ij}]$ is a matrix with elements in $\mathscr{C}$, the matrix $\bar{A} = [\bar{a}_{ij}]$ is called the complex conjugate of A. By (1) and the definition of matrix sums and products, it follows that $\overline{A + B} = \bar{A} + \bar{B}$, $\overline{(kA)} = \bar{k}\bar{A}$, and $\overline{AB} = \bar{A}\bar{B}$ whenever these sums and products are defined.

Also, if A is square, det $\bar{A} = \overline{\det A}$, $\bar{A}$ is nonsingular if and only if A is nonsingular, and in the latter event $(\bar{A})^{-1} = \overline{(A^{-1})}$. Finally, $\bar{A} = A$ if and only if all elements of A are real.

Definition. *If A is any matrix with elements in $\mathscr{C}$, the matrix $\bar{A}^T$ is called the* adjoint *of A and is denoted by A^*.*

Since the adjoint of a matrix plays an important role in what follows, we note its properties as follows.

(1) $(A + B)^* = A^* + B^*$, $(kA)^* = \bar{k}A^*$, $(AB)^* = B^*A^*$.

(2) $A^* = A^T$ if and only if A is real.

(3) $\det A^* = \det \bar{A} = \overline{\det A}$.

Our aim in this chapter is to define an inner product in a vector space over $\mathscr{C}$ that will lead to a development similar to that obtained for real vector spaces in Chapter 4. We begin by defining a dot product in $\mathscr{V}_n(\mathscr{C})$. In order to retain the property that $X \cdot X \geq 0$ we have to modify the definition used in $\mathscr{V}_n(\mathscr{R})$.

Definition. *If $X = [x_1, x_2, \ldots, x_n]$, $Y = [y_1, y_2, \ldots, y_n]$ are any two vectors in $\mathscr{V}_n(\mathscr{C})$, the* standard inner product *or* dot product $X \cdot Y$ *is defined by*

$$X \cdot Y = XY^* = X\bar{Y}^T = x_1\bar{y}_1 + x_2\bar{y}_2 + \cdots + x_n\bar{y}_n.$$

Theorem 8.1. *If $X, Y, Z \in \mathscr{V}_n(\mathscr{C})$, and $k \in \mathscr{C}$, then:*

(a) $X \cdot Y \in \mathscr{C}$, $X \cdot X$ *is real, nonnegative and* $X \cdot X = 0$ *if and only if* $X = O$.
(b) $X \cdot Y = \overline{Y \cdot X}$.
(c) $X \cdot (Y + Z) = X \cdot Y + X \cdot Z$.
(d) $X \cdot (kY) = \bar{k}(X \cdot Y)$.

These four basic properties of the standard inner product follow immediately from the definition and the properties of the complex conjugate. Details are left to the student.

Corollary. *If $X, Y, Z \in \mathscr{V}_n(\mathscr{C})$, and $k \in \mathscr{C}$, then:*

(e) $(X + Y) \cdot Z = X \cdot Z + Y \cdot Z$.
(f) $(kX) \cdot Y = k(X \cdot Y)$.

Proof. We have $(X + Y) \cdot Z = \overline{Z \cdot (X + Y)} = \overline{\bar{Z} \cdot (\bar{X} + \bar{Y})} = \overline{\bar{Z} \cdot \bar{X} + \bar{Z} \cdot \bar{Y}} = \overline{Z \cdot X} + \overline{Z \cdot Y} = X \cdot Z + Y \cdot Z$. This establishes (e). For (f) we have

$$(kX) \cdot Y = \overline{Y \cdot (kX)} = \overline{\bar{k}(Y \cdot X)} = k(\overline{Y \cdot X}) = k(X \cdot Y).$$

Definition. *If X, $Y \in \mathscr{V}_n(\mathscr{C})$ we say that X is orthogonal to Y if $X \cdot Y = 0$.*

By Theorem 8.1(b), $Y \cdot X = 0$ if and only if $X \cdot Y = 0$. Hence, X orthogonal to Y implies Y orthogonal to X and we can say simply that X and Y are orthogonal.

Definition. *If $X \in \mathscr{V}_n(\mathscr{C})$ we define the* length *of X, denoted by $\|X\|$, to be $\sqrt{X \cdot X}$. If $\|X\| = 1$, X is called a* unit vector.

By Theorem 8.1(a), $\|X\|$ is a nonnegative real number and is 0 only if $X = O$. The Cauchy-Schwarz and triangle inequalities can now be proved for the standard inner product in $\mathscr{V}_n(\mathscr{C})$. In fact, the entire theory developed in Sections 4.2, 4.3, and 4.4 can be extended to $\mathscr{V}_n(\mathscr{C})$. To avoid unnecessary repetition, however, we prefer to go directly to the definition of an abstract inner product in any vector space over $\mathscr{C}$. Even then, most of the theorems are so similar to their counterparts in real inner product spaces that, in many cases, detailed proofs can be left to the student to supply.

8.2 Inner Product Spaces over the Complex Field

To define an abstract inner product in an arbitrary vector space $\mathscr{V}$ over $\mathscr{C}$, we adopt as postulates the four properties of the dot product in $\mathscr{V}_n(\mathscr{C})$ listed in Theorem 8.1.

Definition. *A vector space $\mathscr{V}$ over $\mathscr{C}$ is called an* inner product space *if with any two vectors U, V in $\mathscr{V}$ is associated a scalar, denoted by (U, V) and called the* inner product *of U and V, that satisfies the following postulates:*

IP1. *For all $V \in \mathscr{V}$, (V, V) is real and nonnegative and $(V,V) = 0$ if and only if $V = O$.*

IP2. $(U, V) = (\overline{V, U})$.

IP3. $(U, kV) = \bar{k}(U, V)$ *for all k in $\mathscr{C}$.*

IP4. $(U, V + W) = (U, V) + (U, W)$ *for all U, V, W in $\mathscr{V}$.*

As in the corollary to Theorem 8.1 we note that these postulates imply also that $(kU, V) = k(U, V)$ and $(U + V, W) = (U, W) + (V, W)$. Also, by putting $W = O$ in IP4, we get $(U, V) = (U, V) + (U, O)$ so that $(U, O) = 0$ and, by IP2,

$(O, U) = 0$ for all U. These postulates make possible simple computations such as

$$(3U + (2 + i)V, 2iU + 3V) = (3U, 2iU) + (3U, 3V) + ((2 + i)V, 2iU)$$
$$+ ((2 + i)V, 3V)$$
$$= -6i(U, U) + 9(U, V) - 2i(2 + i)(V, U)$$
$$+ 3(2 + i)(V, V).$$

We now give some examples of inner product spaces over $\mathscr{C}$.

Example 1.   By Theorem 8.1 the dot product $X \cdot Y$ in $\mathscr{V}_n(\mathscr{C})$ is an inner product in the sense of our definition.

Example 2.   Let $\mathscr{V}$ be any finite dimensional vector space over $\mathscr{C}$ and let E_1, $E_2, \ldots, E_n$ be any basis of $\mathscr{V}$. Let V_1, V_2 have coordinate vectors $X_1 = \alpha V_1$, $X_2 = \alpha V_2$ relative to the E-basis. Define (V_1, V_2) to be $X_1 \cdot X_2 = \alpha V_1 \cdot \alpha V_2$ and $\mathscr{V}$ becomes an inner product space. The postulates are easily verified. For example,

$$(V_1, V_2 + V_3) = \alpha V_1 \cdot \alpha(V_2 + V_3) = \alpha V_1 \cdot (\alpha V_2 + \alpha V_3)$$
$$= \alpha V_1 \cdot \alpha V_2 + \alpha V_1 \cdot \alpha V_3 = (V_1, V_2) + (V_1, V_3).$$

In this way we find an inner product in $\mathscr{V}$ corresponding to every basis.

Example 3.   Let $\tau : \mathscr{V} \to \mathscr{V}$ be any linear operator in $\mathscr{V}$. If an inner product $(V_1, V_2)_1$ is given, another can be defined by $(V_1, V_2)_2 = (\tau V_1, \tau V_2)_1$. The proof is similar to that in Example 2 with τ replacing α.

Example 4.   Let $\mathscr{V}$ be the space of all complex-valued continuous functions defined on a closed interval $a \leq x \leq b$ of the real line. Then $\mathscr{V}$ becomes an inner product space if, for all $f, g \in \mathscr{V}$, we define the inner product by

$$(f, g) = \int_a^b f(x)\overline{g(x)} \, dx.$$

Verification of the postulates follows from the properties of the integral and the continuity of f and g as in Example 3, Section 4.5.

Definition.   *An inner product space over the field of complex numbers is called a* unitary space.

Definition.   *If $\mathscr{V}$ is a unitary space and V is any vector in $\mathscr{V}$, the length of V is defined by*

$$\|V\| = \sqrt{(V, V)}.$$

Clearly, $\|V\|$ is real and nonnegative.

Theorem 8.2. *If U, V are arbitrary vectors of an inner product space $\mathscr{V}$ over $\mathscr{C}$ and if $k \in \mathscr{C}$, then:*

(a) $\|V\| = 0$ *implies* $V = O$.
(b) $\|kV\| = |k|\,\|V\|$.
(c) $|(U, V)| \leq \|U\|\,\|V\|$ *(Cauchy-Schwarz inequality)*.
(d) $\|U + V\| \leq \|U\| + \|V\|$ *(Triangle inequality)*.

Proof. Part (a) follows from Theorem 8.1(a). Because $\|kV\|^2 = (kV, kV) = k\bar{k}(V, V) = |k|^2\,\|V\|^2$, part (b) follows on taking positive square roots. Since (c) holds trivially if $U = O$, $V = O$ or $(U, V) = 0$ we assume that U and V, and (U, V), are nonzero. For any real numbers h and k,

$$0 \leq (hU - kV, hU - kV)$$
$$= h^2(U, U) - hk[(U, V) + (V, U)] + k^2(V, V)$$
$$= h^2(U, U) - hk[(U, V) + \overline{(U, V)}] + k^2(V, V)$$

and hence for all U, V and all real h, k

(2) $$h^2(U, U) + k^2(V, V) \geq 2hk R(U, V).$$

Now let $h = \|V\|$ and $k = \|U\|$. Since U and V are nonzero, $hk \neq 0$ and (2) gives

(3) $$R(U, V) \leq \|U\|\,\|V\|.$$

Since (3) holds for all U, V we can replace V by $e^{i\theta}V$. Noting that by (b), $\|e^{i\theta}V\| = |e^{i\theta}|\,\|V\| = \|V\|$, we get, for all real θ,

(4) $$R(U, e^{i\theta}V) = R[e^{-i\theta}(U, V)] \leq \|U\|\,\|V\|.$$

Now choose θ to be the amplitude of the complex number (U, V) so that $e^{-i\theta}(U, V)$ is real and positive and $R[e^{-i\theta}(U, V)] = |(U, V)|$. Then (4) becomes

$$|(U, V)| \leq \|U\|\,\|V\|,$$

which proves (c).

To prove (d) we have

$$\|U + V\|^2 = (U + V, U + V) = (U, U) + (U, V) + (V, U) + (V, V)$$
$$= \|U\|^2 + 2R(U, V) + \|V\|^2$$
$$\leq \|U\|^2 + 2\,|U, V| + \|V\|^2$$
$$\leq \|U\|^2 + 2\,\|U\|\,\|V\| + \|V\|^2$$
$$= (\|U\| + \|V\|)^2,$$

and (d) follows on taking positive square roots.

We now define orthogonality in the obvious way: U is orthogonal to V if $(U, V) = 0$. By IP2, V is orthogonal to U if and only if U is orthogonal to V. A vector V in $\mathscr{V}$ is orthogonal to a subspace $\mathscr{S}$ of $\mathscr{V}$ if it is orthogonal to every vector of $\mathscr{S}$ and two subspaces $\mathscr{S}$ and $\mathscr{T}$ are orthogonal if every vector of $\mathscr{S}$ is orthogonal to every vector of $\mathscr{T}$. With very minor modifications the proofs of the theorems in Section 4.6 apply in a unitary space $\mathscr{V}$. We restate the results for easy reference but will not repeat the proofs.

Theorem 8.3. *Let $\mathscr{V}$ be a finite dimensional unitary space.*

(a) *If a vector U in $\mathscr{V}$ is orthogonal to each of the vectors $V_1, V_2, \ldots, V_r$, it is orthogonal to the subspace spanned by $V_1, \ldots, V_r$.*

(b) *A vector U is orthogonal to a subspace $\mathscr{S}$ of $\mathscr{V}$ if and only if $(U, B_i) = 0$ $(i = 1, 2, \ldots, n)$, where $B_1, \ldots, B_n$ is any basis of $\mathscr{V}$.*

(c) *Any set of mutually orthogonal vectors in $\mathscr{V}$ is linearly independent.*

(d) *If $\mathscr{S}$ is a proper subspace of $\mathscr{V}$, then $\mathscr{V}$ contains a nonzero vector V orthogonal to $\mathscr{S}$.*

(e) *If $\mathscr{V} \neq O$, then $\mathscr{V}$ has an orthonormal basis, that is, a basis consisting of mutually orthogonal unit vectors.*

(f) *If $\dim \mathscr{V} = n$ and $E_1, \ldots, E_r$, where $r < n$, are mutually orthogonal unit vectors of $\mathscr{V}$, there exist unit vectors $E_{r+1}, \ldots, E_n$ such that $E_1, E_2, \ldots, E_n$ is an orthonormal basis of $\mathscr{V}$.*

(g) *The Gram-Schmidt orthogonalization process applies exactly as described in Section 4.6.*

Example 5. Using the standard inner product find an orthonormal basis for $\mathscr{V}_3(\mathscr{C})$ of which the first vector is a scalar multiple of $[2, i, -2i]$.

Solution. Let $X_1 = [2, i, -2i]$; to find a vector $X_2 = [x, y, z]$ orthogonal to X_1 we must solve

$$X_2 \cdot X_1 = 2x - iy + 2iz = 0.$$

(a) Choose any solution, for example, $X_2 = [i, -2, -2]$. We now require $X_3 = [x, y, z]$ such that

$$X_3 \cdot X_1 = 2x - iy + 2iz = 0,$$
$$X_3 \cdot X_2 = -ix - 2y - 2z = 0.$$

A solution is $X_3 = [2i, 2, -1]$.

Now $\|X_1\|^2 = (2)(2) + (i)(-i) + (-2i)(2i) = 4 + 1 + 4 = 9$ so that $\|X_1\| = 3$ and similarly $\|x_2\| = \|x_3\| = 3$. An orthonormal basis is therefore

$$E_1 = \left[\frac{2}{3}, \frac{i}{3}, -\frac{2i}{3}\right], \qquad E_2 = \left[\frac{i}{3}, -\frac{2}{3}, -\frac{2}{3}\right], \qquad E_3 = \left[\frac{2i}{3}, \frac{2}{3}, -\frac{1}{3}\right].$$

Exercise 8.1

1. In $\mathcal{V}_3(\mathcal{C})$, with standard inner product, find a unit vector orthogonal to:
   (a) $X_1 = [i, 1 + i, 2 - i]$. (b) $X_2 = [i, 2i, 1 - i]$.
   (c) Both X_1 and X_2.

2. Find an orthonormal basis of $\mathcal{V}_3(\mathcal{C})$ of which the first vector is a scalar multiple of $[1, 2i, 1 + i]$.

3. Show that any orthonormal basis of $\mathcal{V}_n(\mathcal{C})$ that consists of real vectors only is also an orthonormal basis of $\mathcal{V}_n(\mathcal{R})$.

4. Supply proofs for some or all of the statements in Theorem 8.3.

5. Prove that $(A, B) = \text{tr } AB^*$ defines an inner product on the space $\mathcal{M}$ of all $m \times n$ matrices with elements in $\mathcal{C}$. (See Exercise 4.4, Problems 7 and 8.)

6. If $\mathcal{V}$ is an n-dimensional unitary space and $\mathcal{S}$ is an r-dimensional subspace of $\mathcal{V}$, show that
$$\mathcal{S}^\perp = \{V \in \mathcal{V} \mid (V, U) = 0 \text{ for all } U \in \mathcal{S}\}$$
is a subspace, $\dim \mathcal{S}^\perp = n - r$, $\mathcal{V} = \mathcal{S} \oplus \mathcal{S}^\perp$, and $(\mathcal{S}^\perp)^\perp = \mathcal{S}$.

7. Let $\mathcal{S}$ be the subspace of $\mathcal{V}_3(\mathcal{C})$ spanned by the vector $S = [1, 2i, 1 + 2i]$. Find a basis for $\mathcal{S}^\perp$ and find the orthogonal projections of $X = [1 - i, i, 2i]$ onto $\mathcal{S}$ and $\mathcal{S}^\perp$. *Hint.* See Example 2, Section 4.3.

8. Find the orthogonal projection of $X = [17 + 4i, 0, 17 - 4i]$ onto the space $\mathcal{S}$ spanned by $S_1 = [2i, 1 + i, i]$ and $S_2 = [i, 2, -i]$. Also find the projection of X onto $\mathcal{S}^\perp$.

8.3 Orthonormal Bases and Unitary Matrices

Let $E_1, \ldots, E_n$ and $F_1, \ldots, F_n$ be two orthonormal bases of an n-dimensional unitary space $\mathcal{V}$ and let $P = [p_{ij}]$ be the transition matrix from the E-basis to the F-basis. We then have

$$F_1 = p_{11}E_1 + p_{21}E_2 + \cdots + p_{n1}E_n,$$
$$\cdots\cdots\cdots\cdots\cdots\cdots\cdots\cdots\cdots\cdots\cdot$$
$$F_n = p_{1n}E_1 + p_{2n}E_2 + \cdots + p_{nn}E_n,$$

and since $(F_i, F_j) = (E_i, E_j) = \delta_{ij}$

$$(F_i, F_j) = \left(\sum_{k=1}^n p_{ki}E_k, \sum_{k=1}^n p_{kj}E_k\right)$$
$$= \sum_{k=1}^n \sum_{s=1}^n p_{ki}\bar{p}_{sj}(E_k, E_s)$$
$$= p_{1i}\bar{p}_{1j} + p_{2i}\bar{p}_{2j} + \cdots + p_{ni}\bar{p}_{nj} = \delta_{ij}.$$

This states that the columns of the transition matrix P are mutually orthogonal unit vectors of $\mathscr{V}_n(\mathscr{C})$ where as usual orthogonality in $\mathscr{V}_n(\mathscr{C})$ means orthogonality relative to the standard inner product unless otherwise stated. This is equivalent to the matrix equation $P^T \bar{P} = I$ and, taking complex conjugates, this in turn implies $P^*P = I$, or $P^* = P^{-1}$. Thus we also have $PP^* = I$ which states that the rows of P are also mutually orthogonal unit vectors of $\mathscr{V}_n(\mathscr{C})$.

Definition. *A matrix P with complex elements that satisfies $P^*P = PP^* = I$ is called a* unitary *matrix.*

By referring to Theorem 4.19 the student can now easily complete the proof of the following theorem.

Theorem 8.4. *If $E_1, \ldots, E_n$ is an orthonormal basis of a unitary space $\mathscr{V}$ and if P is the transition matrix from the E-basis to a second basis $F_1, \ldots, F_n$, then the F-basis is orthonormal if and only if P is unitary.*

In view of Theorem 8.4, transition from one orthonormal basis to another means a coordinate transformation of the form $X = PX'$ where P is unitary. In the next theorem we state some properties of unitary matrices. The proofs of these are easy and most of them will be left as an exercise.

Theorem 8.5.

(a) *If P is unitary so are P^T, $\bar{P}$, P^*, and P^{-1}.*
(b) *If P and Q are unitary so is PQ.*
(c) *A real matrix is unitary if and only if it is orthogonal.*
(d) *If P is unitary $|\det P| = 1$.*
(e) *All eigenvalues of a unitary matrix have absolute value 1.*

Proof. We prove only (e) and leave the rest as an exercise. If P is unitary with eigenvalue k and corresponding (row) eigenvector X, then $PX^T = kX^T$ and hence $\bar{P}X^* = \bar{k}X^*$. Therefore,

$$k\bar{k}XX^* = (PX^T)^T(\bar{P}X^*) = XP^T\bar{P}X^* = XX^*$$

because P^T is unitary and hence $P^T\bar{P} = I$. Since $XX^* > 0$, it follows that $k\bar{k} = 1$ or $|k| = 1$.

We conclude with the unitary space analogue of Theorem 4.16.

Theorem 8.6. *Let $\mathscr{V}$ be an n-dimensional unitary space and let $E_1, \ldots, E_n$ be any orthonormal basis of $\mathscr{V}$. If $\alpha: \mathscr{V} \to \mathscr{V}_n(\mathscr{C})$ is the isomorphism that maps each vector of $\mathscr{V}$ on its coordinate vector relative to the E-basis, then for all U, V in $\mathscr{V}$,*

$$(U, V) = \alpha U \cdot \alpha V$$

and hence α is an IPS-isomorphism.

The proof of this theorem is the same as that of Theorem 4.16. The equality $(U, V) = \alpha U \cdot \alpha V$ follows from the fact that $(E_i, E_j) = \delta_{ij}$.

Corollary. *If $E_1, \ldots, E_n$ and $F_1, \ldots, F_n$ are two orthonormal bases of $\mathscr{V}$ and V_1, V_2 in $\mathscr{V}$ have E-coordinate vectors X_1, X_2 and F-coordinate vectors Y_1, Y_2, then $Y_1 \cdot Y_2 = X_1 \cdot X_2$.*

The proof of the corollary is trivial because by the theorem, $Y_1 \cdot Y_1 = X_1 \cdot X_2 = (V_1, V_2)$. The result can be stated: the dot product in $\mathscr{V}_n(\mathscr{C})$ is invariant under transformation of coordinates from one orthonormal basis to another.

8.4 Inner Products as Hermitian Bilinear Forms

Let $\mathscr{V}$ be a unitary space of dimension n and let $B_1, \ldots, B_n$ be any basis of $\mathscr{V}$. If $V = \sum_{i=1}^{n} x_i B_i$ and $U = \sum_{j=1}^{n} y_j B_j$ it follows from the postulates for inner products that

$$(U, V) = \sum_{i=1}^{n} \sum_{j=1}^{n} x_i \bar{y}_j (B_i, B_j) = \sum_{i=1}^{n} \sum_{j=1}^{n} b_{ij} x_i \bar{y}_j,$$

where $b_{ij} = (B_i, B_j)$. Because of IP2 we have $b_{ij} = \bar{b}_{ji}$ and if $B = [b_{ij}]$ this can be expressed by the matrix equation $\bar{B}^T = B^* = B$.

Definition. *A matrix A with elements in $\mathscr{C}$ is said to be* Hermitian *or self-adjoint if $A^* = A$.*

If $X = [x_1, x_2, \ldots, x_n]$, $Y = [y_1, y_2, \ldots, y_n]$, and $B = [b_{ij}]$ is an $n \times n$ Hermitian matrix, the expression

$$XBY^* = \sum_{i=1}^{n} \sum_{j=1}^{n} b_{ij} x_i \bar{y}_j$$

is called a *Hermitian bilinear* form in $x_1, \ldots, x_n, y_1, \ldots, y_n$. The expression

$$XBX^* = \sum \sum b_{ij} x_i \bar{x}_j$$

is called a *Hermitian form* in $x_1, \ldots, x_n$.

Let H be any Hermitian matrix. Since XHX^* is a 1×1 matrix, it is symmetric and hence $\overline{XHX^*} = (XHX^*)^* = XH^*X^* = XHX^*$ because H is Hermitian. It follows that if H is Hermitian, XHX^* is real for all X in $\mathscr{V}_n(\mathscr{C})$. In other words, *the range of values of a Hermitian form on $\mathscr{V}_n(\mathscr{C})$ is real.*

Definition. *A Hermitian form XHX^* is said to be* nonnegative definite *if $XHX^* \geq 0$ for all X in $\mathscr{V}_n(\mathscr{C})$. It is* positive definite *if $XHX^* \geq 0$ for all X and in addition $XHX^* = 0$ implies $X = O$. A Hermitian matrix H is called* positive (nonnegative) *definite if XHX^* is positive (nonnegative) definite.*

With these definitions and the foregoing remarks, it is clear that we have the following analogue of Theorem 4.17.

Theorem 8.7. *Let $\mathscr{V}$ be any n-dimensional vector space over $\mathscr{C}$ and let $B_1, \ldots, B_n$ be any basis of $\mathscr{V}$. Every inner product that can be defined on $\mathscr{V}$ has the form $(U, V) = XHY^*$, where X and Y are the coordinate (row) vectors of U and V relative to the B-basis and H is a positive definite Hermitian matrix. Conversely, every positive definite Hermitian bilinear form in the B-coordinates of U and V defines an inner product (U, V).*

Exercise 8.2

1. Prove the properties of unitary matrices stated in Theorem 8.5(a), (b), (c), and (d).
2. Prove that a real matrix is Hermitian if and only if it is symmetric.
3. If H is Hermitian and P is unitary, show that P^*HP is Hermitian.
4. Prove that if H is Hermitian, the elements in the main diagonal of H are real.
5. Prove that if H is Hermitian, det H is real.
6. Prove that the characteristic polynomial of a Hermitian matrix has real coefficients.
7. Find the eigenvalues of the Hermitian matrix

$$H = \begin{bmatrix} a & b + ci \\ b - ci & d \end{bmatrix}$$

and show that they are real. (Here a, b, c, and d are real.)

8.5 Hermitian Matrices

We have seen that Hermitian matrices play the same role relative to inner products in unitary spaces that real symmetric matrices play in Euclidean spaces. We now pursue this analogy further.

Theorem 8.8. *If H is a Hermitian matrix there exists a unitary matrix P such that $P^{-1}HP$ is a diagonal matrix.*

The proof of this theorem is omitted, partly because it can be supplied by obvious modification of the proof of Theorem 6.12 but mainly because the theorem is a special case of Theorem 8.14 which will be proved in Section 8.6.

Corollary 1. *The eigenvalues of a Hermitian matrix are real.*

Proof. In any Hermitian matrix $H = [h_{ij}]$ the fact that $H^* = H$ implies $\bar{h}_{ii} = h_{ii}$ so that the main diagonal elements are real. But since P is unitary, $P^{-1}HP = P^*HP$ is Hermitian because $(P^*HP)^* = P^*H^*P = P^*HP$. Hence the diagonal elements of P^*HP are real. But these diagonal elements are the eigenvalues of H. This corollary includes Theorem 6.11 as a special case.

Corollary 2. *An $n \times n$ Hermitian matrix H has n mutually orthogonal eigenvectors in $\mathscr{V}_n(\mathscr{C})$.*

Proof. Let $P^{-1}HP = D$, where D is diagonal. Then $HP = PD$ which implies, as in Theorem 6.5, that the column vectors of P are eigenvectors of H. Since P is unitary, these column vectors are mutually orthogonal.

Corollary 3. *Every Hermitian form XHX^* is unitarily equivalent (under a transformation $X^T = PY^T$, P unitary) to the form*

(5) $$k_1y_1\bar{y}_1 + k_2y_2\bar{y}_2 + \cdots + k_ny_n\bar{y}_n.$$

Proof. Choose Q unitary such that $Q^*HQ = D$, where D is diagonal with diagonal elements $k_1, k_2, \ldots, k_n$. Let $P = \bar{Q}$ and $X^T = PY^T$. Then $XHX^* = YP^TH\bar{P}\bar{Y}^T = YQ^*HQY^* = YDY^* = \sum_{i=1}^{n} k_iy_i\bar{y}_i$. Since Q is unitary, so is P.

Corollary 4. *A Hermitian matrix H is positive definite if and only if its eigenvalues are all positive and is nonnegative definite if and only if its eigenvalues are nonnegative.*

Proof. Clearly, the equivalent forms XHX^* and YDY^* have the same range of values. By (5), $YDY^* \geq 0$ for all Y if and only if $k_1, \ldots, k_n$ are all ≥ 0. Moreover, if each $k_i > 0$, then $YDY^* = 0$ implies $Y = O$, which in turn implies $X = O$.

Theorem 8.9. *A Hermitian matrix H is positive definite if and only if there exists a nonsingular matrix Q such that $H = QQ^*$.*

Proof. If $H = QQ^*$, $XHX^* = XQQ^*X^* = (XQ)(XQ)^* = Y \cdot Y \geq 0$, where $Y = XQ$. Also, if $XHX^* = 0$, $Y \cdot Y = 0$, and $Y = O$ which implies $X = O$ since Q is nonsingular. The converse is proved as in Theorem 6.16.

Theorem 8.10. *Every nonsingular matrix A with elements in $\mathscr{C}$ can be written in a unique way as a product HP where H is a positive definite Hermitian matrix and P is unitary.*

Proof. Since A is nonsingular, AA^* is positive definite Hermitian. Choose a unitary matrix Q such that $Q^{-1}AA^*Q = D$, a diagonal matrix. Since AA^* is positive definite, the diagonal elements of D are positive. Let D_1 be the diagonal matrix whose diagonal elements are the positive square roots of those of D so that $D_1^2 = D$. We let $H = QD_1Q^*$. Clearly, H is positive definite Hermitian and $H^2 = QDQ^* = AA^*$. Now let $P = H^{-1}A$ and $PP^* = H^{-1}AA^*(H^{-1})^* = H^{-1}H^2(H^*)^{-1} = I$ since $H^* = H$. Hence, P is unitary and $A = HP$.

To prove uniqueness, suppose $A = HP = H_1P_1$, where H, H_1 are positive definite Hermitian and P, P_1 are unitary. Then $AA^* = HPP^*H^* = H^2$ and, similarly, $AA^* = H_1^2$ so that $H^2 = H_1^2$. Now if Q is a unitary matrix that diagonalizes H, then Q also diagonalizes H^2 because $Q^{-1}H^2Q = (Q^{-1}HQ)^2$. Hence the same matrix Q diagonalizes H_1^2. By Theorem 6.8 the eigenvalues of H_1^2 are $k_1^2, \ldots, k_n^2$ where $k_1, \ldots, k_n$ are the eigenvalues of H_1. Thus the eigenvectors of H_1^2 are solutions of

(6) $$(H_1^2 - k_i^2 I)X = (H_1 + k_i I)(H_1 - k_i I)X = O.$$

Since H_1 is positive definite each $k_i > 0$ and $H_1 + k_i I$ is nonsingular. Hence (6) implies $(H_1 - k_i I)X = O$, H_1 and H_1^2 have the same eigenvectors and hence are diagonalized by the same matrix Q. But the diagonal matrices $Q^{-1}HQ$ and $Q^{-1}H_1Q$ are identical because their squares are equal and both are positive definite. It follows that $H_1 = H$ and hence $P_1 = P$.

The above decomposition of a nonsingular matrix A into the product of a positive definite Hermitian matrix and a unitary matrix is called the *polar factorization* of A since, for the case $n = 1$, it reduces to the unique representation of a nonzero complex number $a + bi$ in the polar form $re^{i\theta}$ where $r > 0$. The existence of a polar factorization can be proved for an arbitrary square matrix A but, if A is singular, the Hermitian factor is merely nonnegative definite and the unitary factor is not uniquely determined. The proof for singular matrices is given in [4].

Exercise 8.3

1. Find unitary matrices that will diagonalize

(a) $\begin{bmatrix} 1 & i \\ -i & 1 \end{bmatrix}$.

(b) $\begin{bmatrix} \dfrac{5}{2} & \dfrac{i}{2} \\ -\dfrac{i}{2} & \dfrac{5}{2} \end{bmatrix}$.

2. Prove that for every nonsingular matrix A there exists a unitary matrix Q and a positive definite Hermitian matrix H such that $A = QH$.

3. Prove that every positive definite Hermitian matrix H has a unique positive definite Hermitian "square root" H_1 such that $H_1^2 = H$.

4. Find the positive definite square root of the matrix in Problem 1(b).

5. Prove that every square matrix A with elements in $\mathscr{C}$ can be written in the form $B + iC$ where B and C are Hermitian. *Hint.* Let $B = \frac{1}{2}(A + A^*)$.

6. A matrix A such that $A^T = -A$ is called skew symmetric. Prove that if A is skew symmetric with real elements, then:
   (a) The main diagonal elements of A are all zero.
   (b) The matrix iA is Hermitian.
   (c) There exists a unitary matrix P such that $P^{-1}AP$ is a diagonal matrix with pure imaginary diagonal elements.

7. If $X = [x_1, x_2]$, $Y = [y_1, y_2]$ are any two vectors of $\mathscr{V}_2(\mathscr{C})$ and $(X, Y) = 2x_1\bar{y}_1 + (1 - i)x_2\bar{y}_1 + (1 + i)x_1\bar{y}_2 + 3x_2\bar{y}_2$:
   (a) Prove that (X, Y) is an inner product on $\mathscr{V}_2(\mathscr{C})$.
   (b) Find a vector orthogonal to $[2, i]$ relative to this inner product.

8. Show that if a, b, c, d are real,

$$(X, Y) = ax_1\bar{y}_1 + (b + ci)x_2\bar{y}_1 + (b - ci)x_1\bar{y}_2 + dx_2\bar{y}_2$$

is an inner product on $\mathscr{V}_2(\mathscr{C})$ if and only if $b^2 + c^2 < ad$ and $a > 0$.

9. Find an orthonormal basis of $\mathscr{V}_2(\mathscr{C})$ relative to the inner product in Problem 7.

10. Show that

$$H = \begin{bmatrix} 2 & 1 + 2i \\ 1 - 2i & 3 \end{bmatrix}$$

is positive definite Hermitian and find an orthonormal basis of $\mathscr{V}_2(\mathscr{C})$ relative to the inner product $(X, Y) = XHY^*$.

11. Show that

$$H = \begin{bmatrix} 2 & 0 & 2+i \\ 0 & 3 & 0 \\ 2-i & 0 & 3 \end{bmatrix}$$

is positive definite Hermitian and find an orthonormal basis of $\mathscr{V}_3(\mathscr{C})$ relative to the inner product $(X, Y) = XHY^*$.

12. Let $X_1, X_2, \ldots, X_r$ be any r vectors in $\mathscr{V}_n(\mathscr{C})$ and let $G = [X_i \cdot X_j]$ be the $r \times r$ matrix with $X_i \cdot X_j$ in the ith row and jth column.
(a) Prove that G is Hermitian and hence $\overline{G} = G^T$ and det G is real.
(b) Prove that det $G \geq 0$.
(c) Prove that det $G = 0$ if and only if $X_1, X_2, \ldots, X_r$ are linearly dependent.

13. Let $V_1, V_2, \ldots, V_r$ be any r vectors in an n-dimensional unitary space $\mathscr{V}$. Assume $r \leq n$ and let $G = [g_{ij}]$ where $g_{ij} = (V_i, V_j)$. Prove that (a) det G is real, (b) det $G \geq 0$, and (c) det $G = 0$ if and only if $V_1, \ldots, V_r$ are linearly dependent. *Hint.* For (b) use Theorem 8.6 and Problem 11(b).

8.6 Normal Matrices

A square matrix A is said to be *unitarily* similar to a matrix B if $B = P^{-1}AP$ where P is unitary. In this section we determine exactly what matrices are unitarily similar to diagonal matrices. We begin with a stronger form of the triangularization theorem.

Theorem 8.11. *If A is any matrix with elements in $\mathscr{C}$ there exists a unitary matrix P such that $P^{-1}AP$ is a triangular matrix with the eigenvalues of A in the main diagonal.*

Proof. The proof is similar to that of Theorem 6.7. Let $k_1, k_2, \ldots, k_n$ be the eigenvalues of A and let Q_1 be a unit eigenvector of A corresponding to k_1. By Theorem 8.3(f) there exists a unitary matrix Q with Q_1 as first column. Just as in the proof of Theorem 6.7, we have

$$Q^{-1}AQ = \begin{bmatrix} k_1 & B_1 \\ O & A_1 \end{bmatrix}$$

where A_1 is of order $n - 1$ and has eigenvalues $k_2, \ldots, k_n$. The theorem is therefore proved for $n = 2$. Assuming the result for matrices of order $n - 1$, there exists a unitary matrix R_1 of order $n - 1$ such that $R_1^{-1}A_1R_1$ is triangular with diagonal elements $k_2, \ldots, k_n$. Now if we let

$$R = \begin{bmatrix} 1 & O \\ O & R_1 \end{bmatrix}$$

the matrix $P = QR$ is unitary because Q and R are and $P^{-1}AP$ is triangular with diagonal elements $k_1, k_2, \ldots, k_n$.

Definition. *A matrix A is said to be* normal *if $AA^* = A^*A$.*

Theorem 8.12. *If A is normal and P is unitary, then $P^{-1}AP$ is normal.*

Proof. Since P is unitary, $P^{-1} = P^*$ and

$$(P^*AP)(P^*AP)^* = P^*APP^*A^*P = P^*AA^*P = P^*A^*AP = (P^*A^*P)(P^*AP)$$
$$= (P^*AP)^*(P^*AP).$$

Theorem 8.13. *A triangular matrix is normal if and only if it is diagonal.*

Proof. A diagonal matrix is clearly normal. Let

$$A = \begin{bmatrix} a_{11} & a_{12} & \cdots & a_{1n} \\ 0 & a_{22} & \cdots & a_{2n} \\ \cdots\cdots\cdots\cdots\cdots \\ 0 & 0 & \cdots & a_{nn} \end{bmatrix}$$

and assume that A is normal. The element in the first row and first column of AA^* is $a_{11}\bar{a}_{11} + a_{12}\bar{a}_{12} + \cdots + a_{1n}\bar{a}_{1n}$ and the corresponding element of A^*A is $a_{11}\bar{a}_{11}$. Hence, $AA^* = A^*A$ implies that

$$a_{12}\bar{a}_{12} + a_{13}\bar{a}_{13} + \cdots + a_{1n}\bar{a}_{1n} = 0.$$

Since $a_{1i}\bar{a}_{1i} \geq 0$, we have $a_{12} = a_{13} = \cdots a_{1n} = 0$ and

$$A = \begin{bmatrix} a_{11} & O \\ O & A_1 \end{bmatrix},$$

where A_1 is clearly normal and triangular. This proves the theorem for $n = 2$. Moreover, if we assume the theorem true for matrices of order $n - 1$, then A_1 is diagonal and hence A is diagonal and the proof is complete.

Theorem 8.14. *A matrix A with elements in $\mathscr{C}$ is unitarily similar to a diagonal matrix if and only if A is normal.*

Proof. If A is normal, by Theorem 8.11 there is a unitary matrix P such that $P^{-1}AP$ is triangular. By Theorem 8.12, $P^{-1}AP$ is normal and hence by Theorem 8.13, $P^{-1}AP$ is diagonal. Conversely, if $P^{-1}AP = D$, a diagonal matrix, and P is unitary, then $A = PDP^{-1}$ is normal because D is diagonal and therefore normal.

Corollary. *An $n \times n$ matrix A has n mutually orthogonal eigenvectors in $\mathscr{V}_n(\mathscr{C})$ if and only if A is normal.*

Exercise 8.4

1. Prove that real orthogonal, real symmetric, unitary, and Hermitian matrices are normal.

2. If A is any matrix with elements in $\mathscr{C}$ and $A = B + iC$ where B and C are Hermitian (see Problem 5, Exercise 8.3), prove that A is normal if and only if $BC = CB$.

3. If $A = HP$ is the polar factorization of a nonsingular matrix A, show that A is normal if and only if $HP = PH$.

4. Prove that a matrix A, elements in $\mathscr{C}$, is similar to a diagonal matrix if and only if there exists a positive definite Hermitian matrix H such that $H^{-1}AH$ is normal.

6. Prove that if A is normal and all eigenvalues of A are real, then A is Hermitian.

7. Prove that if A is normal and its eigenvalues all have absolute value 1, then A is unitary.

8. Prove that any diagonal matrix with elements in $\mathscr{C}$ is normal.

9. Use Problems 6, 7, and 8 and Theorem 8.12 to construct a 2×2 normal matrix that is neither diagonal, unitary, nor Hermitian.

8.7 The Real Canonical Form of an Orthogonal Matrix

Let P be any (real) orthogonal matrix. Since P is also unitary, its eigenvalues have absolute value 1 by Theorem 8.5. The real eigenvalues are therefore 1 or -1 and the complex eigenvalues occur in conjugate pairs because the characteristic equation is real. Suppose r eigenvalues are equal to 1, s are equal to -1, and that there are q conjugate pairs $e^{i\varphi_1}, e^{-i\varphi_1}, \ldots, e^{i\varphi_q}, e^{-i\varphi_q}$. Since P is normal, we can choose a unitary matrix U such that $U^*PU = D$ is diagonal. Assume the first r diagonal elements of D are 1's, the next s diagonal elements -1's and these are followed in order by the conjugate pairs $e^{i\varphi_j}, e^{-i\varphi_j}, j = 1, 2, \ldots, q$. Now if we let

$$
S_1 = \begin{bmatrix} \dfrac{1}{\sqrt{2}} & \dfrac{i}{\sqrt{2}} \\[2ex] \dfrac{1}{\sqrt{2}} & -\dfrac{i}{\sqrt{2}} \end{bmatrix}, \qquad D_j = \begin{bmatrix} e^{i\varphi_j} & 0 \\ 0 & e^{-i\varphi_j} \end{bmatrix},
$$

it is easy to verify that S_1 is unitary and that

$$R_j = S_1^* D_j S_1 = \begin{bmatrix} \cos \varphi_j & -\sin \varphi_j \\ \sin \varphi_j & \cos \varphi_j \end{bmatrix}.$$

We now let

$$S = \begin{bmatrix} 1 & & & & & & & & & O \\ & \ddots & & & & & & & & \\ & & 1 & & & & & & & \\ & & & -1 & & & & & & \\ & & & & \ddots & & & & & \\ & & & & & -1 & & & & \\ & & & & & & S_1 & & & \\ & & & & & & & \ddots & & \\ O & & & & & & & & S_1 \end{bmatrix}$$

be the matrix with the first r diagonal elements equal to 1, the next s equal to -1, then q diagonal 2×2 blocks equal to S_1 and all other elements 0. It is clear that S is unitary and that if we put $Q = US$, Q is unitary and

$$(7) \qquad Q^*PQ = S^*DS = \begin{bmatrix} I_r & & & & O \\ & -I_s & & & \\ & & R_1 & & \\ & & & \ddots & \\ O & & & & R_q \end{bmatrix},$$

where I_r, I_s are identity matrices of order r and s and $R_j = S_1^* D_j S_1$.

Now the columns of U are eigenvectors of P and hence the first $r + s$ columns (corresponding to the real eigenvalues) are real. The subsequent columns are conjugate pairs of vectors in $\mathscr{V}_n(\mathscr{C})$ (corresponding to conjugate pairs of eigenvalues). The first $r + s$ columns of US are therefore clearly real. The next two columns are

UY_1 and UY_2, where

$$Y_1 = \begin{bmatrix} 0 \\ \cdot \\ \cdot \\ \cdot \\ 0 \\ \dfrac{1}{\sqrt{2}} \\ \dfrac{1}{\sqrt{2}} \\ \cdot \\ \cdot \\ \cdot \\ 0 \end{bmatrix}, \qquad Y_2 = \begin{bmatrix} 0 \\ \cdot \\ \cdot \\ \cdot \\ 0 \\ \dfrac{i}{\sqrt{2}} \\ -\dfrac{i}{\sqrt{2}} \\ \cdot \\ \cdot \\ \cdot \\ 0 \end{bmatrix}.$$

Thus UY_1 is $\dfrac{1}{\sqrt{2}}$ times the sum of two conjugate columns of U and UY_2 is $\dfrac{i}{\sqrt{2}}$ times the difference of these conjugate columns. Hence, both UY_1 and UY_2 are real. Similarly, all columns of US are real. Thus, $Q = US$ is real and unitary and hence orthogonal and every orthogonal matrix P is orthogonally similar to a matrix in real canonical form (7) characterized by the invariants r, s, and the q real numbers $\varphi_1, \varphi_2, \ldots, \varphi_q$.

8.8 Spectral Decomposition of a Diagonalizable Matrix

The following theorem gives a decomposition of any diagonalizable matrix which is useful for many purposes.

Theorem 8.15. *Let A be any diagonalizable $n \times n$ matrix with elements in $\mathscr{C}$. Let $k_1, k_2, \ldots, k_s$ be the distinct eigenvalues of A. Then A can be written in the form*

(8) $$A = k_1P_1 + k_2P_2 + \cdots + k_sP_s,$$

where the P_i are $n \times n$ matrices having the following properties:

(a) *Each P_i is idempotent, that is, $P_i^2 = P_i$.*
(b) *$P_iP_j = O$ for $i \neq j$.*
(c) *$P_1 + P_2 + \cdots + P_s = I$.*
(d) *Each P_i commutes with A, that is, $AP_i = P_iA$.*

Proof. Since A is diagonalizable, let $P^{-1}AP = D$ be the diagonal form of A. Let E_i be the $n \times n$ diagonal matrix with 1's in those places in the main diagonal in which k_i occurs in D and 0's elsewhere. We then have

(9) $$D = P^{-1}AP = k_1E_1 + k_2E_2 + \cdots + k_sE_s$$

and $E_i^2 = E_i$, $E_iE_j = O$ if $i \neq j$ and $E_1 + E_2 + \cdots + E_s = I$. From (9) we see that

(10) $$A = PDP^{-1} = k_1P_1 + k_2P_2 + \cdots + k_n\check{P}_s,$$

where $P_i = PE_iP^{-1}$. It is clear that $P_1, \ldots, P_s$ satisfy (a), (b), and (c) because $E_1, \ldots, E_s$ do. Finally, by (10) and (a) and (b), $AP_i = k_iP_i$ and $P_iA = k_iP_i$ so $AP_i = P_iA$, $i = 1, 2, \ldots, s$.

The decomposition (8) is called the *spectral decomposition* of A.

Corollary. *If A is diagonalizable and has spectral decomposition (8) and if $f(x)$ is any polynomial, then*

$$f(A) = f(k_1)P_1 + f(k_2)P_2 + \cdots + f(k_s)P_s.$$

Proof. Forming successive powers of A from equation (8) and using (a), (b), and (c), we can see that for any integer $r \geq 0$,

$$A^r = k_1^r P_1 + k_2^r P_2 + \cdots + k_s^r P_s$$

and the result for arbitrary polynomials follows.

8.9 Sequences and Series of Matrices

A sequence $\{S_r\}$ of $n \times n$ matrices in which $S_r = [s_{ij}^{(r)}]$, $(r = 1, 2, 3, \ldots)$ is said to be convergent with limit $S = [s_{ij}]$ if each of the n^2 sequences $\{s_{ij}^{(r)}\}$ is convergent with limit s_{ij}. We then write $\lim_{r \to \infty} S_r = S$.

Lemma 8.16. *If $\{S_r\}$ is a convergent sequence of $n \times n$ matrices with limit S and P is any $n \times n$ matrix, then $\{PS_r\}$ is convergent with limit PS.*

Proof. It is clear that $\lim_{r \to \infty} S_r = S$ if and only if $\lim_{r \to \infty} (S - S_r) = O$. Let $S - S_r = [b_{ij}^{(r)}]$ and let $P = [p_{ij}]$. Now if $PS - PS_r = P(S - S_r) = [c_{ij}^{(r)}]$ we have

$$c_{ij}^{(r)} = p_{i1}b_{1j}^{(r)} + \cdots + p_{in}b_{nj}^{(r)},$$

and since limit $b_{ij}^{(r)} = 0$ for all i, j it follows that limit $c_{ij}^{(r)} = 0$ for all i, j. Hence,

$_{r\to\infty}$

limit $(PS - PS_r) = O$ and limit $PS_r = PS$.

$_{r\to\infty}_{r\to\infty}$

Corollary 1. *If $\{S_r\}$ is convergent with limit S and P is nonsingular, then the sequence $\{P^{-1}S_rP\}$ is convergent with limit $P^{-1}SP$.*

Proof. The theorem can be proved for left multiplication just as for right and the corollary then follows.

An infinite series

(11)
$$\sum_{r=1}^{\infty} A_r,$$

of which the terms are $n \times n$ matrices $A_r = [a_{ij}^{(r)}]$, is said to be convergent with sum $S = [s_{ij}]$ if the sequence $\{S_m\}$ of partial sums

$$S_m = \sum_{r=1}^{m} A_r$$

is convergent with limit S. Thus (11) converges to S if and only if each of the n^2 series $\sum_{r=1}^{\infty} a_{ij}^{(r)}$ is convergent with sum s_{ij}, $(i, j = 1, 2, \ldots, n)$. The following additional corollary to Lemma 8.16 is an immediate consequence of Corollary 1.

Corollary 2. *If the series (11) is convergent with sum S and P is nonsingular, the series*

$$\sum_{r=1}^{\infty} P^{-1}A_rP$$

is convergent with sum $P^{-1}SP$.

Theorem 8.17. *If A is any $n \times n$ matrix with elements in $\mathscr{C}$, the series*

(12)
$$I + A + \frac{1}{2!}A^2 + \cdots + \frac{1}{r!}A^r + \cdots$$

converges.

Proof. Let $A = [a_{ij}]$ and choose a positive real number $M > \max_{i,j} \{n, |a_{ij}|\}$. We then have $|a_{ij}| < M$, and if $A^2 = [b_{ij}] = [\sum a_{ik}a_{kj}]$ we have

$$|b_{ij}| \leq \sum_{k=1}^{n} |a_{ik}| |a_{kj}| < nM^2 < M^4.$$

Now suppose it has been established that $A^r = [c_{ij}]$ where, for all i, j, $|c_{ij}| < M^{2r}$. Then $A^{r+1} = [d_{ij}]$ where

$$|d_{ij}| = \left| \sum_{k=1}^{n} a_{ik}c_{kj} \right| \le \sum_{k=1}^{n} |a_{ik}|\,|c_{kj}| < nM^{2r+1} < M^{2r+2}.$$

Hence if we let $A^r = [a_{ij}^{(r)}]$ each of the n^2 series

(13) $$\sum_{r=1}^{\infty} \frac{1}{r!} |a_{ij}^{(r)}|$$

is dominated by the series

(14) $$\sum_{r=1}^{\infty} \frac{M^{2r}}{r!}.$$

Since (14) converges to the sum e^{M^2}, each of the series (13) converges and therefore (12) converges.

The series (12) is called the matrix exponential series and its sum is denoted by e^A.

Theorem 8.18. *Let A be a diagonalizable $n \times n$ matrix and let $k_1, k_2, \ldots, k_s$ be its distinct eigenvalues. The matrix power series*

(15) $$\sum_{r=0}^{\infty} a_r A^r$$

converges if and only if each of the series

(16) $$\sum_{r=0}^{\infty} a_r k_i^r \qquad\qquad (i = 1, 2, \ldots, s)$$

converges.

Proof. Choose P so that $P^{-1}AP = D$ is diagonal. Then for each r, $P^{-1}A^rP = D^r$.

By Corollary 2, Lemma 8.16, the series (15) converges if and only if $\sum_{r=0}^{\infty} a_r D^r$ converges and hence if and only if each of the series (16) converges.

Theorem 8.19. *Let*

$$p(x) = a_0 + a_1 x + a_2 x^2 + \cdots$$

be any power series with coefficients in $\mathscr{C}$ and let A be a diagonalizable matrix with spectral decomposition

$$A = k_1 P_1 + k_2 P_2 + \cdots + k_s P_s.$$

If the series

$$p(A) = a_0 I + a_1 A + a_2 A^2 + \cdots$$

converges its sum is

$$p(A) = p(k_1)P_1 + p(k_2)P_2 + \cdots + p(k_s)P_s.$$

Proof. As in the proof of Theorem 8.15, let the diagonal form of A be

$$D = P^{-1}AP = k_1 E_1 + k_2 E_2 + \cdots + k_s E_s,$$

so that $P_i = PE_i P^{-1}$, $(i = 1, 2, \ldots, s)$. Obviously, $p(D) = p(k_1)E_1 + \cdots + p(k_s)E_s$ and by Lemma 8.16, Corollary 2,

$$p(A) = p(PDP^{-1}) = P^{-1}p(D)P = p(k_1)P_1 + \cdots + p(k_s)P_s.$$

If we apply Theorem 8.19 to the exponential series we get for any diagonalizable matrix A whose distinct eigenvalues are $k_1, \ldots, k_s$,

$$e^A = e^{k_1}P_1 + e^{k_2}P_2 + \cdots e^{k_s}P_s.$$

We use this to prove the following theorem.

Theorem 8.20. *Every unitary matrix Q can be written in the form $Q = e^{iH}$ where H is Hermitian. Conversely, if H is Hermitian e^{iH} is unitary.*

Proof. Because Q is unitary its eigenvalues have absolute value 1 by Theorem 8.5. Therefore we can denote the distinct eigenvalues of Q by $e^{ik_1}, e^{ik_2}, \ldots, e^{ik_s}$, where $k_1, \ldots, k_s$ are real. Since Q is normal, there is a unitary matrix P such that $P^{-1}QP$ is diagonal and hence

$$P^{-1}QP = e^{ik_1}E_1 + e^{ik_2}E_2 + \cdots + e^{ik_s}E_s,$$

where $E_1, \ldots, E_s$ are the idempotent diagonal matrices used in the proof of Theorem 8.15. It follows from Theorem 8.19 that $P^{-1}QP = e^{iD}$, where

$$D = k_1 E_1 + k_2 E_2 + \cdots + k_s E_s.$$

Hence, by Lemma 8.16, Corollary 2,

$$Q = Pe^{iD}P^{-1} = e^{i(PDP^{-1})} = e^{iH},$$

where $H = PDP^{-1}$. Since D is a real diagonal matrix, it is Hermitian and, since P is unitary, $H = PDP^*$ is also Hermitian. Conversely, suppose H is Hermitian, $k_1, \ldots, k_s$ its distinct eigenvalues, and P a unitary matrix that diagonalizes H. Then

$$P^{-1}HP = k_1 E_1 + k_2 E_2 + \cdots + k_s E_s$$

and

$$P^{-1}e^{iH}P = e^{i(P^{-1}HP)} = e^{ik_1}E_1 + e^{ik_2}E_2 + \cdots + e^{ik_s}E_s.$$

The right-hand side is diagonal and clearly unitary so $P^{-1}e^{iH}P$ is unitary and hence e^{iH} is unitary since P is.

In view of Theorem 8.20, the polar factorization of a nonsingular matrix A becomes $He^{iH'}$, where H and H' are both Hermitian and H is positive definite. This generalizes the polar form $re^{i\theta}$ for a complex number in which $r > 0$ and hence $[r]$ is a 1×1 positive definite Hermitian matrix and $e^{i\theta}$, θ real, is a 1×1 unitary matrix.

The discussion of convergence of matrix series given in this section was introduced mainly to justify the use of the matrix exponential e^A and it is by no means complete. For a fuller treatment of this topic and its many applications it is necessary to introduce a *matrix norm* $\|A\|$ which is a nonnegative real number associated with the matrix A that plays the same role in convergence problems that the absolute value of a complex number plays in the discussion of convergence of series with complex terms. This is an extensive field with many applications, particularly to systems of linear differential equations and to iterative methods for approximating solutions to systems of linear equations or for finding eigenvalues of a matrix. For further information on these topics the student is referred to [10] and [3].

Exercise 8.5

1. If A is nonsingular and has spectral decomposition (8), show that A^{-1} has spectral decomposition

$$A^{-1} = k_1^{-1}P_1 + k_2^{-1}P_2 + \cdots + k_s^{-1}P_s.$$

2. Let A be any $n \times n$ matrix and let P be chosen so that $P^{-1}AP$ is triangular.
   (a) Show that $P^{-1}e^AP$ is triangular.
   (b) Show that if $k_1, \ldots, k_n$ are the eigenvalues of A, then $e^{k_1}, \ldots, e^{k_n}$ are the eigenvalues of e^A.
   (c) Show that e^A is nonsingular.

3. Show that if $AB = BA$, then $e^Ae^B = e^{A+B}$.

4. Show that $(e^A)^{-1} = e^{-A}$.

5. Prove that if A is an $n \times n$ matrix and the series

   (17) $$I + A + A^2 + \cdots + A^r + \cdots$$

   converges, all eigenvalues of A must have absolute values less than 1. (This condition is actually also sufficient for the convergence of this series.)

6. Prove that if (17) converges $I - A$ is nonsingular. *Hint.* Use the triangularization theorem and Problem 5.

7. Prove that if (17) converges its sum is $(I - A)^{-1}$.

8. Prove that every proper orthogonal matrix P has the form $P = e^S$, where S is a real skew symmetric matrix (i.e., $S^T = -S$), and that every matrix of this form is proper orthogonal. *Hint.* Use Theorem 8.20.

9. Given

$$ S = \begin{bmatrix} 0 & \varphi \\ -\varphi & 0 \end{bmatrix}, $$

calculate e^S by forming successive powers of S and substituting in the series for e^S.

8.10 Linear Operators in a Unitary Space

In this section we shall study linear operators in a finite dimensional unitary space $\mathscr{V}$. Because of the basic IPS-isomorphism between $\mathscr{V}$ and $\mathscr{V}_n(\mathscr{C})$ and the algebra isomorphism between $\mathscr{L}(\mathscr{V}, \mathscr{V})$ and $\mathscr{C}_n$, theorems about operators can be derived from theorems about matrices and vice versa. Since we have developed the matrix theory first, we shall use it extensively although some of our results will be proved directly from properties of operators when this is more convenient. We shall define unitary, Hermitian, and normal operators corresponding to unitary, Hermitian, and normal matrices. For this purpose we introduce the *adjoint* of an operator which will be defined after the next theorem.

Theorem 8.21. *If $\mathscr{V}$ is an n-dimensional unitary space and $\tau \in \mathscr{L}(\mathscr{V}, \mathscr{V})$, there exists a uniquely determined linear operator τ^* in $\mathscr{L}(\mathscr{V}, \mathscr{V})$ such that for all U, V in $\mathscr{V}$, $(\tau U, V) = (U, \tau^* V)$.*

Proof.

1. *Existence of τ^*.* Let $E_1, E_2, \ldots, E_n$ be any orthonormal basis of $\mathscr{V}$. Let $A_\tau = [a_{ij}]$ be the matrix of τ relative to the E-basis and define τ^* to be the linear operator in $\mathscr{V}$ defined by

$$ \tau^* E_i = \bar{a}_{i1} E_1 + \bar{a}_{i2} E_2 + \cdots + \bar{a}_{in} E_n, $$

so that the matrix of τ^* relative to the E-basis is A_τ^*, the adjoint of A_τ. Now let U, V be arbitrary vectors in $\mathscr{V}$ and let X, Y be their coordinate (column) vectors relative to the E-basis. By Theorem 8.6,

$$ (\tau U, V) = (A_\tau X) \cdot Y = X^T A_\tau^T \bar{Y} = X^T \overline{(A_\tau^* Y)} = X \cdot (A_\tau^* Y) = (U, \tau^* V) $$

and hence τ^* satisfies the requirement of the theorem.

2. *Uniqueness of* τ^*. If, given $\tau \in \mathscr{L}(\mathscr{V}, \mathscr{V})$, we have $(\tau U, V) = (U, \sigma_1 V) = (U, \sigma_2 V)$ for all U, V in $\mathscr{V}$, then $(U, (\sigma_1 - \sigma_2)V) = (U, \sigma_1 V - \sigma_2 V) = (U, \sigma_1 V) - (U, \sigma_2 V) = 0$. However, $[U, (\sigma_1 - \sigma_2)V] = 0$ for all U implies $(\sigma_1 - \sigma_2)V = O$, which holds, therefore, for all V in $\mathscr{V}$. Hence, $\sigma_1 - \sigma_2$ is the zero operator and $\sigma_1 = \sigma_2$. Thus the condition $(\tau U, V) = (U, \tau^* V)$ determines τ^* uniquely.

Definition. *If* $\tau \in \mathscr{L}(\mathscr{V}, \mathscr{V})$ *the uniquely determined operator* τ^* *such that* $(\tau U, V) = (U, \tau^* V)$ *for all* U, V *in* $\mathscr{V}$ *is called the* adjoint *of* τ.

Corollary 1. *If* $\tau \in \mathscr{L}(\mathscr{V}, \mathscr{V})$ *and* $E_1, \ldots, E_n$ *is any orthonormal basis of* V, *then* $A_{\tau^*}^{(E)} = (A_\tau^{(E)})^*$.

Corollary 2. *If* $\tau \in \mathscr{L}(\mathscr{V}, \mathscr{V})$, *then* $\tau^{**} = \tau$.

Proof. This follows from Corollary 1 or can be proved directly from the definition Thus $(\tau^* U, V) = \overline{(V, \tau^* U)} = \overline{(\tau V, U)} = (U, \tau V)$ and hence, by the uniqueness proved in the Theorem, $\tau^{**} = \tau$.

Definition. *If* $\tau \in \mathscr{L}(\mathscr{V}, \mathscr{V})$, *then* τ *is* unitary *if* $\tau^* \tau = I$, Hermitian *(or* self-adjoint) *if* $\tau^* = \tau$, *and* normal *if* $\tau \tau^* = \tau^* \tau$.

Theorem 8.22. *A linear operator* $\tau: \mathscr{V} \to \mathscr{V}$ *is unitary, Hermitian, or normal if and only if its matrix* $A_\tau^{(E)}$ *relative to any orthonormal basis* $E_1, \ldots, E_n$ *is unitary, Hermitian, or normal, respectively.*

Proof. This follows immediately from Theorem 8.21, Corollary 1. For example, if $\tau^* \tau = I$, then $A_\tau^* A_\tau = A_{\tau^*} A_\tau = A_{\tau^* \tau} = I$ because $\tau^* \tau = I$. Hence, if τ is unitary A_τ is unitary. The proof for Hermitian and normal operators is similar.

Theorem 8.23. *A linear operator* $\tau: \mathscr{V} \to \mathscr{V}$ *is unitary if and only if it preserves inner products, that is, if and only if*

$$(\tau U, \tau V) = (U, V) \quad \text{for all } U, V \text{ in } \mathscr{V}.$$

Proof. Since $(\tau U, \tau V) = (U, \tau^* \tau V)$, clearly inner products are preserved if and only if $\tau^* \tau = I$.

Corollary. *A linear operator is unitary if and only if it preserves lengths, that is,* $\|\tau V\| = \|V\|$ *for all* V *in* $\mathscr{V}$.

Proof. Clearly, if τ preserves inner products it preserves lengths. The converse follows from the identity

(18) $4(U, V) = \|U + V\|^2 - \|U - V\|^2 + i \|U + iV\|^2 - i \|U - iV\|^2.$

We note that both unitary and Hermitian operators are normal. Theorems 8.14 (Corollary), 6.18, and 8.6 immediately imply the following.

Theorem 8.24. *If $\mathscr{V}$ is an n-dimensional unitary space and $\tau \in \mathscr{L}(\mathscr{V}, \mathscr{V})$, a necessary and sufficient condition that τ be normal is that τ have n mutually orthogonal eigenvectors.*

Let $\mathscr{V}$ be an n-dimensional unitary space, let τ be a Hermitian operator in $\mathscr{V}$, and let A_τ be its matrix relative to an orthonormal basis. Since the eigenvalues of τ are the eigenvalues of A_τ, they are real by Theorem 8.8, Corollary 1. A Hermitian operator τ is said to be *positive* if all its eigenvalues are positive. Thus, τ is positive if A_τ is positive definite. By Theorem 8.24 τ has n mutually orthogonal eigenvectors, and hence a positive operator is a magnification in n mutually orthogonal directions.

Now let σ be any nonsingular linear operator on $\mathscr{V}$ and A_σ its matrix relative to an orthonormal basis $E_1, \ldots, E_n$. By Theorem 8.10 $A_\sigma = HP$ where H is positive definite Hermitian and P is unitary. We can therefore define a positive operator τ by $A_\tau^{(E)} = H$ and a unitary operator ρ by $A_\rho^{(E)} = P$. Then $A_\sigma^{(E)} = A_\tau^{(E)} A_\rho^{(E)} = A_{\tau\rho}^{(E)}$ and hence $\sigma = \tau\rho$. This *polar factorization* of σ is unique because the polar factorization of A_σ is unique. We have therefore proved the following.

Theorem 8.25. *Every nonsingular operator $\sigma \in \mathscr{L}(\mathscr{V}, \mathscr{V})$ has a unique representation as a product $\tau\rho$, where τ is a positive operator and ρ is a unitary operator.*

We recall that if $\mathscr{V}$ is the direct sum of two subspaces $\mathscr{S}$ and $\mathscr{T}$, then every vector V of $\mathscr{V}$ has a unique representation $V = S + T$, where $S \in \mathscr{S}$ and $T \in \mathscr{T}$, and that the mapping $\pi : \mathscr{V} \to \mathscr{V}$ defined by $\pi V = S$ is called a projection of $\mathscr{V}$ on $\mathscr{S}$. This projection π depends on the subspace $\mathscr{T}$, which is not uniquely determined by $\mathscr{S}$, and should strictly be called the projection of $\mathscr{V}$ on $\mathscr{S}$ *in the direction of $\mathscr{T}$.* In particular, if $\mathscr{T} = \mathscr{S}^\perp$, π is called the *orthogonal projection* of $\mathscr{V}$ on $\mathscr{S}$.

Theorem 8.26. *A linear operator $\pi : \mathscr{V} \to \mathscr{V}$ is a projection if and only if $\pi^2 = \pi$.*

Proof. Suppose $\pi \in \mathscr{L}(\mathscr{V}, \mathscr{V})$ and $\pi^2 = \pi$. Let $\mathscr{S} = \text{im } \pi$ and $\mathscr{T} = \text{im } (I - \pi)$. Since every vector V in $\mathscr{V}$ can be written

(19) $V = \pi V + (V - \pi V) = \pi V + (I - \pi)V,$

it is clear that $\mathcal{V} = \mathcal{S} + \mathcal{T}$. If $V \in \mathcal{S} \cap \mathcal{T}$, then $V = \pi V_1 = (I - \pi)V_2$. Applying π again, we get $\pi^2 V_1 = (\pi - \pi^2)V_2 = O$, since $\pi^2 = \pi$, and hence $V = \pi V_1 = O$. Therefore, $\mathcal{S} \cap \mathcal{T} = O$ and $\mathcal{V} = \mathcal{S} \oplus \mathcal{T}$. Equation (19) now shows that π is the projection of $\mathcal{V}$ on $\mathcal{S}$ in the direction $\mathcal{T}$. The converse, that if π is a projection $\pi^2 = \pi$, is clear from the definition.

Corollary. *If π is a projection the space $\mathcal{T} = \mathrm{im}\,(I - \pi)$ is the null space of π.*

Proof. If $U = (I - \pi)V$, then $\pi U = (\pi - \pi^2)V = O$ and $\mathrm{im}\,(I - \pi) \subset$ null space of π. Conversely, if $\pi U = O$, then $U = \pi U + (I - \pi)U = (I - \pi)U$ and $U \in \mathrm{im}\,(I - \pi)$.

Theorem 8.27. *(Spectral theorem for normal operators) Let $\mathcal{V}$ be a finite dimensional unitary space and τ a normal operator on $\mathcal{V}$. If the distinct eigenvalues of τ are $k_1, k_2, \ldots, k_s$, then*

$$\tau = k_1 \pi_1 + k_2 \pi_2 + \cdots + k_s \pi_s,$$

where π_i is the orthogonal projection of $\mathcal{V}$ onto a subspace $\mathcal{W}_i = \mathrm{im}\,\pi_i$. Also, $\pi_i \pi_j = O$ for $i \neq j$, and $\pi_1 + \pi_2 + \cdots + \pi_s = I$. Moreover, $\mathcal{V} = \mathcal{W}_1 \oplus \mathcal{W}_2 \oplus \cdots \oplus \mathcal{W}_s$ and the subspaces $\mathcal{W}_i$ are mutually orthogonal.

Proof. Choose any orthonormal basis $B_1, \ldots, B_n$ of $\mathcal{V}$ and let A_τ be the matrix of τ relative to the B-basis. The distinct eigenvalues of A_τ are (Theorem 6.18) $k_1, k_2, \ldots, k_s$ and by Theorem 8.15

$$A_\tau = k_1 P_1 + k_2 P_2 + \cdots + k_s P_s,$$

where $P_i^2 = P_i$, $P_i P_j = O$ for $i \neq j$, and $P_1 + P_2 + \cdots + P_s = I$. Moreover, since A_τ is normal and hence unitarily similar to a diagonal matrix, it follows (see proof of Theorem 8.15) that each P_i has the form $Q^* E_i Q$, where E_i is a real diagonal matrix and Q is unitary. Hence, each P_i is Hermitian and $P_i^* = P_i$. If we now let $\pi_i : \mathcal{V} \to \mathcal{V}$ be the linear operator defined by $A_{\pi_i}^{(B)} = P_i$, it is clear that $\pi_i^* = \pi_i$ and from the basic isomorphism $\sigma \to A_\sigma^{(E)}$, we have

(20) $$\pi_i^2 = \pi_i \qquad (i = 1, 2, \ldots, s),$$

(21) $$\pi_i \pi_j = O \qquad \text{for } i \neq j,$$

(22) $$\pi_1 + \pi_2 + \cdots + \pi_s = I.$$

By (20) and Theorem 8.26 π_i is a projection of $\mathcal{V}$ onto the subspace $\mathcal{W}_i = \mathrm{im}\,\pi_i$.

Let $\mathcal{N}_i$ be the null space of π_i which, by the corollary to Theorem 8.26, is equal to im $(I - \pi_i)$. Thus, $\mathcal{V} = \mathcal{W}_i \oplus \mathcal{N}_i$. Now if $W \in \mathcal{W}_i$ and $N \in \mathcal{N}_i$, we have

$$(W, N) = (\pi_i V_1, (I - \pi_i) V_2)$$

$$= (V_1, (\pi_i - \pi_i^2) V_2), \qquad \text{since } \pi_i^* = \pi_i$$

$$= (V_1, O), \qquad \text{since } \pi_i^2 = \pi_i$$

$$= 0.$$

Thus, $\mathcal{N}_i \subset \mathcal{W}_i^\perp$, and because dim $\mathcal{N}_i = n - \dim \mathcal{W}_i$, $\mathcal{N}_i = \mathcal{W}_i^\perp$. Hence π_i is the orthogonal projection of $\mathcal{V}$ onto $\mathcal{W}_i$. Finally, if $V \in \mathcal{V}$, we have by (22)

$$V = \pi_1 V + \pi_2 V + \cdots + \pi_s V$$

and hence $\mathcal{V} = \sum \mathcal{W}_i$. To show the sum is direct suppose, for example, that for $W_i \in \mathcal{W}_i$ $(i = 1, 2, \ldots, s)$ we have

(23) $$W_1 = W_2 + W_3 + \cdots + W_s.$$

Since for each i, $W_i = \pi_i V_i$ for some $V_i \in \mathcal{V}$, we have by (20) and (21), $\pi_i W_i = W_i$ and $\pi_i W_j = O$ if $j \neq i$. Applying π_1 to (23), we get $W_1 = \pi_1 W_1 = \sum_{i=2}^{s} \pi_1 W_i = O$. Hence, $\mathcal{W}_1 \cap (\mathcal{W}_2 + \cdots + \mathcal{W}_s) = O$ and, similarly, each $\mathcal{W}_i$ has O intersection with the sum of the $\mathcal{W}_j$'s with $j \neq i$. Hence,

$$\mathcal{V} = \mathcal{W}_1 \oplus \mathcal{W}_2 \oplus \cdots \oplus \mathcal{W}_s$$

and because for $j \neq i$ $\mathcal{W}_j \subset \mathcal{N}_i = \mathcal{W}_i^\perp$, the subspaces $\mathcal{W}_i$ are mutually orthogonal.

Exercise 8.6

1. Prove that if $\tau: \mathcal{V} \to \mathcal{V}$ is unitary, then τ maps orthogonal vectors onto orthogonal vectors.

2. Prove that $\tau: \mathcal{V} \to \mathcal{V}$ is unitary if and only if τ maps an orthonormal basis of $\mathcal{V}$ onto an orthonormal basis.

3. An operator $\tau: \mathcal{V} \to \mathcal{V}$ is said to be diagonalizable if its eigenvectors span $\mathcal{V}$. Prove that τ is diagonalizable if and only if its matrix A relative to some basis is diagonalizable.

4. If τ is diagonalizable and its distinct eigenvalues are $k_1, \ldots, k_s$, prove that there exist (not necessarily orthogonal) projections $\pi_1, \ldots, \pi_s$ such that:
   (a) $\tau = k_1 \pi_1 + k_2 \pi_2 + \cdots + k_s \pi_s$.
   (b) $\pi_i^2 = \pi_i$ and $\pi_i \pi_j = O$ if $i \neq j$.
   (c) $\pi_1 + \pi_2 + \cdots + \pi_s = I$.
   (d) $\mathcal{V} = \mathcal{W}_1 \oplus \mathcal{W}_2 \oplus \cdots \oplus \mathcal{W}_s$ where $\mathcal{W}_i = \text{im } \pi_i$.

(e) $\mathscr{W}_i$ is the null space of $\tau - k_i I$, $i = 1, 2, \ldots, s$.

(f) Each subspace $\mathscr{W}_i$ is invariant under τ.

(g) $\tau \pi_i = \pi_i \tau$, $i = 1, 2, \ldots, s$.

5. Translate Problems 1 through 7. Exercise 8.4 into statements about linear operators.

6. Prove that a projection is self-adjoint (i.e., Hermitian) if and only if it is an orthogonal projection.

7. Prove that if τ is a normal operator in $\mathscr{V}$, then $\|\tau V\| = \|\tau^* V\|$ for all V in $\mathscr{V}$.

8. Verify the identity (18).

REFERENCES

1. Baer, Reinhold, *Linear Algebra and Projective Geometry*, Academic Press Inc., New York, 1952.
2. Bellman, Richard, *Introduction to Matrix Analysis*, McGraw-Hill, New York, 1960.
3. Fadeev D. K. and V. N. Fadeeva, *Computational Methods of Linear Algebra*, W. H. Freeman and Co., San Francisco, 1963.
4. Halmos, Paul R., *Finite-dimensional Vector Spaces*, Second Edition, Van Nostrand, Princeton, N.J., 1958.
5. Hoffman, Kenneth and Ray Kunze, *Linear Algebra*, Prentice-Hall Inc., Englewood Cliffs, N.J., 1961.
6. Jacobson, Nathan, *Lectures in Abstract Algebra*, Volume 1, Van Nostrand, New York, 1951.
7. Marcus, Marvin, and Henryk Minc, *Introduction to Linear Algebra*, Macmillan, New York, 1965.
8. Milne, W. E., *Numerical Solution of Differential Equations*, Wiley, New York, 1953.
9. Murdoch, David C., *Analytic Geometry with an Introduction to Vectors and Matrices*, Wiley, New York, 1966.
10. Varga, Richard S., *Matrix Iterative Analysis*, Prentice-Hall, Englewood Cliffs, N.J., 1962.
11. Whittaker, E. T., *A Treatise on the Analytical Dynamics of Particles and Rigid Bodies*, Fourth Edition, Dover Publications, New York, 1944.

ANSWERS TO PROBLEMS

Exercise 0.1

1. (a) $OA = 3\sqrt{6}$, $AB = \sqrt{118}$, $BC = 2\sqrt{22}$, $CA = \sqrt{62}$.

   (b) $\overrightarrow{OA} = [2, -1, -7]$, $\overrightarrow{BO} = [-3, -5, -2]$, $\overrightarrow{AB} = [1, 6, 9]$,

   $\overrightarrow{BA} = [-1, -6, -9]$, $\overrightarrow{BC} = [6, -4, -6]$, $\overrightarrow{CA} = [-7, -2, -3]$.

   (c) $P(1, 6, 9)$, $Q(6, -4, -6)$, $R(-7, -2, -3)$.

   (d) $M(8, -5, -13)$.

2. $(8, 0, 3)$, $(-4, 6, 7)$, $(2, -2, 1)$.

3. $x = x' + 2$, $y = y' - 3$, $z = z' + 4$. (a) $A(3, 6, -5)$. (b) $B(0, 3, -1)$. (c) $C(0, 0, 0)$.

   (d) $D(-2, 3, -4)$.

4. Use the distance formula.

Exercise 0.2

1. (a) $AB = \sqrt{102}$, $BC = \sqrt{83}$, $CA = 3\sqrt{5}$.

   (b) $\overrightarrow{AB} = [-7, 7, 2]$, $\overrightarrow{BC} = [5, -3, -7]$, $\overrightarrow{CA} = [2, -4, -5]$.

   (c) $[-2, 4, 5]$ and $[0, 0, 0]$.

   (d) $M(\frac{1}{2}, \frac{5}{2}, -\frac{1}{2})$, $N(\frac{3}{2}, \frac{1}{2}, 2)$, $Q(4, -1, -\frac{3}{2})$.

   (e) $(2, \frac{2}{3}, 0)$.

4. $C(0, 11, -14)$, $D(6, -7, 16)$.

6. $(-8, 9, 0)$, $(\frac{13}{4}, 0, \frac{9}{2})$, $(0, \frac{13}{5}, \frac{16}{5})$.

8. (a) $(-10, -7, 14)$.   (b) $(-8, -14, 14)$.   (c) $(-\frac{4}{3}, \frac{5}{3}, \frac{19}{6})$.

Exercise 0.3

1. (a) $[\frac{2}{7}, \frac{6}{7}, -\frac{3}{7}]$.   (b) $[\frac{2}{3}, \frac{2}{3}, \frac{1}{3}]$.   (c) $\left[\dfrac{4}{\sqrt{26}}, \dfrac{1}{\sqrt{26}}, -\dfrac{9}{\sqrt{26}}\right]$.

2. The vectors in Problem 1 multiplied by -1.

5. $\pm\dfrac{1}{\sqrt{195}} [11, -5, -7]$.   6. $\dfrac{16}{\sqrt{17}\,\sqrt{37}}$.

7. $\dfrac{1}{\sqrt{374}}$ and $\dfrac{21}{\sqrt{814}}$.

8. $k[2, -15, 12]$, k any scalar.

9. (a) $[-1, -1, -1]$.

(b) $[2, -1, 2]$.

(c) $[7, 0, 0]$, $[0, 2, 0]$, $[0, 0, 4]$.

Exercise 0.4

1. (a) $3x - y + 5z + 8 = 0$.

(b) $3x - y - 4z + 10 = 0$.

(c) $y = 3$.

(d) $16x - 18y - 5z = 32$.

(e) $x + y + z = 1$.

(f) $5y - z = 0$.

(g) $x - 2z = 0$.

(h) $2x + y = 7$.

(i) $x - 20y - 7z + 3 = 0$.

(j) $2x + 11y - 5z + 21 = 0$.

2. (Answers to this question are not unique.)

(a) $x = 1 + 2t, y = 5 - 7t, z = 4t$.

(b) $x = 1 - 3t, y = 4t, z = 9 - 5t$.

(c) $x = 1 + 4t, y = 7 - 3t, z = 6 + 3t$.

(d) $x = 4 + 2t, y = 1 - t, z = -2 + 3t$.

(e) $x = 4t, y = 5 - 2t, z = 6 + 5t$.

(f) $x = 7 - t, y = -1 + t, z = 4 + t$.

(g) $x = 7 + 7t, y = -1 - 3t, z = 4 - 11t$.

3. (a) $(\frac{1}{2}, \frac{11}{2}, \frac{23}{2})$.

(b) $(0, 7, 13)$, $(\frac{7}{3}, 0, 6)$, $(\frac{13}{3}, -6, 0)$.

4. $(\frac{27}{28}, -\frac{39}{28}, \frac{1}{4})$.

7. (a) $[-21, 2, 11]$.

(b) $[-11, 9, 13]$.

9. $\frac{48}{7}$.

10. $\dfrac{7}{\sqrt{6}}$.

11. $\left| \dfrac{ax_1 + by_1 + cz_1 - d}{\sqrt{a^2 + b^2 + c^2}} \right|$.

Exercise 0.5

1. (a) $4x^2 - y^2 - z^2 = 0$.

(b) $(x - 2)^2 + (z + 1)^2 = 9$.

(c) $4x^2 + y^2 + 4z^2 = 16$.

(d) $4x^2 + y^2 + z^2 = 16$.

(e) $z^2 = 8y$.

(f) $2x - y = 0$.

2. (a) $x^2 + 2xy + 2y^2 = 4x + 4y$.

(b) $x^4 + 2x^2y^2 + y^4 = 16x^2 + 4y^2$.

(c) $x^2 + xy + y^2 - 2x - 2y = 0$.

(d) $3x - 7y + 4 = 0$.

3. $147(x^2 + y^2 + z^2) = 4(6x - 2y + 3z)^2$.

Exercise 0.6

1. (a) Hyperboloid of 2 sheets.

(b) Quadric cone.

(c) Hyperboloid of one sheet.

(d) Parabolic cylinder.

(e) Elliptic cylinder.

(f) Two intersecting planes.

(g) Hyperbolic paraboloid.

(h) Ellipsoid.

(i) Elliptic paraboloid.

(j) Ellipsoid of revolution.

(k) Sphere.

2. (a) New origin $(1, -\frac{1}{2}, -3)$; $4(x^2 + y^2 + z^2) = 121$; sphere, radius $\frac{11}{2}$.

   (b) New origin $(\frac{5}{2}, -1, -\frac{21}{32})$; $x^2 + y^2 = 8z$; elliptic paraboloid of revolution.

   (c) New origin $(-\frac{3}{2}, -1, -\frac{1}{8})$; $16(2x^2 - y^2 + 4z^2) = 249$; hyperboloid of one sheet.

   (d) New origin $(-4, 3, -\frac{5}{2})$; $16x^2 + 4y^2 + 36z^2 = 517$; ellipsoid.

   (e) New origin $(0, \frac{5}{4}, \frac{3}{2})$; $x^2 - 4z^2 + 16y = 0$; hyperbolic paraboloid.

3. (a) No real graph.

   (b) Single point $(1, -2, 5)$.

   (c) The straight line $x = 0$, $y = 1$, parallel to the z-axis.

   (d) The point $(0, 0, 1)$ and the line $x = 1$, $z = 2$.

   (e) The intersection of the plane $x + y + z = 0$ and the elliptic cylinder $x^2 + 4y^2 = 4$.

   (f) The infinite set of points $(0, 0, n\pi)$, n any integer.

   (g) The infinite set of straight lines $x = 0$, $z = n\pi$, n any integer.

   (h) A parabolic cylinder, generating lines parallel to the z-axis.

   (i) A plane containing the z-axis.

   (j) The two planes $x - y = 0$ and $x + y = 0$.

Exercise 1.1

1. (a) Domain $\mathscr{R}$, range $\mathscr{L}$, $\operatorname{im} f_1 = \mathscr{L}$. (b) $\operatorname{Im} f_2 = \mathscr{L}$.

   (c) $\operatorname{Im} f_3 = \{x \in \mathscr{L} \mid x \geq 2\}$. (d) $\operatorname{Im} f_4 = \{x \in \mathscr{L} \mid x \geq 0\}$.

   (e) $\operatorname{Im} f_5 = \mathscr{R}_0$. (f) $\operatorname{Im} f_6 = \mathscr{R}_0$.

   (g) $\operatorname{Im} f_7 = \mathscr{R}_0$. (h) $\operatorname{Im} f_8 = \mathscr{E}_2$.

   (i) $\operatorname{Im} f_9 = \mathscr{E}_2$. (j) $\operatorname{Im} f_{10} = \mathscr{R}$.

2. (a) f_2, f_3, f_5, f_6, f_8. (d) $f_2^{-1}: \mathscr{L} \to \mathscr{L}$ with

   (b) $f_1, f_2, f_5, f_6, f_7, f_8, f_9, f_{10}$. $f_2^{-1}(x) = x - 1$;

   (c) f_2, f_5, f_6, f_8. $f_5^{-1} = f_6; f_6^{-1} = f_5; f_8^{-1} = f_8$.

3. (a) Domain $\mathscr{A}$, range $\mathscr{R}$, $\operatorname{im} f_1 = \mathscr{R}$. (b) $\operatorname{Im} f_2 = \mathscr{D}$.

   (c) $\operatorname{Im} f_3 = \mathscr{D}$. (d) $\operatorname{Im} f_4 = \mathscr{F}$.

   (e) $\operatorname{Im} f_5 = \mathscr{F}$. Only f_1, f_3 are invertible.

4. $f_2 f_1: \mathscr{R} \to \mathscr{L}$ and $f_2 f_1(x) = [x] + 1$.

   $f_2 f_4: \mathscr{L} \to \mathscr{L}$ and $f_2 f_4(x) = x^2 + 1$.

   $f_4 f_2: \mathscr{L} \to \mathscr{L}$ and $f_4 f_2(x) = (x + 1)^2$.

   $f_6 f_7: \mathscr{R} \to \mathscr{R}_0$ and $f_6 f_7(x) = |x|$.

   $f_8 f_9: \mathscr{E}_3 \to \mathscr{E}_2$ and $f_8 f_9[(x, y, z)] = (x + y, x + 3y)$.

   $f_5 f_6 = f_6 f_5 = I_{\mathscr{R}_0}$.

Exercise 1.2

1. (d), (f), and (g) are fields.   3. No.

Exercise 1.3

6. No.   7. Yes.   8. No.   9. No.

10. If and only if $c = 0$.

11. The space consisting of all vectors lying in a plane through the origin perpendicular to the vector $[1, -2, 3]$.

12. All vectors lying in the line of intersection of the two planes whose equations are given.

13. $k[-9, 1, 4]$ for any $k \neq 0$.

17. (a) $k[-1, -2, 1, 1]$, $k \in \mathscr{F}$.

   (b) $k_1[-\frac{7}{3}, -2, -\frac{11}{3}, 3, -2] + k_2[-1, -2, -4, 2, -3]$.

   (c) $k\left[-\dfrac{1+i}{2}, 2+2i, 1+6i \right]$.

   (d) $k\left[-2, 1, \dfrac{-(2+4i)}{3}, 1 \right]$.

Exercise 1.4

1. (a) $X_1 - 2X_2 + X_3 = O$ where X_1, X_2, X_3 are the three given vectors.

   (b) $-2X_1 + X_3 = O$.　　　　　　　　(c) $-5X_1 + 2X_2 - X_3 = O$.

10. Write all three functions in terms of $\sin x$ and $\cos x$ and then show that a nontrivial linear relation can be found.

Exercise 1.5

1. (a) $3X - 2Y - Z = O$.　　　(c) No.　　　(d) Yes, e.g. $X - 2Y$.

   (e) dim $\mathscr{T} = 2$, dim $\mathscr{S} = 2$, dim $\mathscr{S} \cap \mathscr{T} = 1$, dim $\mathscr{S} + \mathscr{T} = 3$.

2. $s + t \leq n$.　　　　　　　　**3.** $k[5, -2, 3]$, k any scalar $\neq 0$.

10. $[\frac{1}{2}, 1, 0, 0]$, $[-\frac{3}{2}, 0, 1, 0]$, $[\frac{5}{2}, 0, 0, 1]$. Dim $\mathscr{S} = 3$.

Exercise 1.6

1. $[\frac{1}{8}, \frac{5}{8}]$.

2. Choose basis $F_1 = \left[\dfrac{1}{\sqrt{5}}, \dfrac{2}{\sqrt{5}} \right]$, $F_2 = \left[\dfrac{1}{\sqrt{5}}, -\dfrac{2}{\sqrt{5}} \right]$; then coordinates are $\left(\dfrac{5\sqrt{5}}{2}, \dfrac{\sqrt{5}}{2} \right)$.

3. $[-\frac{7}{8}, -\frac{15}{8}, \frac{17}{4}]$.

4. $x = a_1 x' + a_2 y' + a_3 z'$, $y = b_1 x' + b_2 y' + b_3 z'$, $z = c_1 x' + c_2 y' + c_3 z'$.

5. $x'^2 + y'^2 + 2x'y' \cos(\alpha - \beta) = r^2$.　　　　**6.** $4x'y' = a^2 + b^2$.

Exercise 2.1

1. AB, BC, CB^T, AA.

2.
$$AB = \begin{bmatrix} 34 & 3 \\ 5 & 7 \\ 20 & 16 \end{bmatrix}, \quad BC = \begin{bmatrix} 10 & 1 \\ 4 & 13 \\ 28 & 0 \end{bmatrix}, \quad CB^T = \begin{bmatrix} 7 & -8 & 22 \\ 7 & 10 & 18 \end{bmatrix},$$

$$AA = \begin{bmatrix} 5 & 6 & 20 \\ 6 & -5 & 15 \\ 14 & 4 & 19 \end{bmatrix}.$$

3.

$$XX^T = [30], \quad X^T X = \begin{bmatrix} 9 & 3 & 6 & 12 \\ 3 & 1 & 2 & 4 \\ 6 & 2 & 4 & 8 \\ 12 & 4 & 8 & 16 \end{bmatrix}.$$

4.

$$AB = \begin{bmatrix} 2 & 9 \\ 4 & -2 \end{bmatrix}, \quad BA = \begin{bmatrix} 4 & 3 \\ 8 & -4 \end{bmatrix}.$$

5.

$$AB = \begin{bmatrix} 0 & 0 \\ 0 & 0 \end{bmatrix}, \quad BA = \begin{bmatrix} 10 & -20 \\ 5 & -10 \end{bmatrix}.$$

9.

$$\begin{bmatrix} -5 & -14 \\ 42 & 23 \end{bmatrix}.$$

11. (a) No. (b) Yes. (c) Yes.

13. (a)

$$AB = \begin{bmatrix} 20 & -2 & 1 & 1 \\ 7 & 4 & 10 & 19 \\ 17 & 17 & 11 & 23 \\ 18 & 2 & 3 & 10 \end{bmatrix}.$$

14. A basis is $\begin{bmatrix} 1 & 0 \\ 0 & 0 \end{bmatrix}, \begin{bmatrix} 0 & 1 \\ 0 & 0 \end{bmatrix}, \begin{bmatrix} 0 & 0 \\ 1 & 0 \end{bmatrix}, \begin{bmatrix} 0 & 0 \\ 0 & 1 \end{bmatrix}.$

15. Choose basis E_{ij}, $i = 1, 2, \ldots, m$, $j = 1, 2, \ldots, n$, where E_{ij} is the matrix with 1 in the ith row, jth column and all other elements 0.

16. Dim $\mathcal{M} = mn - 1$.

Exercise 2.2

1. $[z + \frac{20}{3}, -\frac{2}{3}, z]$, z arbitrary.   **2.** $x = y = 0$.   **3.** No solution.

4. $k_1[0, 1, 1, 1] + k_2[1, -1, 0, 0]$, k_1, k_2 arbitrary.   **5.** $x = 3, y = 0, z = -1$.

6. $x = \frac{5}{2} - \frac{3}{2}z, y = -\frac{3}{2} + \frac{1}{2}z$, z arbitrary.

7. $[-19, 8, -3]$.   **8.** $x = y = z = 0$.   **9.** No solution.

10. $[9, -1, 0] + k[-2, 1, 1]$.   **11.** $x = 14, y = -4$.

12. No solution.   **13.** $[2, 0] + k[3, 1]$.

14. $x = \frac{7}{5}$.   **15.** No solution.   **16.** x arbitrary.

17. $x = y = z = 0$.   **18.** $[\frac{6}{7}, -\frac{3}{7}, 1, \frac{5}{7}]$.

19. $k_1[2, 1, 0, -4] + k_2[-10, 13, 4, 0]$.

20. No solution.   **22.** $[9, -8, 3]$ is in the space.

Exercise 2.3

1. (b) and (d).

3. (a) $\begin{bmatrix} 1 & 0 & \frac{26}{7} & 0 \\ 0 & 1 & -\frac{3}{7} & 0 \\ 0 & 0 & 0 & 1 \end{bmatrix}.$ (b) $\begin{bmatrix} 1 & 0 & \frac{23}{7} \\ 0 & 1 & \frac{3}{7} \\ 0 & 0 & 0 \end{bmatrix}.$

(c) $\begin{bmatrix} 1 & 0 & \frac{13}{7} & \frac{1}{7} \\ 0 & 1 & -\frac{2}{7} & -\frac{5}{7} \\ 0 & 0 & 0 & 0 \\ 0 & 0 & 0 & 0 \end{bmatrix}.$

Exercise 2.4

1. (Answers not unique but obtained from the appropriate reduced row-echelon matrix.)

(a) $[1, 0, 1]$, $[0, 1, 1]$.   (b) $[1, 0, -1]$, $[0, 1, 2]$.

(c) $[1, 0, -2, 1]$, $[0, 1, 1, 1]$.   (d) $[1, 0, \frac{3}{2}, 0]$, $[0, 1, -\frac{1}{2}, 0]$, $[0, 0, 0, 1]$.

2. (a) 3.   (b) 2.   (c) 2.   (d) 3.   (e) 2.

4. No.   **5.** $3c = 7a + b$ and $3d = 2a - 7b$.

6. Basis for $\mathscr{S}$: $[11, 0, -3, 18]$, $[0, 11, 5, -8]$.
Basis for $\mathscr{T}$: $[2, 0, 5, -1]$, $[0, 1, 6, -5]$.
Basis for $\mathscr{S} + \mathscr{T}$: $[61, 0, 0, 87]$, $[0, 61, 0, -23]$, $[0, 0, 61, -47]$.
Basis for $\mathscr{S} \cap \mathscr{T}$: $[-2, 1, 1, -4]$.

7. Dim $\mathscr{T} = $ dim $\mathscr{S} = 2$, dim $\mathscr{S} + \mathscr{T} = 3$. Dim $\mathscr{S} \cap \mathscr{T} = 1$.
Basis for $\mathscr{S} + \mathscr{T}$: $[5, 0, 0, 6]$, $[0, 2, 0, -1]$, $[0, 0, 5, -8]$.
Basis for $\mathscr{S} \cap \mathscr{T}$: $[1, 8, 2, -6]$.

8. (a) $d = 3a + 4b + 2c$ and $e = 2a - 3b + c$.
(b) $c = a - b$, $d = a + b$, $e = 20a - 15b$.
(c) $4c = 5a + 3b$, $d = a + b$, $2e = 5a + b$.

Exercise 2.5

1. (a) $[\frac{1}{3}, 1, 0, 0]$, $[-\frac{1}{3}, 0, 1, 0]$, $[-\frac{4}{3}, 0, 0, 1]$.

(b) $[-1, 1, 1, 0, 0]$, $[-7, 3, 0, 1, 0]$, $[12, -5, 0, 0, 1]$.

(c) $[2, 1, 0]$, $[-1, 0, 1]$.   (d) The zero space. No basis.

(e) $[0, 0]$.   (f) $[-1, 1, 2]$.

(g) The zero space. No basis.

(h) $[5, -24, 25, 11, 0]$, $[-13, 14, -10, 0, 11]$.

Exercise 2.6

1. (a) Row space: $[1, 0, 2]$, $[0, 1, 1]$.
Column space: $[1, 0]$, $[0, 1]$. Null space $[2, 1, -1]$.

(b) Row space: $[1, 0, 2]$, $[0, 1, 1]$.
Column space: $[7, 0, 12]$, $[0, 7, 1]$.
Null space: $[2, 1, -1]$.

(c) Row space: $[5, 0, 8, -19]$, $[0, 5, 6, 2]$.
Column space: $[1, 0, 3]$, $[0, 1, -2]$.
Null space: $[-8, -6, 5, 0]$, $[19, -2, 0, 5]$.

(d) Row space (and column space): $[1, 0, 0]$, $[0, 1, 0]$.
Null space: $[0, 0, 1]$.

2. No. (See Problem 3.)

Exercise 2.7

(a) $[\frac{20}{3}, -\frac{2}{3}, 0] + k[1, 0, 1]$.   (b) $[1, 1]$.   (c) No solution.

(d) $[10, 0, 0, -6] + k_1[-1, 1, 0, 0] + k_2[1, 0, 1, 1]$.

(e) $[3, 0, -1]$.   (f) $[-2, 0, 3] + k[-3, 1, 2]$.

3. (a) Every vector $\overrightarrow{OP}$, P on a fixed line l, is equal to the sum of a fixed vector $\overrightarrow{OP_1}$, P_1 on l, and a vector $\overrightarrow{OQ}$ where Q is on the line through O parallel to l.

5. (a) $\begin{bmatrix} -1 & 3 \\ 2 & -5 \end{bmatrix}$.

(b) $\begin{bmatrix} \frac{2}{3} & -\frac{1}{3} \\ -\frac{5}{3} & \frac{4}{3} \end{bmatrix}$.

(c) $\begin{bmatrix} -1 & -6 & 4 \\ -1 & -13 & 8 \\ 1 & 8 & -5 \end{bmatrix}$.

6. $\begin{bmatrix} \frac{9}{2} & -\frac{3}{2} & -\frac{1}{2} \\ -2 & 1 & 0 \\ \frac{1}{2} & -\frac{1}{2} & \frac{1}{2} \end{bmatrix}$.

(a) $[9, -6, 13]$.

(b) $[\frac{15}{2}, -5, \frac{9}{2})$.

7. $[a_{ij}]$ nonsingular.

12. Intersection is a point in case (a), a line in case (c), and a plane in case (e). No intersection in cases (b) and (d). Case (f) cannot occur.

Exercise 2.8

2. $R = \begin{bmatrix} 1 & 0 & 2 \\ 0 & 1 & -\frac{1}{3} \\ 0 & 0 & 0 \end{bmatrix}$. $P = \begin{bmatrix} 1 & 0 & 0 \\ 0 & \frac{1}{3} & 0 \\ -2 & -1 & 1 \end{bmatrix}$.

3. (a) $\begin{bmatrix} 1 & 0 \\ 2 & 1 \end{bmatrix}\begin{bmatrix} 1 & 0 \\ 0 & 2 \end{bmatrix}\begin{bmatrix} 1 & 3 \\ 0 & 1 \end{bmatrix}$.

(b) $\begin{bmatrix} 2 & 0 \\ 0 & 1 \end{bmatrix}\begin{bmatrix} 1 & 0 \\ 3 & 1 \end{bmatrix}\begin{bmatrix} 1 & 0 \\ 0 & \frac{7}{2} \end{bmatrix}\begin{bmatrix} 1 & -\frac{1}{2} \\ 0 & 1 \end{bmatrix}$.

These answers are not unique.

Exercise 2.9

(a) $\begin{bmatrix} 8 & 35 \\ 1 & 19 \\ \hline 13 & 14 \end{bmatrix}$.

(b) $\begin{bmatrix} 6 & 9 & 12 & 17 \\ 7 & 12 & 11 & 17 \\ \hline 8 & 14 & 12 & 40 \\ 9 & 13 & 19 & 4 \end{bmatrix}$.

(c) $\begin{bmatrix} 4 & 0 & 0 & 0 & 0 \\ \hline 11 & 23 & 16 & 0 & 0 \\ 9 & -8 & 7 & 0 & 0 \\ \hline 8 & 28 & 0 & 8 & -9 \\ 2 & 57 & 27 & 9 & 35 \end{bmatrix}$.

Exercise 3.1

1.
$$\sigma\tau = \begin{pmatrix} 1 & 2 & 3 & 4 & 5 & 6 \\ 1 & 4 & 6 & 5 & 2 & 3 \end{pmatrix}, \quad \tau\sigma = \begin{pmatrix} 1 & 2 & 3 & 4 & 5 & 6 \\ 5 & 4 & 3 & 2 & 6 & 1 \end{pmatrix},$$

$$\sigma^2 = \begin{pmatrix} 1 & 2 & 3 & 4 & 5 & 6 \\ 3 & 2 & 4 & 1 & 5 & 6 \end{pmatrix}, \quad \tau^2 = \begin{pmatrix} 1 & 2 & 3 & 4 & 5 & 6 \\ 2 & 3 & 1 & 6 & 4 & 5 \end{pmatrix},$$

$$\tau^3 = I, \quad \sigma^{-1} = \begin{pmatrix} 1 & 2 & 3 & 4 & 5 & 6 \\ 3 & 6 & 4 & 1 & 5 & 2 \end{pmatrix},$$

$$(\sigma\tau)^{-1} = \begin{pmatrix} 1 & 2 & 3 & 4 & 5 & 6 \\ 1 & 5 & 6 & 2 & 4 & 3 \end{pmatrix}.$$

Exercise 3.2

1. (a) 60.   (b) 9.   (c) 264.

(d) 270.   (e) -2.   (f) -312.

6. $\det A = (-1)^{\frac{n(n-1)}{2}}$.

Exercise 3.3

1. (i) -45.   (ii) 240.

8. (a) $k[1, 1, 1]$.   (b) $k[22, -16, -1, -20]$.

Exercise 4.1

Answers to Problems 1–5 are not unique and the student should check his own answers.

6. $k[-4, 1, 5, 2]$, k any nonzero scalar.

7. Use Theorem 1.1.

Exercise 4.2

1. (a) $[-7, 4, 0]$, $[1, 0, 2]$. (Answer not unique.)

2. (a) $k[-9, 7, 5]$.   (b) $k[-43, 3, 16]$.
(k any nonzero scalar.)

3. $[\frac{8}{3}, \frac{8}{3}, \frac{8}{3}]$ and $[\frac{1}{3}, \frac{4}{3}, -\frac{5}{3}]$.

4. $[\frac{61}{14}, \frac{32}{14}, \frac{1}{14}]$ and $[\frac{9}{14}, -\frac{18}{14}, \frac{27}{14}]$.

5. $[1, 2, 0, -2] = \frac{1}{42}[71, 60, 11, -74] + \frac{1}{42}[-29, 24, -11, -10]$.

6. $[\frac{52}{19}, -\frac{2}{19}, -\frac{22}{19}]$.   **7.** $[\frac{9}{11}, -\frac{9}{11}, \frac{27}{11}]$.

8. Solution space. $[-\frac{1}{7}, \frac{4}{7}, 1, 0]$, $[-1, 1, 0, 1]$.
Orthog. compl. $[-7, -7, 3, 0]$, $[4, 1, 0, 3]$.
(Bases not unique.)

9. $[-\frac{247}{173}, \frac{433}{173}, \frac{434}{173}, \frac{185}{173}]$ and $[\frac{420}{173}, \frac{259}{173}, -\frac{88}{173}, \frac{161}{173}]$.

Exercise 4.3

1. (a) 7.   (b) 7.   (c) 10.   (d) $4\sqrt{65}$.   (e) $3\sqrt{6}$.

2. (a) 26.   (b) 2.

5. (a) $\sqrt{4359}$.   (b) 12.   **6.** $\sqrt{1049}$.   **7.** 27.

Exercise 4.4

3. $X_2 = [-\frac{11}{5}, \frac{4}{5}]$.

4. (a) and (e) are inner products, (b), (c), (d) are not.

5. $\dfrac{1}{\sqrt{5}} [2, 0, 1], \dfrac{1}{\sqrt{270}} [-7, -5, 14], \dfrac{1}{3\sqrt{6}} [-1, 7, 2]$.

6. (b) $1, \sqrt{3}(2x - 1), \sqrt{5}(6x^2 - 6x + 1)$.

(c) $\left[\dfrac{a}{3} + \dfrac{b}{2} + c, \dfrac{a + b}{\sqrt{12}}, \dfrac{a}{\sqrt{180}}\right]$.

(d) $\sigma(ax^2 + bx + c) = \left[\frac{1}{6}(2a + 3b + 6c), \dfrac{1}{\sqrt{12}}(a + b), \dfrac{a}{\sqrt{180}}\right]$.

9. (a) $\begin{bmatrix} 1 & 0 & 0 \\ 0 & 1 & -5 \\ 0 & 2 & 0 \end{bmatrix}$.   (b) $\begin{bmatrix} 2 & 3 & 5 \\ 0 & -2 & 0 \\ 7 & 0 & 4 \end{bmatrix}$,

10. (a) $\begin{bmatrix} 1 & 0 & 0 \\ 0 & 5 & 0 \\ 0 & 0 & -7 \end{bmatrix}$.   (b) $\begin{bmatrix} 0 & 1 & 3 \\ 1 & 0 & -2 \\ 3 & -2 & 0 \end{bmatrix}$.   (c) $\begin{bmatrix} 1 & -\frac{1}{2} & -\frac{3}{2} \\ -\frac{1}{2} & 2 & 2 \\ -\frac{3}{2} & 2 & -5 \end{bmatrix}$.

11. $XAX^T = 2x_1^2 + 3x_2^2 + 4x_3^2 + 2x_1x_2 + 10x_1x_3 - 4x_2x_3$.

$XAY = 2x_1y_1 + 3x_2y_2 + 4x_3y_3 + x_1y_2 + x_2y_1 + 5x_1y_3 + 5x_3y_1 - 2x_2y_3 - 2x_3y_2$.

13. Basis of $\mathscr{S}_1^\perp$ is $x - \frac{1}{2}, x^2 - x + \frac{1}{6}$; of $\mathscr{S}_2^\perp, -\frac{1}{3}x^2 + \frac{1}{4}x, \frac{3}{2}x - 1$; of $\mathscr{S}_3^\perp, x^2 - x + \frac{1}{6}$.

14. (a) $1 - 2x + 3x^2 = 1 + [3(x^2 - x + \frac{1}{6}) + (x - \frac{1}{2})]$.

(b) $1 - 2x + 3x^2 = \frac{7}{4}x - [9(-\frac{1}{3}x^2 + \frac{1}{4}x) + (\frac{3}{2}x - 1)]$.

Exercise 4.5

1. (a) $\begin{bmatrix} 2 & 3 \\ 1 & 2 \end{bmatrix}$.   (b) $\begin{bmatrix} \frac{4}{7} & -\frac{9}{7} \\ \frac{13}{7} & \frac{4}{7} \end{bmatrix}$.   (c) $\begin{bmatrix} 0 & 0 & 1 \\ 1 & 0 & 0 \\ 0 & 1 & 0 \end{bmatrix}$.

(d) $\begin{bmatrix} 2 & 1 & -3 \\ -1 & 1 & 0 \\ 0 & 1 & 4 \end{bmatrix}$.   (e) $\begin{bmatrix} \frac{4}{15} & -\frac{7}{15} & \frac{1}{5} \\ \frac{4}{15} & \frac{8}{15} & \frac{1}{5} \\ -\frac{1}{15} & -\frac{2}{15} & \frac{1}{5} \end{bmatrix}$.   (f) $\begin{bmatrix} -\frac{1}{5} & \frac{7}{15} & -\frac{4}{15} \\ \frac{4}{5} & \frac{7}{15} & \frac{11}{15} \\ -\frac{1}{5} & \frac{2}{15} & \frac{1}{15} \end{bmatrix}$.

2.
$$F_1 = \left[-\frac{2}{\sqrt{5}}, \frac{1}{\sqrt{5}}\right], \quad F_2 = \left[\frac{2}{\sqrt{5}}, \frac{1}{\sqrt{5}}\right].$$

(a)
$$\begin{bmatrix} -\dfrac{2}{\sqrt{5}} & \dfrac{2}{\sqrt{5}} \\[2mm] \dfrac{1}{\sqrt{5}} & \dfrac{1}{\sqrt{5}} \end{bmatrix}.$$

(b) $\left(-\dfrac{\sqrt{5}}{4}, \dfrac{\sqrt{5}}{2}\right), \left(\dfrac{\sqrt{5}}{2}, \dfrac{\sqrt{5}}{2}\right), (2\sqrt{5}, 3\sqrt{5}).$

$[F_1, F_2$ may be interchanged with corresponding changes in (a) and (b).].

3. $4x'y' + a^2 + b^2 = 0$ or $4x'y' - a^2 - b^2 = 0$ depending on which asymptote is chosen for the x'-axis.

4. $[2, 1, 3, 4] = F_1 + \frac{1}{2}F_3 + \frac{3}{2}F_4.$

5. $x'y' = 12.$

6. Basis not unique. The three basis vectors are the columns of the transition matrix.

8. (a) $\begin{bmatrix} \frac{2}{3} & -\frac{2}{3} & \frac{1}{3} \\ \frac{2}{3} & \frac{1}{3} & -\frac{2}{3} \\ \frac{1}{3} & \frac{2}{3} & \frac{2}{3} \end{bmatrix}.$

9. (a) only.

10. (a) $\begin{bmatrix} \frac{2}{7} & \frac{3}{7} & \frac{6}{7} \\ \frac{6}{7} & \frac{2}{7} & -\frac{3}{7} \\ \frac{3}{7} & -\frac{6}{7} & \frac{2}{7} \end{bmatrix}.$

(b) $\begin{bmatrix} \frac{3}{5} & \frac{4}{5} & 0 \\ 0 & 0 & 1 \\ -\frac{4}{5} & \frac{3}{5} & 0 \end{bmatrix}.$

(c) $\begin{bmatrix} -\frac{18}{35} & \frac{26}{35} & \frac{3}{7} \\ \frac{6}{7} & \frac{3}{7} & \frac{2}{7} \\ \frac{1}{35} & \frac{18}{35} & -\frac{6}{7} \end{bmatrix}.$

(d) $\begin{bmatrix} \frac{16}{21} & \frac{13}{21} & -\frac{4}{21} \\ \frac{11}{21} & -\frac{16}{21} & -\frac{8}{21} \\ \frac{8}{21} & -\frac{4}{21} & \frac{19}{21} \end{bmatrix}.$

11. (c) only.

13. $-x_1^2 + 5x_2^2 + 2x_3^2 = 10.$

Exercise 5.1

1. (a) $A_r = \begin{bmatrix} 1 & 0 & 0 \\ 0 & 1 & 0 \end{bmatrix}, \quad y_1 = x_1, \quad y_2 = x_2.$

(b) $A_r = \begin{bmatrix} 0 & 1 & 1 \\ 1 & 0 & 1 \end{bmatrix}, \quad y_1 = x_2 + x_3, \quad y_2 = x_1 + x_3.$

(c) $A_r = \begin{bmatrix} 1 & -5 & 7 \\ 2 & 1 & 3 \end{bmatrix}, \quad y_1 = x_1 - 5x_2 + 7x_3, \quad y_2 = 2x_1 + x_2 + 3x_3.$

(d) $A_\tau = \begin{bmatrix} 0 & 0 & 0 \\ 0 & 0 & 0 \end{bmatrix}$, $y_1 = 0$, $y_2 = 0$.

(e) $A_\tau = \begin{bmatrix} 4 & 4 & 4 \\ -3 & -3 & -3 \end{bmatrix}$, $y_1 = 4(x_1 + x_2 + x_3)$, $y_2 = -3(x_1 + x_2 + x_3)$.

2. (a) $A_\sigma = \begin{bmatrix} 1 & 0 \\ 0 & 0 \\ 0 & 1 \end{bmatrix}$.

(b) $A_\sigma = \begin{bmatrix} 1 & 2 \\ 2 & -1 \\ 3 & 4 \end{bmatrix}$.

(c) $A_\sigma = \begin{bmatrix} 0 & 1 \\ 0 & 0 \\ 0 & 0 \end{bmatrix}$.

(d) $A_\sigma = \begin{bmatrix} 1 & 1 \\ 2 & 2 \\ -6 & -6 \end{bmatrix}$.

(e) $A_\sigma = \begin{bmatrix} 0 & 0 \\ 0 & 0 \\ 0 & 0 \end{bmatrix}$.

3. $A_{\sigma\tau} = \begin{bmatrix} 2 & 1 & 3 \\ -1 & 2 & 1 \\ 4 & 3 & 7 \end{bmatrix} = \begin{bmatrix} 1 & 2 \\ 2 & -1 \\ 3 & 4 \end{bmatrix} \begin{bmatrix} 0 & 1 & 1 \\ 1 & 0 & 1 \end{bmatrix}$.

4. (a) $A_{\sigma\tau} = \begin{bmatrix} 5 & -3 & 13 \\ 0 & -11 & 11 \\ 11 & -11 & 33 \end{bmatrix}$.

(b) $A_{\sigma\tau} = \begin{bmatrix} -3 & -3 & -3 \\ 0 & 0 & 0 \\ 0 & 0 & 0 \end{bmatrix}$.

5. $A_{\tau\sigma} = \begin{bmatrix} 5 & 3 \\ 4 & 6 \end{bmatrix} = \begin{bmatrix} 0 & 1 & 1 \\ 1 & 0 & 1 \end{bmatrix} \begin{bmatrix} 1 & 2 \\ 2 & -1 \\ 3 & 4 \end{bmatrix}$.

6. (a) $A_{\tau\sigma} = \begin{bmatrix} 12 & 35 \\ 13 & 15 \end{bmatrix}$.

(b) $A_{\tau\sigma} = \begin{bmatrix} 0 & 4 \\ 0 & -3 \end{bmatrix}$.

9. (b) $A_\tau = \begin{bmatrix} \frac{8}{9} & \frac{2}{9} & -\frac{2}{9} \\ \frac{2}{9} & \frac{5}{9} & \frac{4}{9} \\ -\frac{2}{9} & \frac{4}{9} & \frac{5}{9} \end{bmatrix}$.

(c) Choose $F_3 = [1, -2, 2]$ and F_1, F_2 orthogonal to F_3.

(d) $A_\tau^{(F)} = \begin{bmatrix} 1 & 0 & 0 \\ 0 & 1 & 0 \\ 0 & 0 & 0 \end{bmatrix}$.

Exercise 5.2

1. (a) $A_\sigma^{(E)} = \begin{bmatrix} 1 & 1 \\ -1 & 1 \end{bmatrix}$, $A_\tau^{(E)} = \begin{bmatrix} 0 & 2 \\ 0 & 1 \end{bmatrix}$, $A_\varphi^{(E)} = \begin{bmatrix} -2 & 0 \\ 0 & 5 \end{bmatrix}$.

(b) $A_\sigma^{(F)} = \begin{bmatrix} 1 & 1 \\ -1 & -1 \end{bmatrix}$, $A_\tau^{(F)} = \begin{bmatrix} -\frac{1}{2} & \frac{1}{2} \\ -\frac{3}{2} & \frac{3}{2} \end{bmatrix}$, $A_\varphi^{(F)} = \begin{bmatrix} \frac{3}{2} & -\frac{7}{2} \\ -\frac{7}{2} & \frac{3}{2} \end{bmatrix}$.

2. (b) $\begin{bmatrix} 1 & 0 \\ 0 & 0 \end{bmatrix}$.

(c) Projection onto the subspace spanned by F_1.

(d) Rotation (counterclockwise) through $60°$.

(e) $\varphi F_1 = [1, -2]$, $\varphi F_2 = [6, -10]$.

(f) $A_\varphi^{(F)} = \begin{bmatrix} 1 & 0 \\ 0 & 2 \end{bmatrix}$.

(g) φ leaves vectors parallel to F_1 invariant, multiplies vectors parallel to F_2 by 2. A magnification by factor 2 in the direction of F_2.

3. $A_{\sigma\tau} = \begin{bmatrix} \frac{9}{5} & -\frac{12}{5} \\ \frac{4}{5} & \frac{3}{5} \end{bmatrix}$. $\qquad A_{\sigma\tau} = \begin{bmatrix} \frac{9}{5} & -\frac{4}{5} \\ \frac{12}{5} & \frac{3}{5} \end{bmatrix}$.

4. $A_\tau^{(E)} = \begin{bmatrix} 2 & -1 \\ 5 & 6 \end{bmatrix} = A_\tau^{(F)}$. $\qquad A_{\tau^{-1}}^{(E)} = \begin{bmatrix} \frac{6}{17} & \frac{1}{17} \\ -\frac{5}{17} & \frac{2}{17} \end{bmatrix} = A_{\tau^{-1}}^{(F)}$.

5. (a) $A_\tau^{(F)} = \begin{bmatrix} \frac{9}{7} & -\frac{1}{7} \\ \frac{10}{7} & -\frac{5}{7} \end{bmatrix}$. $\qquad$ (b) $A_\tau^{(G)} = \begin{bmatrix} \frac{9}{7} & -\frac{1}{7} \\ \frac{10}{7} & -\frac{5}{7} \end{bmatrix}$.

(c) $A_\tau^{(E)} = \begin{bmatrix} \frac{11}{7} & -\frac{6}{7} \\ 1 & -1 \end{bmatrix}$.

6. $A_\tau = \begin{bmatrix} -3 & 0 \\ 0 & 2 \end{bmatrix}$, a magnification followed by a reflection.

8. $A_\tau = \begin{bmatrix} 3 & -\frac{2}{3} & -\frac{2}{3} \\ -\frac{2}{3} & \frac{10}{3} & 0 \\ -\frac{2}{3} & 0 & \frac{8}{3} \end{bmatrix}$.

10. $x' = 4x$, $y' = 3y$. $\qquad$ 11. $x = 2(x' - y')$, $y = 3(x' + y')$.

15. $\pi ab = \pi \begin{vmatrix} a & 0 \\ 0 & b \end{vmatrix}$. $\qquad$ 17. $\frac{4}{3}\pi abc = \frac{4}{3}\pi \begin{vmatrix} a & 0 & 0 \\ 0 & b & 0 \\ 0 & 0 & c \end{vmatrix}$. $\qquad$ 18. $|\det \tau| = 1$.

20. Choose a basis such that the first r vectors are a basis of $\mathscr{S}$ and the last $n - r$ a basis of $\mathscr{T}$.

Exercise 5.3

1. Axis of rotation: $x = z$, $y = 0$; angle: $\cos \theta = \frac{1}{3}$, $0 < \theta < \frac{\pi}{2}$.

3. (a) $\begin{bmatrix} (8 + 5\sqrt{2})/18 & (8 - 7\sqrt{2})/18 & (2 + 2\sqrt{2})/9 \\ (8 - \sqrt{2})/18 & (8 + 5\sqrt{2})/18 & (2 - 4\sqrt{2})/9 \\ (2 - 4\sqrt{2})/9 & (2 + 2\sqrt{2})/9 & (1 + 4\sqrt{2})/9 \end{bmatrix}$.

(b) $\begin{bmatrix} \frac{7}{9} & \frac{4}{9} & -\frac{4}{9} \\ \frac{4}{9} & \frac{1}{9} & \frac{8}{9} \\ -\frac{4}{9} & \frac{8}{9} & \frac{1}{9} \end{bmatrix}$.

4. (a) $\begin{bmatrix} -1 & 0 & 0 \\ 0 & 1 & 0 \\ 0 & 0 & 1 \end{bmatrix}$.
   (b) $\begin{bmatrix} 0 & 1 & 0 \\ 1 & 0 & 0 \\ 0 & 0 & 1 \end{bmatrix}$.
   (c) $\begin{bmatrix} -\frac{1}{3} & -\frac{2}{3} & \frac{2}{3} \\ -\frac{2}{3} & \frac{2}{3} & \frac{1}{3} \\ \frac{2}{3} & \frac{1}{3} & \frac{2}{3} \end{bmatrix}$.

(d) $\begin{bmatrix} 1 & 0 & 0 \\ 0 & \dfrac{\sqrt{2}}{2} & -\dfrac{\sqrt{2}}{2} \\ 0 & \dfrac{\sqrt{2}}{2} & \dfrac{\sqrt{2}}{2} \end{bmatrix}$.
   (e) $\begin{bmatrix} 0 & 0 & 1 \\ 1 & 0 & 0 \\ 0 & 1 & 0 \end{bmatrix}$.

7.
$$A_{\tau\sigma} = \begin{bmatrix} -\frac{2}{3} & \frac{2}{3} & \frac{1}{3} \\ -\frac{1}{3} & -\frac{2}{3} & \frac{2}{3} \\ \frac{2}{3} & \frac{1}{3} & \frac{2}{3} \end{bmatrix}; \text{ axis along the vector } [1, 1, 3].$$

Exercise 6.1

1. (a) 14, 1; [1, 3], [4, −1].

(b) $\frac{3}{2} \pm \frac{1}{2}\sqrt{37}$; $[6, 1 - \sqrt{37}]$, $[6, 1 + \sqrt{37}]$.

(c) $2 \pm \sqrt{3}i$; $[2, 1 - \sqrt{3}i]$, $[2, 1 + \sqrt{3}i]$.

(d) 8, −1, −1; linearly independent eigenvectors [2, 1, 2], [0, 2, −1], [1, 0, −1]. (Not unique.)

(e) $1, 1 \pm 2\sqrt{2}i$; $[1, 0, 1]$, $[\sqrt{2} \pm i, -\sqrt{2} \pm 2i, -\sqrt{2} \mp i]$.

2. All eigenvalues are 1. Every nonzero vector is an eigenvector.

3. (a) $D = \begin{bmatrix} 1 & 0 \\ 0 & -1 \end{bmatrix}$, $P = \begin{bmatrix} 2 & 3 \\ 5 & 4 \end{bmatrix}$.

(b) $D = \begin{bmatrix} 2 & 0 \\ 0 & -7 \end{bmatrix}$, $P = \begin{bmatrix} 1 & -3 \\ -1 & 2 \end{bmatrix}$.

(c) $D = \begin{bmatrix} 3 + 4i & 0 \\ 0 & 3 - 4i \end{bmatrix}$, $P = \begin{bmatrix} 1 & -1 \\ i & i \end{bmatrix}$.

(d) $D = \begin{bmatrix} e^{i\varphi} & 0 \\ 0 & e^{-i\varphi} \end{bmatrix}$, $P = \begin{bmatrix} 1 & 1 \\ -i & i \end{bmatrix}$.

(f) $D = \begin{bmatrix} 1 & 0 & 0 \\ 0 & 2 & 0 \\ 0 & 0 & 5 \end{bmatrix}$, $P = \begin{bmatrix} 2 & 1 & 0 \\ 1 & 1 & 1 \\ 4 & 2 & 1 \end{bmatrix}$.

(g) $D = \begin{bmatrix} 1 & 0 & 0 & 0 \\ 0 & -1 & 0 & 0 \\ 0 & 0 & 2 & 0 \\ 0 & 0 & 0 & 2 \end{bmatrix}$, $P = \begin{bmatrix} 1 & 0 & 0 & -5 \\ 0 & 1 & 0 & 2 \\ 0 & 0 & 1 & 0 \\ 0 & 0 & 0 & 1 \end{bmatrix}$.

8. $Q^{-1}AQ = \begin{bmatrix} d & c \\ b & a \end{bmatrix}$.

Exercise 6.2

1. $\begin{bmatrix} 1 & 0 & 0 \\ 0 & 3 & 2 \\ 0 & 1 & 1 \end{bmatrix}.$

2. $P = \begin{bmatrix} 1 & 2 & 0 \\ 2 & 2 & 0 \\ 3 & 1 & 1 \end{bmatrix}$ (third column arbitrary provided P is nonsingular).

Exercise 6.3

1. (a) $y_1 = c_1 e^x + c_2 e^{-x} + c_3 e^{-3x}.$
   (b) $y_1 = c_1 e^x + c_2 \sin 2x + c_3 \cos 2x.$
   (c) $y_1 = c_1 e^{3x} \sin x + c_2 e^{3x} \cos x, \; y_2 = -c_1 e^{3x} \cos x + c_2 e^{3x} \sin x.$
   (d) $y_1 = \frac{1}{3} x e^x - \frac{4}{9} e^x + \frac{1}{3} e^{-x} - 1 + c_1 e^x - c_2 e^{2x} + c_3 e^{-2x}.$
   $y_2 = \frac{1}{3} x e^x - \frac{1}{9} e^x - e^{-x} + c_1 e^x - 2c_3 e^{-2x}.$
   $y_3 = \frac{1}{3} x e^x + \frac{5}{9} e^x - \frac{1}{3} e^{-x} - 1 + c_1 e^x + c_2 e^{2x} + c_3 e^{-2x}.$

2. $y_1 = c_1 e^x + 2c_2 x e^x, \quad y_2 = 3c_2 e^x + 2c_3 e^{2x}, \quad y_3 = c_2 e^x + c_3 e^{2x}.$

4. $y = c_1 e^x + c_2 e^{-x} + c_3 \cos x + c_4 \sin x.$

Exercise 6.4

1. (a) $\begin{bmatrix} \dfrac{3}{\sqrt{10}} & -\dfrac{1}{\sqrt{10}} \\ \dfrac{1}{\sqrt{10}} & \dfrac{3}{\sqrt{10}} \end{bmatrix}.$
   (b) $\begin{bmatrix} \frac{3}{5} & \frac{4}{5} \\ -\frac{4}{5} & \frac{3}{5} \end{bmatrix}.$
   (c) $\begin{bmatrix} \dfrac{\sqrt{3}}{2} & \dfrac{1}{2} \\ -\dfrac{1}{2} & \dfrac{\sqrt{3}}{2} \end{bmatrix}.$

   (d) $\begin{bmatrix} \frac{1}{3} & \frac{2}{3} & \frac{2}{3} \\ \frac{2}{3} & \frac{1}{3} & -\frac{2}{3} \\ \frac{2}{3} & -\frac{2}{3} & \frac{1}{3} \end{bmatrix}.$
   (e) $\begin{bmatrix} \dfrac{2}{\sqrt{5}} & \dfrac{1}{\sqrt{6}} & -\dfrac{1}{\sqrt{30}} \\ 0 & \dfrac{1}{\sqrt{6}} & \dfrac{5}{\sqrt{30}} \\ \dfrac{1}{\sqrt{5}} & -\dfrac{2}{\sqrt{6}} & \dfrac{2}{\sqrt{30}} \end{bmatrix}.$

2. (a) $\begin{bmatrix} \frac{3}{5} & -\frac{4}{5} \\ \frac{4}{5} & \frac{3}{5} \end{bmatrix} \begin{bmatrix} \frac{5}{2} & -\frac{3}{2} \\ -\frac{3}{2} & \frac{5}{2} \end{bmatrix}.$
   (b) $\begin{bmatrix} \dfrac{1}{\sqrt{2}} & \dfrac{1}{\sqrt{2}} \\ -\dfrac{1}{\sqrt{2}} & \dfrac{1}{\sqrt{2}} \end{bmatrix} \begin{bmatrix} 5 & 3 \\ 3 & 5 \end{bmatrix}.$

   (c) $\begin{bmatrix} \frac{2}{3} & -\frac{2}{3} & -\frac{1}{3} \\ \frac{1}{3} & \frac{2}{3} & -\frac{2}{3} \\ \frac{2}{3} & \frac{1}{3} & \frac{2}{3} \end{bmatrix} \begin{bmatrix} \frac{5}{3} & \frac{1}{3} & -\frac{1}{3} \\ \frac{1}{3} & \frac{5}{3} & \frac{1}{3} \\ -\frac{1}{3} & \frac{1}{3} & \frac{5}{3} \end{bmatrix}.$

Exercise 7.1

1. (a) $\begin{bmatrix} 1 & -1 \\ -1 & 2 \end{bmatrix}$, rank 2.

(b) $\begin{bmatrix} 0 & \frac{1}{2} \\ \frac{1}{2} & -1 \end{bmatrix}$, rank 2.

(c) $\begin{bmatrix} 9 & -6 \\ -6 & 4 \end{bmatrix}$, rank 1.

(d) $\begin{bmatrix} 4 & 2 & -2 \\ 2 & 1 & 4 \\ -2 & 4 & -8 \end{bmatrix}$, rank 3.

(e) $\begin{bmatrix} 1 & 2 & 0 \\ 2 & 0 & 2 \\ 0 & 2 & -1 \end{bmatrix}$, rank 2.

(f) $\begin{bmatrix} 0 & \frac{1}{2} & 0 & 0 \\ \frac{1}{2} & 0 & 0 & 0 \\ 0 & 0 & 0 & -\frac{1}{2} \\ 0 & 0 & -\frac{1}{2} & 0 \end{bmatrix}$, rank 4.

2. (a), (b), (c), (e).

3. (a) $y_1^2 + y_2^2$. (b) $y_1^2 - y_2^2$. (c) y_1^2. (d) $y_1^2 + y_2^2 - y_3^2$.

(e) $y_1^2 - y_2^2$;

$$P = \begin{bmatrix} \dfrac{2}{3\sqrt{3}} & \dfrac{1}{3\sqrt{3}} & \dfrac{2}{3} \\[2mm] \dfrac{2}{3\sqrt{3}} & -\dfrac{2}{3\sqrt{3}} & -\dfrac{1}{3} \\[2mm] \dfrac{1}{3\sqrt{3}} & \dfrac{2}{3\sqrt{3}} & -\dfrac{2}{3} \end{bmatrix}.$$

(f) $y_1^2 + y_2^2 - y_3^2 - y_4^2$;

$$P = \frac{1}{\sqrt{2}} \begin{bmatrix} 1 & 1 & 1 & -1 \\ 1 & 1 & -1 & 1 \\ 1 & -1 & 1 & 1 \\ -1 & 1 & 1 & 1 \end{bmatrix}.$$

4. (a) $[(\sqrt{2}+1)x_1 - (\sqrt{2}+1)x_2 - (2-\sqrt{2})x_3]$
$$\times [(\sqrt{2}-1)x_1 - (\sqrt{2}-1)x_2 - (2+\sqrt{2})x_3].$$

(b) $(x_1 + x_2 + x_3)(x_1 - x_2 - 5x_3)$.

Exercise 7.2

1. (a) Matrix of rotation:

$$P = \begin{bmatrix} \dfrac{1}{\sqrt{3}} & \dfrac{1}{\sqrt{6}} & \dfrac{1}{\sqrt{2}} \\[2mm] \dfrac{1}{\sqrt{3}} & -\dfrac{2}{\sqrt{6}} & 0 \\[2mm] \dfrac{1}{\sqrt{3}} & \dfrac{1}{\sqrt{6}} & -\dfrac{1}{\sqrt{2}} \end{bmatrix};$$

standard form: $2x^2 - y^2 - z^2 = 1$; hyperboloid of 2 sheets.

(b) $\left(\dfrac{3}{2} + \dfrac{\sqrt{15}}{2}\right)x^2 + \left(\dfrac{3}{2} - \dfrac{\sqrt{15}}{2}\right)y^2 = 2$; hyperbolic cylinder.

(c)
$$P = \begin{bmatrix} 1 & 0 & 0 \\ 0 & \dfrac{\sqrt{2}}{2} & \dfrac{\sqrt{2}}{2} \\ 0 & -\dfrac{\sqrt{2}}{2} & \dfrac{\sqrt{2}}{2} \end{bmatrix}; \quad \text{new origin } \left(-1, \dfrac{7\sqrt{2}}{2}, -\dfrac{\sqrt{2}}{2}\right);$$

$2x^2 + y^2 - z^2 = 50$; hyperboloid of one sheet.

2. (a)
$$P = \begin{bmatrix} \dfrac{2}{\sqrt{5}} & -\dfrac{1}{\sqrt{5}} \\ \dfrac{1}{\sqrt{5}} & \dfrac{2}{\sqrt{5}} \end{bmatrix}; \quad \text{parabola.}$$

(b)
$$P = \begin{bmatrix} \dfrac{2}{\sqrt{5}} & -\dfrac{1}{\sqrt{5}} \\ \dfrac{1}{\sqrt{5}} & \dfrac{2}{\sqrt{5}} \end{bmatrix}; \quad x^2 + 6y^2 = 36; \quad \text{ellipse.}$$

(c)
$$P = \begin{bmatrix} \dfrac{3}{\sqrt{10}} & -\dfrac{1}{\sqrt{10}} \\ \dfrac{1}{\sqrt{10}} & \dfrac{3}{\sqrt{10}} \end{bmatrix}; \quad x^2 - y^2 = 4; \quad \text{hyperbola.}$$

Exercise 7.3

1. (a) $\det D = -512$; eigenvalues 2, 4, 10; ellipse; $10x^2 + 20y^2 + 50z^2 = 32$.
$$\left[\det H = (2)(4)(10) = 80; \frac{\det D}{\det H} = -\frac{32}{5}\right]$$

(b) $\det D = 0$; eigenvalues $0, \frac{3}{2}(1 \pm \sqrt{3})$; pair of lines;
$(\sqrt{3} + 1)x^2 - (\sqrt{3} - 1)y^2 = 0$.

(c) $\det D = 0$; eigenvalues $3 \pm 4\sqrt{2}, -1$; cone;
$(3 + 4\sqrt{2})x^2 + (3 - 4\sqrt{2})y^2 - z^2 = 0$.

5. $\det D > 0$, $\det H = 0$, $j < 0$.

Exercise 8.1

1. (a) and (b). There are infinitely many solutions. Verify your own.

(c) $\pm \dfrac{1}{\sqrt{19}} [4i, -i, -1 + i]$.

7. $\left[\dfrac{7 + i}{10}, \dfrac{-1 + 7i}{5}, \dfrac{1 + 3i}{2}\right]$ and $\left[\dfrac{3 - 11i}{10}, \dfrac{1 - 2i}{5}, \dfrac{-1 + i}{2}\right]$.

8. $[15 - 2i, 6, 15 + 2i]$ and $[2 + 6i, -6, 2 - 6i]$.

Exercise 8.3

1. (a) $\begin{bmatrix} -\dfrac{1}{\sqrt{2}} & \dfrac{1}{\sqrt{2}} \\[2ex] \dfrac{i}{\sqrt{2}} & \dfrac{i}{\sqrt{2}} \end{bmatrix}$.   (b) $\begin{bmatrix} \dfrac{1}{\sqrt{2}} & \dfrac{i}{\sqrt{2}} \\[2ex] \dfrac{i}{\sqrt{2}} & \dfrac{1}{\sqrt{2}} \end{bmatrix}$.

4. $\begin{bmatrix} \dfrac{\sqrt{3} + \sqrt{2}}{2} & \left(\dfrac{\sqrt{3} - \sqrt{2}}{2}\right)i \\[3ex] \left(\dfrac{\sqrt{2} - \sqrt{3}}{2}\right)i & \dfrac{\sqrt{3} + \sqrt{2}}{2} \end{bmatrix}$.

7. (b) Any scalar multiple of $[2 - 5i, -5 + i]$.

INDEX